IET HISTORY OF TECHNOLOGY SERIES 26

Series Editors: Dr B. Bowers
Dr C. Hempstead

History of Telegraphy

Other volumes in this series:

Volume 4   **The history of electric wires and cables** R.M. Black
Volume 6   **Technical history of the beginnings of radar** S.S. Swords
Volume 7   **British television: the formative years** R.W. Burns
Volume 9   **Vintage telephones of the world** P.J. Povey and R. Earl
Volume 10   **The GEC research laboratories 1919–1984** R.J. Clayton and J. Algar
Volume 11   **Metres to microwaves** E.B. Callick
Volume 12   **A history of the world semiconductor industry** P.R. Morris
Volume 13   **Wireless: the crucial decade 1924–34** G. Bussey
Volume 14   **A scientists war – the diary of Sir Clifford Paterson 1939–45** R.J. Clayton and J. Algar (Editors)
Volume 15   **Electrical technology in mining: the dawn of a new age** A.V. Jones and R.P. Tarkenter
Volume 16   **Curiosity perfectly satisfied: Faraday's travels in Europe 1813-1815** B. Bowers and L. Symonds (Editors)
Volume 17   **Michael Faraday's 'Chemical Notes, Hints, Suggestions and Objects of Pursuit' of 1822** R.D. Tweney and D. Gooding (Editors)
Volume 18   **Lord Kelvin: his influence on electrical measurements and units** P. Tunbridge
Volume 19   **History of international broadcasting, volume 1** J. Wood
Volume 20   **The early history of radio: from Faraday to Marconi** G.R.M. Garratt
Volume 21   **Exhibiting electricity** K.G. Beauchamp
Volume 22   **Television: an international history of the formative years** R.W. Burns
Volume 23   **History of international broadcasting, volume 2** J. Wood
Volume 24   **Life and times of Alan Dower Blumlein** R.W. Burns
Volume 26   **A history of telegraphy: its technology and application** K.G. Beauchamp
Volume 27   **Restoring Baird's image** D.F. McLean
Volume 28   **John Logie Baird: television pioneer** R.W. Burns
Volume 29   **Sir Charles Wheatstone, 2nd edition** B. Bowers
Volume 30   **Radio man: the remarkable rise and fall of C.O. Stanley** M. Frankland
Volume 31   **Electric railways, 1880–1990** M.C. Duffy
Volume 32   **Communications: an international history of the formative years** R.W. Burns
Volume 33   **Spacecraft technology: the early years** M. Williamson

History of Telegraphy

Ken Beauchamp

The Institution of Engineering and Technology

Published by The Institution of Engineering and Technology, London, United Kingdom

First edition © 2001 The Institution of Electrical Engineers
New cover © 2008 The Institution of Engineering and Technology

First published 2001

The Institution of Engineering and Technology
Michael Faraday House
Six Hills Way, Stevenage
Herts, SG1 2AY, United Kingdom

www.theiet.org

British Library Cataloguing in Publication Data
Beauchamp, K.G. (Kenneth George), 1923–1999
A history of telegraphy: its technology and application
(IET history of technology series no. 26)
1. Telegraph – History
I. Title
621.3'83'09

ISBN (10 digit) 0 85296 792 6
ISBN (13 digit) 978-0-85296-792-8

Typeset by RefineCatch Ltd, Bungay, Suffolk
First printed in the UK by Short Run Press Ltd, Exeter, Devon
Reprinted in the UK by Lightning Source UK Ltd, Milton Keynes

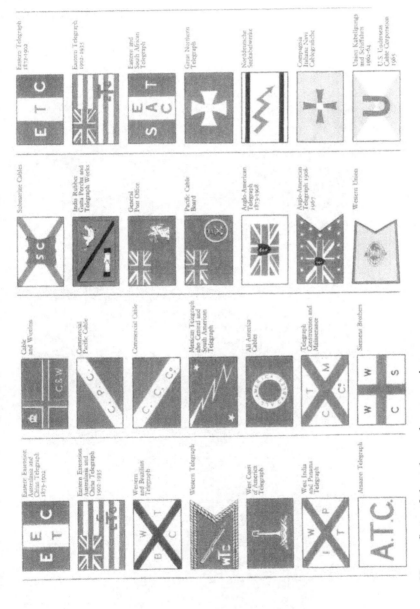

House flags of the telegraph companies
Source: Haigh, 'Cableships and Cables', 1979

Contents

List of figures   xiii
List of tables   xvii
Preface   xviii
Abbreviations   xxi

PART 1 – *TERRESTRIAL TELEGRAPHY*   1

1 Things mechanical   3
    1.1   Shutter systems   4
    1.2   Chappé's telegraph   6
    1.3   Popham and the Admiralty installation   8
    1.4   Some semaphore systems in Europe   14
    1.5   Semaphore in the United States   16
    1.6   Operations   17
        References   18

2 Early electrical ideas   20
    2.1   Electrostatic devices   20
    2.2   Electrochemical devices   22
    2.3   The 'needle' telegraphs   24
    2.4   Cooke and Wheatstone   30
    2.5   Telegraphy on the railways   34
    2.6   Dial telegraphs   40
    2.7   Codes and ciphers   47
        References   49

3 Commercial telegraphy   51
    3.1   Morse and single-line working   51
    3.2   Telegraph companies in the United States 57
    3.3   Development in Britain   69
    3.4   The Telegraph Acts of 1868–69   73
        3.4.1 The Electric & International Telegraph
              Company   74

	3.4.2 The British & Irish Magnetic Telegraph Company	77
	3.4.3 The London & District Telegraph Company	77
	3.4.4 The United Kingdom Electric Telegraph Company	78
	3.4.5 The Universal Private Telegraph Company	79
	3.4.6 The Reuter's Telegram Company	80
	3.4.7 The Exchange Telegraph Company	80
3.5	Faster, cheaper telegrams	81
3.6	Recording and printing	85
3.7	Overhead or underground?	90
3.8	Telegraphy in British India	94
	References	99

4 Military operations — 102
4.1	War in the Crimea	103
4.2	The Indian Mutiny	108
4.3	The American Civil War	110
4.4	European conflicts	115
4.5	African colonial wars	117
	4.5.1 The Nile and Egyptian campaigns	119
	4.5.2 The Boer War	122
4.6	Early British Army telegraph training	126
4.7	The Telegraph Battalion	126
4.8	The Society of Telegraph Engineers	127
4.9	British Army signalling	128
	References	132

5 Submarine cables — 134
5.1	Leaving the land	134
5.2	Gutta-percha	135
5.3	Crossing the Channel	138
5.4	The Siberian Telegraph	142
5.5	Oceanic cables	147
5.6	Theory and techniques	148
	5.6.1 Loading	151
	5.6.2 Sensitive detectors	152
5.7	Atlantic crossing	154
5.8	Links to South America	156
5.9	Cable-laying technology	158
5.10	A Committee of Inquiry	160
5.11	A Cable to India . . .	162
5.12	. . . and further East	168
5.13	The Australian connection	171
5.14	The world encompassed	174
	References	178

PART 2 – *AERIAL TELEGRAPHY*   181

6 Marconi and the experimenters   183
   6.1   Beginnings   184
   6.2   Marconi   186
   6.3   Transatlantic attempt   190
   6.4   Spark and arc   192
         6.4.1 'Short spark' operation   194
         6.4.2 The electric arc   200
         6.4.3 The high-frequency alternator   201
         6.4.4 Frequency multiplication   202
   6.5   Production and power   204
         6.5.1 The 'Marconi system'   206
         6.5.2 Marconi high-power stations   208
         6.5.3 Duplex working   210
         6.5.4 Telefunken and Siemens   210
         6.5.5 Keying at high power   215
         6.5.6 Continuous waves in the United States   215
               References   221

7 Telegraphy for peace . . .   224
   7.1   The advent of thermionics   225
         7.1.1 Detection   225
         7.1.2 Amplification and oscillation   227
         7.1.3 Transmission   229
   7.2   Linking the Empire   231
         7.2.1 Cable and wireless   238
         7.2.2 Reuters and the news service   240
   7.3   Maritime communication   240
         7.3.1 Wireless training in the merchant navy   244
         7.3.2 Codes, telegrams and newspapers at sea   246
   7.4   Life-saving at sea   248
   7.5   International agreements   254
   7.6   Civil aviation   257
   7.7   The role of amateurs   260
         References   262

8 . . . and at war   266
   8.1   Army wireless before 1914   267
   8.2   War on the ground, 1914–18   268
         8.2.1 Wireless direction-finding   269
         8.2.2 Trench warfare   272
         8.2.3 Wireless at the front   276
         8.2.4 Two military engagements   281
   8.3   The inter-war years   284

8.4	War on the ground, 1939–45	285
	8.4.1 Line working	286
	8.4.2 The African campaigns	288
	8.4.3 Communication systems	289
	8.4.4 Across the Channel	292
8.5	Army wireless in the Second World War	294
	8.5.1 Allied wireless equipment	296
	8.5.2 German Army wireless	301
8.6	British Army training and recruitment	304
	References	305
9	**Military telegraphy at sea**	**308**
9.1	Wireless experiments at sea	308
9.2	War at sea, 1914–18	314
	9.2.1 Shipboard wireless equipment	317
	9.2.2 The Naval wireless telegraph network	322
	9.2.3 Cable operations	323
9.3	The shore stations	324
9.4	The inter-war years	327
9.5	War at sea, 1939–45	329
	9.5.1 Allied wireless equipment	333
	9.5.2 German wireless equipment	337
9.6	Cable ships and cables	340
9.7	British naval wireless training	342
	References	345
10	**Military telegraphy in the air**	**348**
10.1	The dirigible	348
10.2	War in the air, 1914–18	350
	10.2.1 British airborne equipment	353
	10.2.2 Training telegraphists for air operations	359
	10.2.3 American airborne equipment	360
	10.2.4 German airborne equipment	362
10.3	The inter-war years	365
	10.3.1 Use of shorter wavelengths	367
	10.3.2 The 1929 development programme	369
10.4	War in the air, 1939–45	370
	10.4.1 British airborne equipment	371
	10.4.2 American airborne equipment	376
	10.4.3 German airborne equipment	377
10.5	RAF wireless training	383
	References	386
11	**Epilogue**	**389**
11.1	The demise of Morse	389
11.2	High-speed telegraphy	390

11.3   Baudot and the new codes   391
        11.3.1 Keyboard machines   396
        11.3.2 The teleprinter   397
        11.3.3 Military use of the teleprinter   398
11.4   Telegram, telex and the telephone   399
        11.4.1 The telex service   399
        11.4.2 Telephony by submarine cable and satellite   401
11.5   The digital revolution   403
        References   404

Index   405

List of figures

Frontispiece	House flags of the telegraph companies	v
1.1	Murray's shutter system, devised for the Admiralty in 1796	5
1.2	A station in Claude Chappé's telegraph system	7
1.3	Chappé's semaphore network in France, 1850	9
1.4	Popham's semaphore line to Portsmouth	10
1.5	A London–Portsmouth semaphore station	11
1.6	Gamble's mobile semaphore system, used for the Army	12
1.7	Watson's semaphore station on the South Downs, 1841	13
1.8	The Prussian semaphore alphabet	15
1.9	A Prussian station in the Potsdam–Koblenz semaphore line	16
2.1	Sömmerring's electrochemical telegraph	23
2.2	The indicator for Schilling's telegraph system	27
2.3	Principle of the astatic galvanometer	27
2.4	Circuit diagram for Schilling's six-needle telegraph	29
2.5	Alexander's electric telegraph	33
2.6	Wheatstone's five-needle 'hatchment telegraph', first used on the Euston–Camden railway line in 1837	34
2.7	Cooke's two-needle telegraph	38
2.8	Robinson's 'Wireless Electric Signals'	39
2.9	An early telegraph advertisement, 1845	41
2.10	Wheatstone and Cooke's ABC Telegraph	42
2.11	Post Office rules for operating a Wheatstone ABC instrument	43
2.12	Fardeln's dial telegraph	45
2.13	Siemens dial telegraph	45
2.14	Operation of the Siemens dial telegraph	46
3.1	Morse's recording instrument, 1837	53
3.2	Reproduction from Morse's notebook, showing his original code	54
3.3	(a) The original (American) Morse code. (b) The International Morse code	56
3.4	Inductive communication with a moving train	60
3.5	The House printing telegraph	61
3.6	A Morse sounder	63
3.7	A Western Union station	66

3.8   Two variations on the Morse key: (a) an International Morse key
       and (b) an American 'bug' key   68
3.9   The Central Telegraph Station, London, 1849   70
3.10 Some of the equipment in use at the Central Telegraph Office in
       1868   76
3.11 The differential relay   82
3.12 The bridge duplex method   83
3.13 The Prussian underground telegraph network in 1890   92
3.14 Street telegraph communication in the United States   93
3.15 An Indian railway training diagram   98
4.1   The Russian telegraph network to the Crimea in 1855   104
4.2   Siemens' Crimea telegraph equipment, consisting of an auto-
       matic writing system (left) and a three-key perforator (right)   105
4.3   Latimer Clark's cable plough   106
4.4   The Crimean Peninsula, showing the position of the submarine
       cable   107
4.5   Portable keyboard Morse instrument   113
4.6   'Fixing the cable' by the Royal Corps of Signals in the first
       Ashanti War   118
4.7   The Nile   121
4.8   Southern Africa   123
4.9   A Boer War railway blockhouse   124
4.10 Army signalling: time coding   129
4.11 Army signalling: code form   130
4.12 Cartoon from *The Military Telegraph Bulletin*   131
5.1   'Tapping the tree'   136
5.2   Laying the cross-Channel cable   139
5.3   Submarine telegraph cable laid between Britain and France, 1851   140
5.4   Cables laid between Britain, Ireland and the Continent in 1854   141
5.5   The Siberian Telegraph line   143
5.6   Telegraph arrival curves   149
5.7   The telegraph 'cable code', first used for the transatlantic cable in
       the 1860s   150
5.8   Thomson's mirror galvanometer   153
5.9   Cable routes to India   163
5.10 Australian telegraph connections   173
5.11 The first world-encircling telegram, sent from Sir Sandford
       Fleming to Lord Minton, the Foreign Secretary, in 1902   176
6.1   Lodge's coherer   185
6.2   Marconi's 'syntonic' tuned transmitter   189
6.3   Marconi's reliable magnetic detector   191
6.4   Various forms of quenched spark gap   193
6.5   The Marconi timing disc   194
6.6   Performance of Wien's short spark gap   196
6.7   Arrangement of Wien's short spark circuit   196

6.8	Galletti's multiple spark gap transmitter	199
6.9	Duddell's arc generator	200
6.10	Schematic diagram for the Marconi 0.5 kW Type T17 set	207
6.11	Marconi short-range multiple tuner, the Mk III	208
6.12	Telefunken quenched spark ship installation	212
6.13	Simplified schematic diagram of the Nauen HF alternator transmitter, 1919	214
7.1	Early thermionic valves: (a) Fleming's diode valve, (b) De Forest's triode valve	226
7.2	The Caernarvon valve transmitter	230
7.3	Empire beam transmitting aerial	238
7.4	Advertisement for telegraphy training, 1919	245
7.5	Marconi uniforms available for purchase in 1919	246
7.6	The *The North Atlantic Times* – the first ship's newspaper	249
7.7	Schematic diagram of the Marconi auto-alarm receiver	253
8.1	Bellini–Tosi directional system: aerial and crossed coil connections	270
8.2	Routes of German Zeppelins raiding over the east coast of England in 1916	271
8.3	The Fullerphone line instrument	275
8.4	Standardised grid for the front line cables on the Western Front, 1917	276
8.5	German quenched spark field unit, 1911	278
8.6	The wireless telegraphy trench set, Mk III, 1918	280
8.7	Wireless equipment distributed over the Allied front in August 1918	282
8.8	The No. 10 army set	291
8.9	Invasion communications across the English Channel	293
9.1	Admiral Rozhestvensky's voyage to the Far East	313
9.2	Schematic diagram of the submarine Type 10 set	318
9.3	Brown's sensitive relay	319
9.4	Schematic diagram of the naval Type 31 transmitter	320
9.5	Post Office coastal stations, 1919	326
9.6	Position of W/T offices in capital ships	330
9.7	The Marconi CR100 receiver (Navy Type B28)	331
9.8	Notice appealing for young volunteers for telegraph operator training with the Post Office (*Wireless World* July 1941)	344
10.1	Schematic diagram of the French Rouzet transmitter	352
10.2	The Sterling spark transmitter	353
10.3	Four First World War spark transmitters: from left to right, Type 52m, battery and rotary of Type 54b, Rouzet set, Type 54b, Type 52	357
10.4	RAF Type Tf valve receiver, 1917	358
10.5	RAF Type T21 transmitter, 1921	359
10.6	A Morse sending tablet	363
10.7	RAF Type T1154/R1155 HF-MF transmitter and receiver: (a) T1154, (b) R1155	373
10.8	Lifeboat Type T1333 transmitter	377

10.9   Schematic diagram of the German FuG 10 transmitter   380
10.10   Assembly of units forming a complete FuG 10 aircraft installation   381
10.11   Interchangeable HF Pentode, type RV12P2000, used in
        Luftwaffe airborne receivers   382
11.1   The Baudot transmission system   394
11.2   The five-unit Baudot code   395
11.3   The International Alphabet No. 1   397

List of tables

2.1	Electric telegraphs to the year 1850	26
2.2	Francis Bacon's *Exemplum alphabeti biliterarii* (1605)	28
2.3	Schilling's six-needle code table	30
2.4	Steinheil's two-needle code table	36
2.5	Cooke and Wheatstone's code table for the 'hatchment telegraph'	37
3.1	The main North American telegraph companies in 1852, and their principal developers	58
3.2	Companies named in the Telegraph Acts (Great Britain) of 1868–69	74
3.3	Underground telegraph lines in Europe in 1881 (from statistics issued by the World Telegraph Union, Bern, 1881)	92
5.1	Principal submarine cables to 1903	144
7.1	The international Q code as defined by the International Radio-telegraphic Convention, London, 1912	248
8.1	Allied army wireless sets, 1939–45	297
8.2	German Army wireless sets, 1936–45	303
9.1	Ships fitted with wireless telegraphy apparatus in 1906	310
9.2	Royal Navy transmitting and receiving sets, 1914–18	321
9.3	Admiralty W/T chain of transmitting stations, 1915–16	323
9.4	Royal Navy transmitting and receiving sets, 1930	328
9.5	Allied navy transmitting and receiving sets, 1939–45	332
9.6	German Navy transmitting and receiving sets, 1935–45	338
10.1	RFC/RAF wireless equipment to 1918	356
10.2	Luftwaffe airborne equipment, 1933–45	379
11.1	Systems considered by the High-speed Telegraph Enquiry of 1913	392

Preface

A dominating theme of the twenty-first century is going to be the transmission and processing of information. The continuing development and maintenance of a global network of communication channels is vital, for without it business and much of human endeavour would be atrophied. But how did all this come about? It is to telegraphy that we owe many of the ideas that have come to the fore in the present-day digital world. Complex methods of data compression, error recovery, flow control, encryption and even the computer techniques of the stored program and priority tagging can be found in the ingenuity of the telegraph pioneers. The arrival of cheap, long-distance communication stimulated the initial development of coded transmissions, primarily through the use of telegraphy and the Morse code, and had a tremendous influence on the development of communications for almost two hundred years. It stimulated the design and production of technical equipment in a pre-electronic age; influenced business and public communications through the ubiquitous telegram; permitted safe travel on railways; made possible navigation at sea and in the air; and saved countless lives. A major technological infrastructure was created and a new industry was born. The role of the military telegraph has had a dominant effect on the funding and technology of the process, from the Crimean War, through minor engagements, to the tremendous organisational problems presented by two major world wars, in which it played a unique part; and has culminated in today's widespread activities using the Internet, whose beginnings are directly related to the development of the military-funded Arpanet, itself influenced by the coding pioneers of the telegraph.

This book presents a brief record of the growth of the techniques of telegraphy over two centuries of development. While a major purpose is to depict the discoveries and ingenuity of the experimenters, engineers and scientists involved and the equipment they designed and built, the organisation, applications and effects of its use on society are also considered – a society which was as fascinated by its new ability to reach out and contact people by electric telegraphy across the world as the present generation is by today's parallel development of the Internet.

We can clearly define two major phases in the growth of the electric telegraph, from its beginnings in the nineteenth century, in its initial use via landlines and submarine cables and the later development through the medium of wireless transmission. The first phase is considered in Part 1, Terrestrial Telegraphy, which commences in Chapter 1 with the early experiments in communication by pre-electric means. It is salutary to note that well before an electrical network was established in Europe and the United States, many countries had fully operational nationwide communication systems, each consisting of hundreds of intercommunicating stations, operated by purely mechanical means. This both intensified the search for faster and more efficient techniques and laid the foundation for the organisational techniques which were later to be developed. Early electrical ideas are considered in Chapter 2, leading to the expanding commercial and military use which are discussed in Chapters 3 and 4. The military played a significant role in the application of line telegraphy, not only in providing funds for development but also in testing the new ideas in the rigorous environment of military campaigns. Chapter 5 considers the ability to cross the world's oceans with linking cables, the crowning achievement of line telegraphy, which came in the second half of the nineteenth century.

The second phase of growth is considered in Part 2, Aerial Telegraphy. The tremendous fillip that wireless communication gave to the transmission of telegraphic information is described in Chapters 6 and 7. Here the early work of Marconi, Lodge, Fleming, von Arco, Fessingden, Tesla and many others is considered, together with its extension to naval operations by Jackson and Popov, culminating in the widespread use of wireless telegraphy in linking empires, controlling business, and military communications at sea and in the air. Wireless communication, and particularly wireless telegraphy, played a vital role in both world wars and its use on land, sea and air is considered in Chapters 8 to 10, which make use of much previously unpublished material from the Public Record Office and military establishments. An epilogue forming the final Chapter 11 takes a look at the successors to this telegraphic enterprise, using new methods of coding, which have led to its modern descendant, the global communication network.

During the gestation period for this work the author had reason to thank very many organisations and individuals who were the source of much information on telegraphy from widely scattered sources. Chief among these are the libraries: the British Library, the Public Record Office, the National Maritime Library, the Library of Congress, the library of the Deutsches Museum in Munich, the libraries of the Institution of Electrical Engineers and the Institution of Mechanical Engineers, and the Science Museum Library. Particular thanks for much help and advice are due to Susan Bennett of the Library of the Royal Society of Arts; Jean Hort of the US Navy Department Library, Washington;

Stefanie Bennett, Information Division of the National Maritime Museum Library; Melvyn Rees of the Patent Office; Lenore Symons, archivist of the Institution of Electrical Engineers; Robert Colburn, archivist of the IEEE History Center at Rutgers University; Gertrude Gerstle, archivist of Cable & Wireless plc; David Hay, archivist of British Telecom plc; Dr Thwaites, archivist of the Royal Corps of Signals; Lieutenant-Colonel W.E. Legg, museum curator of HMS *Collingwood*; Dr Graham Winbolt, curator of the Winbolt Collection of Military Communications; Dr Wittendorfer and Herbert Böhnen of Siemens AG; Peter Singleton of the Air Historical Branch of the RAF; Douglas Robb of the RAF Air Defence Radar Museum, Neatishead; Yap Siow Hong of Singapore Botanical Gardens; Arthur Bauer of the Foundation Centre for German Communication and related Technology in the Netherlands; John Beavis of the School of Conservation Sciences, Bournemouth University; Brian Cooke of the Crimean War Research Society; Professor Patrick Parker of the US Naval Postgraduate School; Professor Ron Williams and the Shelburne Museum of Vermont; Jack Howlett, Jack Bridge and Gordon Woodward of the IEE Professional Group S7; and Brian Bowers of the Science Museum.

Ken Beauchamp
Lancaster, September 1999

Abbreviations

A	ampere
a.c.	alternating current
ACT	Apparatus Carrier Telephone
ARC	Automatic error correction
ASCII	American Standard Code for Information Exchange
AEG	Allgemeine Elektrizitäts Gesellschaft
AF	audio frequency
AG	Aktien Gesellschaft (limited company)
AM	amplitude modulation
ARQ	automatic repeat request
AT&T	American Telegraph & Telephone Co.
BFO	beat frequency oscillator
BWG	Birmingham wire gauge
CCITT	Consultative Committee for Telegraph Communication
CTO	Central Telegraph Office
CW	continuous wave
d.c.	direct current
DCCC	double current cable code
D/F	direction-finding
DLH	Deutsches Lufthansa
DTN	Defence Teleprinter Network
e.m.f.	electromotive force
FDM	frequency-division multiplex
FM	frequency modulation
FuG	Funk Gerat (wireless apparatus)
GE	General Electric
GHQ	General Headquarters
GPO	General Post Office
HF	high frequency
HT	high tension (volts)
Hz	hertz (cycles per second)
ICW	interrupted continuous wave
IEE	Institution of Electrical Engineers

IEEE	Institution of Electrical & Electronic Engineers (USA)
ITU	International Telecommunications Union
kHz	kilohertz
km	kilometre
kVA	kilovolt-ampere
kW	kilowatt
LF	Low frequency
LC	inductance/capacitance
LT	Low tension (volts)
m	metre
MCW	modulated continuous wave
MF	medium frequency
MHz	megahertz
	modem modulator/demodulator
MO–PA	master oscillator–power amplifier
NESCO	National Electric Signal Company
RA	Royal Artillery
RAAF	Royal Australian Air Force
RAE	Royal Aeronautical Establishment
RAF	Royal Air Force
RCA	Radio Corporation of America
RE	Royal Engineers
RF	radio frequency
RFC	Royal Flying Corps
RN	Royal Navy
RNAS	Royal Naval Air Service
RNV(W)R	Royal Naval Volunteer (Wireless) Reserve
RSGB	Radio Society of Great Britain
RST	readability, signal strength and tone (the Braaten code)
R/T	radio telegraphy (speech)
SWG	standard wire gauge
TDM	time-division multiplex
TRF	tuned radio frequency
UHF	ultra-high frequency
USAF	United States Air Force
V	volt
VHF	very high frequency
W	watt
WM	wireless mechanic
WO	War Office
WOAG	wireless operator/air gunner
WOM	wireless operator/mechanic
W/Op	wireless operator
W/T	wireless telegraphy (Morse)
wpm	words per minute

Kenneth Beauchamp

1923–1999

Dr Ken Beauchamp was an electrical engineer with a keen interest in the history of his chosen field. He played an active part in the historical activities of the Institution of Electrical Engineers. He was the author of *Exhibiting Electricity*, published by the IEE in 1997, which provided an account of the many exhibitions where electricity and electrical devices have been displayed over the last two centuries.

Dr Beauchamp completed the manuscript and chose the illustrations, but did not live to see his work published. The Series Editors and Publisher are grateful to Mrs Margaret Beauchamp for allowing them to publish this book as a tribute to the author and as a further contribution to the historical studies which he encouraged and enjoyed.

PART 1

TERRESTRIAL TELEGRAPHY

Chapter 1

Things mechanical

'Communication at a distance' is an idea which has been present in one form or another since people have had any information worth communicating. Early attempts made use of smoke signals, low-frequency drum beats, wind instruments and, as the technology improved, reflecting mirrors, flags and beacon fires. The last of these is interesting as it makes use of a number of useful ideas capable of further development. In its usual form it consisted of a number of 'line of sight' fire beacons, situated at high, fixed locations, which enabled simple messages, using a previously agreed coding repertoire, to be sent along a chain of such beacons.[1] The information content was low, however, and the satisfactory transmission of information by such visual means, conveying an acceptable range of meaning over long distances in a reasonable time, was not to occur until the end of the eighteenth century, when a variety of ingenious attempts to establish long lines of communicating devices were made.

This was prompted by the requirements of the military: the navy in England and the army in France. As first developed in Europe, the signalling systems adopted were mechanical structures, and significant changes in the appearance of the mechanism could be observed at a distance. This development depended for its practical implementation on the advent of the optical telescope, which although invented in Holland in about 1608 was not manufactured to a useful accuracy until 1757, when John Dollond's achromatic object lens became available. The combination of a refracting telescope and a clearly recognisable signalling structure, mounted on a hill or tall building and within telescopic sight of a similar structure some distance away, constituted a workable system, of which very many were erected in most of the developed regions of the

---

[1] Such as the system passed by an Act of Scottish Parliament in 1455, which stipulated 'That one bale or fagot [of straw set alight] shall be a warning of the approach of the English, two bales that they are actually coming, and four bales, blazing side by side, shall note that the enemy is in great force.'

world in the eighteenth and early nineteenth centuries [1]. The two systems which were in use, the shutter system and the semaphore system, corresponding in modern technology to digital and analogue systems, both established a firm foundation of telegraphic and organisational techniques for the more versatile electrical systems that were to follow.

1.1 Shutter systems

As early as 1684, Robert Hooke, Secretary to the Royal Society, proposed a scheme in which large boards or shutters of different shapes could be hung in a wooden frame to convey by their shape different letters of the alphabet, and viewed from a distance by using a telescope [2]. Others suggested similar schemes, but it was over a hundred years later, in 1796, that a practical working system was brought into use. That year the Admiralty was persuaded to fund Lord George Murray [3] to set up a six-shutter scheme to link the Admiralty in Whitehall with its bases in Portsmouth and, later, Plymouth, in order to communicate with ships at sea [4]. Other contenders for this task included the Reverend John Gamble, Chaplain General to the Forces, who devised a system of five vertical shutters similar to the Murray system [5]. The positions of the shutters (open or closed) in the Murray system, used to represent the first seven letters of the alphabet, are shown in Figure 1.1. It is likely that the Murray system was chosen by the Admiralty for its number of shutters, which allowed for 63 different combinations; after the letters of the alphabet had been allocated, the remaining combinations could be used to represent the numerals 0 to 9 and various coded words. The shutters were controlled by pulling on ropes from inside a specially designed building surmounted by the large and heavy wooden shutter system. Each building was over 10 m high, including 5 m for the shutter system, and located on a suitable hill, having visual sight of neighbouring signalling stations in the communication chain. Many of these locations, now bereft of signalling systems, are known to this day as 'Telegraph Hill'. In good weather the system was surprisingly effective: a message between London and Portsmouth might take about 15 min to pass through the ten stations en route, with a shorter time required for acknowledgement. While Lord Murray carried out the planning of the shutter system, it was George Roebuck, a surveyor from London later appointed an Inspector of Telegraphs, who was responsible for constructing the Portsmouth line, together with a later set of 15 stations from London to Deal and Sheerness. Roebuck's final shutter line was constructed in 1807 to connect the Admiralty in London with Yarmouth, then an important naval station. Murray and Roebuck's systems worked on the principle of sending one letter at a time. The suggestion of using a cipher system to transmit numbers to represent

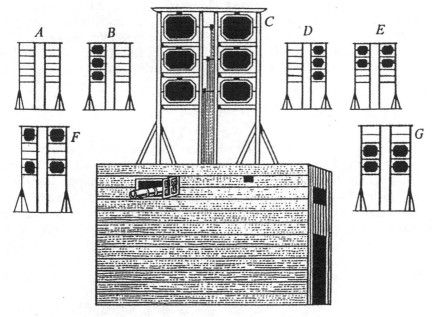

Figure 1.1   Murray's shutter system, devised for the Admiralty in 1796
Source: *Admiralty Archives, London*

individual words was made by Lieutenant-Colonel Macdonald in a book published in 1808 [6], but declined by the Admiralty on the grounds that this could lead to serious errors in the event of faulty transmission of only a single number.

The only other European country to adopt the shutter system was Sweden. In 1795 Abraham Edelcrantz, private secretary to the King of Sweden, inaugurated a chain of shutter telegraphs from Stockholm to Fredricksborg, later extended by the addition of further lines from Grissleham to Signliskär and Eckerö on Åland. His system consisted of a matrix of nine shutters, with a tenth larger one mounted on top. This gave $2^{10} = 1024$ different signals, many of which were used in a number coding system (actually a cipher system) to improve the speed of communication. By 1809 the Swedish network comprised about fifty stations distributed over a distance of 200 km. It was manned by a skilled team of telegraph operators wearing a distinctive uniform approved by the Swedish King, Carl XIV Johan [7], the first uniformed signal cadre to appear in Europe.

Lord Murray's shutter telegraph systems for the military were intended only for temporary use during the Napoleonic Wars, and were brought to an abrupt closure at the conclusion of the Peninsular War in May 1814, only to be opened again, somewhat hurriedly, after Napoleon escaped from Elba to recommence the war with England a year later. After the final victory at Waterloo in June 1815, and the closure and dismantling of the shutter systems, the Admiralty declared its intention to establish a permanent system of signalling stations to Portsmouth, but this time using a quite different semaphore plan.

1.2 Chappé's telegraph

Shutter systems were a peculiarly British solution to the communications problem. They were cumbersome and prone to mechanical defects, and after 1815 were gradually replaced by a semaphore system, already in use in France, in which movable wooden arms conveyed the information through the network. This soon came to be accepted as the standard mechanical telegraph in the first half of the nineteenth century, copied and replicated in many nations, and not only in Europe.

Whereas in England it was the Navy that was building 'line of sight' shutter systems, in France Napoleon's army was busily constructing semaphore lines, based on a design by Claude Chappé, who had previously demonstrated his method to the Committee of Public Instruction in Paris in 1793 [8]. France was then in turmoil, assailed on every side by most of the armies of Europe. A secure and effective method of coordination between Napoleon's scattered forces was vital if they were to survive the onslaught of the opposing armies. Chappé was fortunate to be active in carrying out experiments with his semaphore at a time when the French military and civil authorities were anxious to establish a good system of intercommunication for governmental and military purposes. Once he had convinced the Committee that his system would work, he was given extraordinary support in developing it. His design for a 'telegraph station' is shown in Figure 1.2. It was a substantial structure, consisting of a mast 5 m high carrying a 4.5 m rotatable cross-beam known as the 'regulator'. At each end of the regulator was fixed another arm, 2 m long, referred to as the 'indicators'. Each indicator was capable of taking up seven different positions at intervals of 45°, and the regulator arm carrying the indicators had four possible orientations. This gave a total of $7 \times 7 \times 4 = 196$ usable signals, though in practice only 96 positions were ever used. Almost from the inception of this scheme, Chappé used the signals, not directly to signify letters of the alphabet and numbers, but as a medium for the transmission of coded signals, arranged as a set of ciphers. In 1795 he published an extensive cipher book containing a total of 8464 letters, numbers, words and phrases. Each one was transmitted as a pair of semaphore signals as listed in his cipher book.

Initially, Chappé was empowered to establish a chain of communicat-

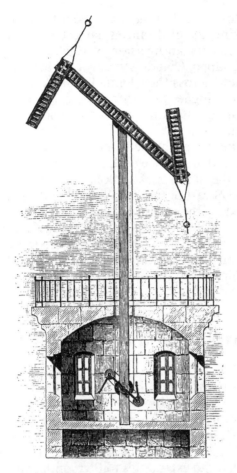

Figure 1.2   A station in Claude Chappés telegraph system
Source: Ernest, 'Claude Chappé', 1893

ing stations to connect Paris to Lille. He was given carte blanche to 'place his machines in any belfry, towers or emplacement that he might choose, and he could cut down any trees that might interfere with the line of vision'. The entire system was working by 1794, with a line of 15 stations over a distance of 230 km, and is considered by many historians to have played a major part in turning the tide of war in favour of the French forces. It was over this line that the first formal telegraph message was sent announcing Napoleon's capture of Le Quesnoy from the Austrians in 1794.[2] In the following decade Claude Chappé, now designated Ingénieur Télégraphe by the government, installed further lines radiating

[2] The word *telegraph* ('far writer') is reputed to have been coined by Chappé, though he had earlier described his invention as a *tachygraphe* (fast writer).

from Paris, London, Strasbourg, Brest, Béhobie (Spain) and numerous branch lines linking 29 of France's largest cities to Paris. This also included preparation for an invasion of England. Napoleon Bonaparte asked Abraham Chappé, Claude's younger brother, to design a telegraph capable of signalling across the Channel. The station on the French side was installed at Boulogne, but no record exists to indicate interest in establishing a similar station on British shores in the event of a successful invasion.

Chappé's growing telegraph network was staffed by an army of operators and superintendents. Since each station had to be within telescopic sight of the next, the number of stations was very great, with over fifty between Paris and Strasbourg alone. The extent of this impressive network as completed by the middle of the century is shown in Figure 1.3, with extensions to Spain, Italy, Belgium, Switzerland and Germany. By 1852 a total of more than 4000 km of lines were in operation, with over 550 stations in France, but by this time the electric telegraph was beginning to replace Chappé's lines, and they gradually fell out of use during the next three decades [9].

1.3 Popham and the Admiralty installation

In England the shutter systems were not developed much further than the Admiralty line to Portsmouth and the line from London to Chatham and Sheerness. However, by the 1820s semaphore development in France caused the Admiralty to consider its use as a replacement line and, among several contenders for this task, Sir Home Riggs Popham was asked to establish his design for a semaphore line linking London to Portsmouth, and later to extend this to Plymouth (which further task was never completed). The map (Figure 1.4) shows his route and further extensions to Dover and Deal, which also carried some private business and shipping traffic. Popham used a simpler system of arms for his semaphore without additional indicator arms at the ends, which had been found to be somewhat difficult to adjust remotely, independently of the main regulator arm. This followed an earlier system of three rotatable arms on a tall mast, invented by Depillon in France in 1801, which had been adopted for communication between ship and shore and was considered easier than the Chappé system to 'read' at a distance at times of poor visibility. The stations erected by Popham on the Portsmouth line consisted of a mast carrying pairs of two arms, each arm capable of being set at a recognised angle to the vertical, as may be seen from Figure 1.5, reproduced from the *Mechanics' Magazine* for 24th September 1825. Each arm could take up to six positions, plus one 'at rest' – in line with the supporting mast. This gave a total of 48 possible signals which were allotted to the letters of the alphabet, the numerals 0 to 9, and a few

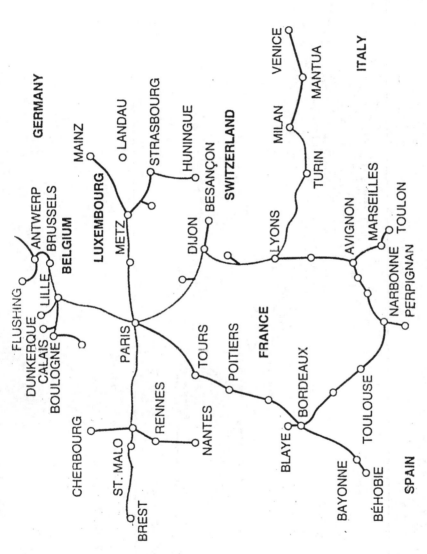

Figure 1.3   Chappés semaphore network in France, 1850

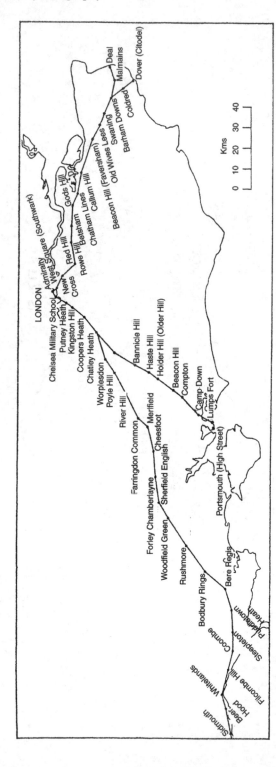

Figure 1.4   Popham's semaphore line to Portsmouth
Source: Wilson, 'The Old Telegraphs', 1976

arbitrary positions for words and phrases such as 'fog' or 'closing down'. The system was successful and attracted the attention of the Royal Society of Arts, which in 1816 awarded Admiral Popham (as he was then) one of their first gold medals for technological achievement. Popham at that time was no stranger to academic accomplishment, having earlier carried out much scientific work for the Admiralty, and in recognition he had been made an FRS in 1799.

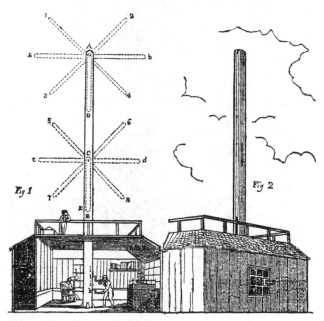

Mechanics' Magazine,

MUSEUM, REGISTER, JOURNAL, AND GAZETTE.

[No. 109.]   SATURDAY, SEPTEMBER 24, 1825.   [Price 3d.]

"To industrious study is to be ascribed the invention and perfection of all those arts whereby human life is civilized, and the world cultivated with numberless accommodations, ornaments, and beauties. All the comely, the stately, the pleasant, and useful works, which we view with delight, or enjoy with comfort, industry did contrive them—industry did frame them."—*Barrow*.

THE SEMAPHORE,

NOW IN USE BETWEEN LONDON AND PORTSMOUTH.

Figure 1.5   A London–Portsmouth semaphore station
Source: Mechanic's Magazine, 24th September 1825

In operation, Popham's semaphore was found to be faster than the shutter system, and a simple message such as the noonday time signal could be sent from Greenwich to Portsmouth, and acknowledged, in a few minutes (weather permitting).[3] Attempts were made to use the system at night by hanging lanterns at the end of the semaphore arms. A Lieutenant-Colonel Pasley, an earlier rival to Popham in telegraphic matters, describes such a system, which he called 'a nocturnal telegraph', in a handwritten paper to the Admiralty in 1822, but there are no records of this being used in the Portsmouth line [10]. Eventually, in 1840, the line of semaphore stations was closed on the grounds of excessive running costs, and replaced by an electric telegraph alongside the South-Western Railway, linking the Admiralty with its port at Plymouth [11].

The use of the mechanical telegraph by the Admiralty prompted the Army to express an interest in a semaphore system, but in their case portability was considered essential. To meet this requirement the Reverend Gamble, who had unsuccessfully bid for the naval shutter system, designed a mobile telegraph in 1797, shown in Figure 1.6. Twelve

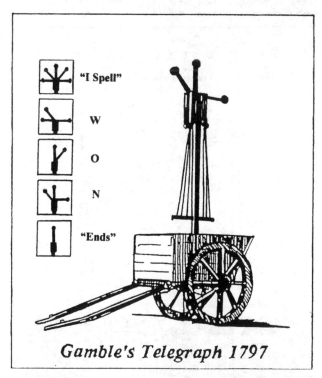

Figure 1.6   Gamble's mobile semaphore system, used by the Army

---

[3] It is recorded that Popham's system was in full operation for some 200 days of the year and in part for a further 60 days. Carrier pigeons were kept in reserve for the 'off days'.

sets of these carts with their semaphore telegraph were ordered, each capable of expressing 31 signal variations by suitable settings of the three signal arms, observed at a distance with the aid of a telescope.

Few commercial semaphore systems were erected in England. The most successful of them was the Holyhead–Liverpool telegraph, built in 1826 and having nine intermediate stations between Holyhead mountain, off Anglesey, and Liverpool, a total distance of 116 km. The line was used to communicate news of ships arriving in the Mersey, replacing an earlier 'ball and flag' system. The Holyhead line was the work of another naval man, Lieutenant Barnard Lindsay Watson. Later, in a civilian capacity, he negotiated for sites for similar commercial stations between London and the South Downs, essentially identical to Popham's stations, but mounted on a substantial wooden tower which accommodated the operating staff (Figure 1.7) [12]. According to an advertisement in the *Shipping and Mercantile Gazette* for 2nd November 1841, 'Watson's General Telegraph Association', responsible for operating the scheme, was prepared to provide information on 'vessels of all nations entered in the Telegraph List to be reported at any of his stations for the payment of 20s [£1] per annum'. His coding structure 'included the selection of

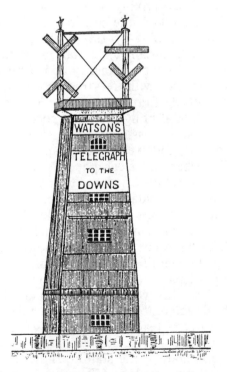

Figure 1.7   Watson's semaphore station on the South Downs, 1841
Source: Wilson, 'The Old Telegraphs', 1976

every word in the language of common occurrence'. Each word was numbered and allocated to one of nine subclasses, each of which contained 999 numbers. In his cipher book he listed all these, including letters of the alphabet, compass points, time signals and several thousand phrases. This complexity implied a far higher expertise in managing a semaphore station than had been achieved by any earlier system in England, including the Navy.

1.4 Some semaphore systems in Europe

As well as France's extensive network of stations in Europe, the turn of the nineteenth century saw Chappé's semaphores widely adopted in neighbouring countries and in America. A Russian line of stations, established between Sevastopol and Moscow, proved invaluable to their army early in the Crimean conflict (until replaced by an electrical system), enabling short messages to be conveyed in about two days [13]. For a while this afforded the Russians an advantage in Crimean communication, until the Allies established an electric telegraph line between Varna and the Crimean peninsula (of which more later). Some lengthy semaphore lines were developed linking St Petersburg, Kronstadt, Pushkin and Galchina, and one 830 km in length from St Petersburg to Warsaw, arranged with generous funding by Tsar Nicholas I. The Warsaw link, staffed by 1320 personnel, contained 220 stations, each incorporating tall brick towers 'constructed for permanency and beauty – neatly painted and the grounds ornamented with trees and flowers' [14].

The first country outside France to adopt Chappé's semaphore was, however, Sweden in 1795, although this was soon replaced by a shutter system, following the same cross-country route described earlier in this chapter. This was quickly followed by a semaphore system in Denmark in 1802. A line of mechanical telegraphs was also established in 1832 in Prussia, under the direction of Major Franz August O'Etzel for use by the military. It linked Potsdam, Brandenburg, Magdelburg, Berlin, Paderborn, Köln, Koblenz and Treves. Initially a single-arm regulator was used with two indicator arms located at each end, as in the original Chappé system. This allowed capitals and lower-case letters of the German alphabet to be identified, together with a few letter combinations, punctuation marks and numerals. The complete set of signs used is shown in Figure 1.8. This was eventually replaced with a multi-arm system based on the design of the English commercial stations erected by Watson, but having three instead of two pairs of signal arms. These were affixed to a 6.5 m mast and erected on a substantial building which housed the officer and staff of the station, shown in Figure 1.9. Since each arm could be set at intervals of 45°, a remarkable total of $4^6 = 4096$ different signals could theoretically be distinguished. A transmission rate

Figure 1.8   The Prussian semaphore alphabet
Source: Deutches Museum Archives, Munich

of 1.5 signals per minute has been claimed for the entire system, which seems rather high [15]. Some 61 stations were built along the 750 km route from Potsdam to Koblenz, and a special corps of the army, the 'Telegraphen Korps', attached to the general headquarters in Berlin, was created to manage the whole enterprise.

Figure 1.9   A Prussian station in the Potsdam–Koblenz semaphore line
Source: Deutches Museum Archives, Munich

1.5 Semaphore in the United States

The first semaphore in the United States, again based on Chappé's system, was built by Jonathan Grout of Belchertown, Massachusetts, in 1801. It was over 100 km in length, linking Martha's Vineyard off the New England coast with Boston, and its purpose was to transmit news about shipping entering the straits. Sixteen stations were established along the route. In the years to 1856 several other semaphore lines were built linking Boston, Hull and New York, all providing shipping information for ship owners and traders [16]. One of these, constructed by John R. Parker in 1824, was used to identify ships approaching Long Island Sound and to telegraph this information to New York. Parker was an enthusiastic advocate of the semaphore, and in his book *The United States Telegraph Vocabulary*, published in Boston in 1832, he lauds it in the following terms:

The Semaphoric Telegraph for land operation, with two Arms and an Indicator at the top, to denote the section of the Alphabet from whence the communication proceeds, is an original invention, combining simplicity of plan, with economy, certainty, and celerity, in execution. By an attentive perusal of the instructions, and a careful examination of the illustrations given in the Plates, together with their explanations, any person may readily become acquainted with the method of making and receiving Signals.

In 1837 he initiated a proposal to Congress for a cross-country line of semaphores linking New York with New Orleans, a distance of 1900 km. This was not carried out, mainly on account of the high cost, but the request did have one unexpected result. One of the objectors to the proposal was Samuel F.B. Morse, who took the opportunity to present his own ideas on telegraphic communication to Congress – with far-reaching effect, as we shall see in a later chapter [17].

1.6 Operations

Semaphore systems were extremely labour-intensive, as consecutive stations were seldom more than 12 km apart. A staff of at least five were required at each: two to operate the ropes or windlass controlling the signalling arms, two to man the forward and rear telescopes, and one to supervise operations (he was generally the one who could read and understand the codes and ciphers used). In Britain in the 1850s this gave employment to several thousand, but led to considerable hardship when the various lines were suddenly closed in favour of the new electric telegraphs, which required a different calibre of staff to maintain and operate them.

However, difficulties were beginning to appear before this date, arising from new methods of working which involved complex transmission codes or ciphers demanded by the users, whether military or commercial. These new codes allowed each individual position of the semaphore arm or arms to represent not just a letter, but a complete word or phrase, to be interpreted by reference to a cipher book, a technique initiated by Chappé in 1795. This meant that a single error in transmission could render a complete message unintelligible, and so imposed a level of accuracy not often reached by the staff manning the telegraph stations. Repetition was used to overcome the problem, but this of course slowed down the transmission of the message. To be certain that the correct combination had been sent, each of Chappé's teams had to confirm that the next station down the line was repeating the message correctly by sending a confirmatory signal, and this further reduced the rate of transmission.

From the beginning use was made of the surplus positions of the shutters or telegraph arms after the alphanumeric positions had been allocated. These were use to indicate instructions such as 'ready' and 'end of word', and certain words such as 'and', 'fog' and 'sail'. As the use of the system developed, more elaborate coding arrangements were introduced, often adapted from the codes already in use by the British Navy for flag signalling in which a combination of letters and numbers could be related to an extensive repertoire of meaning: over 30 000 for Popham's system of flag signalling, later adopted as the International Flag Code [18]. In other codes and ciphers, particularly those used by the military, each of a set of numbers arranged as series of groups of four numbers, related to a word or phrase found in the station codebook. The complexity of these new requirements, together with the semaphore's inability to operate in conditions of poor visibility, rendered the electric systems, when they arrived, extremely attractive to both military and commercial users. Mechanical systems lingered on in certain areas, particularly where used by the military. In Britain the Admiralty continued to use the Portsmouth line of semaphores until 1851, the year of the Great Exhibition, and in Russia the line from Moscow to Sevastopol was still in use in 1856, while the Crimean War was being fought. The organisational methods and ideas developed during over fifty years of semaphore signalling provided a valuable background, enabling the successful electric systems which eventually replaced them to be implemented rapidly. The sites themselves were (and in some cases still are) valuable for many later 'line of sight' communications systems.

References

1   WILSON, G.: 'The old telegraphs' (Phillimore, London, 1976)
2   'Hooke, Dr. R[obert]', *in* 'Dictionary of national biography', vol. 27, pp. 283–6
3   'Murray, Lord George (1761–1803)', *in* 'Dictionary of national biography', vol. 39, p. 361
4   HOLMES, T.W.: 'The semaphore 1796–1847' (Arthur H. Stockwell, Ilfracombe, Devon, 1983)
5   GAMBLE, J.: 'Observations on telegraphic experiments; or the different modes which have been or may be adopted for the purpose of distant communication' (London, 1795)
6   MACDONALD, J.: 'A treatise on telegraphic communication, naval, military and political' (T. Egerton, London, 1808)
7   HOLZMANN, G.J., and POHRSON, B.: 'The first data networks', *Scientific American*, 1994, **270**, pp. 112–17
8   JACQUEZ, ERNEST: 'Claude Chappé' (Alphonse Picard, Paris, 1893)
9   APPLEYARD, R.: 'Pioneers of electrical communication' (Macmillan, London, 1930)
10 PASLEY, Lt Col.: 'Telegraph signal rules' (1822), Institution of Electrical Engineers Library, London

11 '4th Report of the Railway Committee of the House of Commons' (1840), Fiche no. 43.97, vol. 13, p. 181, Public Record Office, Kew, Surrey

12 WATSON, B.L.: 'Watson's code of signals for the use of vessels at sea and adapted for the semaphoric telegraph as established from the Downs to London, Holyhead to Liverpool, St Catherines, Isle of Wight to Southampton, Spurn to Hull, Start to Dartmouth and the several reporting stations upon prominent headlands around the coast' (Henry Causton, London, 1842, 6th edn)

13 DODD, G.: 'The Russian war 1884–6' (Chambers, London, 1856)

14 SCHAFFNER, T.P.: 'The telegraph manual' (Pudney & Russell, New York, 1859)

15 FEYERABEND, E.: 'Der Telegraph von Gauss und Weber im Werden der elektrischer Telegraphie' (Reichspostministerium, Berlin, 1933)

16 O'REILLY, H.: 'Christopher Colles and the first proposal of a telegraph system in the United States', *Historical Magazine*, 1807, **15**, pp. 262–9

17 COE, LEWIS: 'The telegraph – a history of Morse's invention and its predecessors in the United States' (McFarland, Jefferson, Mass., 1911)

18 POPHAM, Adm. Sir HOSIE.: 'Telegraphic signals' (Edgerton, London, 1812)

Chapter 2

Early electrical ideas

While scores of semaphore installations were being established on hill-tops all over the European mainland, enthusiastic technologists were already considering how electricity could be used to achieve the same ends. Indeed, Chappé himself had considered electricity in 1790 and abandoned it in favour of mechanical systems. He was sufficiently concerned, however, to leave behind a record of his ideas [1], which were taken up by Ronalds in 1816.

However, this was not the first proposal for 'signalling at a distance by the use of electricity'. The idea was initially brought to public attention by a letter from the enigmatic 'C.M.' in the *Scots' Magazine*, Edinburgh, on 17th February 1753 [2]; his identity has never been satisfactorily disclosed, although various suggestions have been made. He proposed an electric conducting line (or rather a series of such lines, one for each letter of the alphabet), each terminating in a pith ball suspended from the end of the line. Below each ball are placed pieces of paper, each marked with a letter of the alphabet. At the sending end of a line, contact is made with a source of electrostatic charge. The charge travels along the line to the ball and causes the appropriate piece of paper to rise up and display its letter. This is repeated for each letter of the message. When the charge is removed from the electrified ball, the charged paper falls. Although not recognised by 'C.M.', this scheme makes use of a common earth return circuit.

2.1 Electrostatic devices

This was the first of a number of communication devices, proposed or constructed, to make use of static electricity, widely known to technologists since the publication of *De magnete* by William Gilbert in 1600, and easily generated by rubbing an insulating material, such as a glass or a sulphur ball. While these ideas did not result in any practical equipment, they were important in establishing the value of copper and iron wires for the transmission of electric charge over long distances when suitably insulated against earth contact. Several experimenters demonstrated their equipment, usually by effecting the movement of lightweight pith balls, attracted by an electric charge, at the receiving end of the line. The idea of coding repeated movements of the ball at the end of a single line instead of employing multiple lines and a pre-arranged code to spell out a message was already being considered by M. Lomond and others, despite the appalling slowness inherent in this method of communication. Other workers retained multiple lines, as suggested by 'C.M.', notably Bozolus in 1767 and Le Sage in 1782. Don Salvá of Barcelona put forward a scheme whereby the multiple lines could be activated by Leyden jars, previously charged from an electrostatic generator at the transmitting end, and an operator at the other end would detect the active lines by experiencing shocks through his fingers [3]!

The fullest account we have of the application of static electricity to the communication problem comes from Sir Francis Ronalds, who left a detailed description of the experiments carried out in the garden of his house in Hammersmith in 1823 [4]. This was a realisation of Chappé's electric 'synchronous telegraph' and consisted of a pair of stout wooden frames between which were strung over 12 km of wire – in effect, a simulation of a practical telegraph between two stations. A frictional machine charged one end of the line, and to the other was connected a pair of pith balls which diverged when the line was charged. At the sending station Ronalds arranged a dial capable of rotation marked with the letters of the alphabet, seen through an aperture, one letter at a time. A similar dial was provided at the receiving end. Both were controlled separately by a clockwork mechanism and kept in synchronism. The dials were rotated in a series of steps at 1 s intervals, controlled by a clock escapement. This enabled the dial operator to pause for a second as each letter reached the aperture, and read the letter easily. An operator was required at the sending end in order to charge the line at precisely the moment that the desired letter appeared in the aperture; his opposite number at the receiving end would observe the moment when the pith balls diverged and note the letter appearing in *his* aperture. The system was cumbersome and impracticable for distant communication, but was valuable as a testing ground. With its use, Ronalds was able to confirm that the signal attenuation of an electric charge over long distances was small, and no transmission delay was experienced. He also devised a very sound method of insulation for his buried cables, which he used later. He insulated the copper wire with lengths of glass tube, which he encased in soft pitch in long wooden troughs. Many years later, in 1862, a length of his line was recovered from a Hammersmith garden and found to be in perfect condition. A short specimen is preserved at the Science Museum in London.

Probably the last of the experimenters to make use of electrostatic potential for a working telegraph system was the Honorable Harrison Gray Dyer of New York, who set up a telegraph on Long Island. A record of the US Supreme Court made during one of the many legal actions pursued by Samuel F.B. Morse concerning the recognition of his system notes that,

In 1827 or 1828 he [Dyer] is proved by Cornwall to have constructed a telegraph on Long Island race course by wires on poles, using glass insulators . . . he used common electricity and not electromagnetism, and but one wire which operated a spark, which after going through paper chemically prepared, so as to leave a red mark on it, passed into the ground without a return circuit [5, 6].

2.2 Electrochemical devices

With the availability of convenient sources of electric potential provided by the invention of the electric pile by Alessandro Volta in 1800, experimenters now had an alternative to electrostatic generators and the Leyden jar, which could be used for communication purposes. The first practical outcome was the reappearance of Don Salvá with his electrochemical telegraph in 1804, now using a voltaic pile, as these early versions of the battery were known, as a source of electricity. He proposed to use the visual effects of the electrolysis of water to determine whether an electric current had reached its destination at the end of a pair of signal wires, by immersing the wires in a flask of water and observing the bubbles of hydrogen and oxygen that emerged. In his actual installation he used an earth return and a single transmission wire. The use of an earth return (strictly speaking, a 'water return') had been recorded since 1804 through the work of Giovanni [7], who sent a current through a wire supported on the masts of boats across Calais harbour, with a second wire terminating in a copper plate immersed in the water at either end [8]. Salvá's invention was largely ignored by the European technological fraternity at the time, and it is not known whether the more successful and similar devices constructed by Sömmerring were inspired by Salvá's work. Samuel Thomas Sömmerring was a member of the Munich Academy of Sciences and, according to Dr Hamel [9], was requested by the Bavarian authorities to develop a telegraph system for the military. This was seen as vital by the authorities, all too aware of the contribution made by Chappé's semaphore system in assisting Napoleon's successful campaign to thwart an invasion of Austrian troops across the Isar river into Bavaria in 1809. Sömmerring's practical working system is shown in Figure 2.1. It consisted of two major parts: a transmitter consisting of a source of electricity (a voltaic pile) connected by two leads terminating in pins, one a common return, and the other

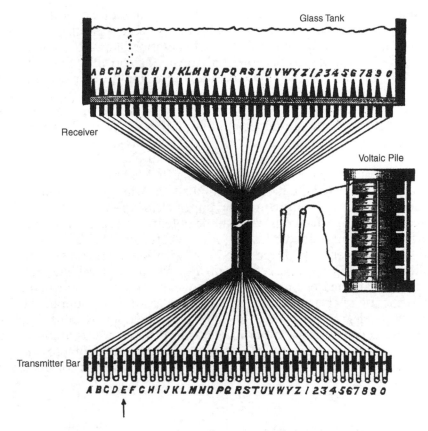

Figure 2.1   Sömmerring's electrochemical telegraph
Source: Hamel, Journal of the Royal Society of Arts 7, 1859

going to one of 35 alphabetically labelled terminals arranged on a wooden bar; and a receiver consisting of a glass tank filled with water and with 35 gold points aligned along its bottom. Each of the points was connected on the outside of the tank to the transmitting wires and labelled alphabetically, as in the transmitter bar. Upon connecting a given named terminal to a voltaic source the circuit is completed and current flows through the correspondingly labelled point in the receiving tank. Electrolysis ensues and is detected by observing bubbles rising from that point, the bubbles are more copious from the negative terminal, where hydrogen is released. This fact could be used to reduce the number of terminals required by reversing the battery connections, and although Sömmerring mentions this he did not make use of it in any of his experimental systems.

At about the time that Sömmerring was busy in Europe with electrolysis as a telegraph detector, Dr J.R. Coxe of Philadelphia was considering a similar scheme as 'a possible means of establishing telegraphic communication with much rapidity, and, perhaps, with less expense than any hitherto employed'. He described how he would transmit, through a key, electric potential to the far end of a line where it would be used for the decomposition of, for example, water or salts, and that with a suitable coding arrangement he could use this phenomenon for the communication of words, sentences or figures from one station to another [10]. He does not appear to have followed this up, however, and no record exists of any further work by him on this subject.

Much later, in 1843, R. Smith, a lecturer in chemistry at Blackford, Scotland, constructed a telegraph in which the electrical potential at the end of a line affected the movement of a stylus, which caused changes in a chemical contained in a moving paper strip upon which the stylus rested. The change in colour when a line potential was present provided a moving record of a coded message sent down the line [11]. The technique had been used earlier by Dyer in 1827, as mentioned previously, when in addition to using the observation of a spark at the receiving end, the spark also affected chemically prepared paper leaving a red mark [6]. While the use of chemically impregnated paper found no support at that time as a telegraph detector (in 1843 the development of the needle telegraph had eclipsed all other methods in Britain), it was applied later by Bain, Davy, Morse and others as a useful method of providing a permanent record of coded electrical signals, developments considered in the next chapter.

2.3 The 'needle' telegraphs

None of the early electrostatic or electrochemical devices resulted in any usable telegraph systems, and remained inferior to the established mechanical semaphore systems already in place. Ronalds attempted to interest the government of the day in his invention, and in 1816 entered into a lengthy correspondence with the Admiralty in which he requested the opportunity to demonstrate his apparatus as a replacement for the mechanical semaphore, still in use between London and the south coast ports. Eventually he received a brief reply from Sir John Barrow, Secretary to the Admiralty, informing him that 'telegraphs of any kind are now wholly unnecessary and no other than the one now in use will be adopted' [12]. The Admiralty did not replace its visual telegraphs until 1851, when the electric telegraph was well advanced. Despite this rebuff, Ronalds continued to take an interest in telegraph matters, and contributed to the work of the Society of Telegraph Engineers and various journals over the next few decades. Personal recognition came in 1870, when he was knighted for his work on the telegraph.

The first glimmerings of a viable electrical system came with Oersted's discovery in 1819. He found that a magnetic needle is deflected at right angles to a wire carrying an electric current when the wire is located close to the needle. Ampère in the following year proposed an electromagnetic telegraph based on the use of many pairs of conductors (he did not consider an earth return) and magnetised needles, with a pair of wires for each letter of the alphabet [13]. He did not pursue the idea further – but others did, and a flood of applications followed (a fairly complete list is given in Table 2.1). Considerable impetus was given to these new developments by J.S.C. Schweigger's invention of the galvanometer or multiplier, which increased the sensitivity of the process by enclosing the needle in a coil of many turns of wire through which a current is passed [14].

Possibly the first to combine Schweigger's multiplier with Oersted's discovery was the Russian Baron Pawel Lwowitsch Schilling, an attaché at the Russian Embassy in Munich in about 1820. He had become acquainted with Sömmerring during a previous visit to Munich in 1805 and was well aware of the latter's work on telegraphy. Schilling's experience with 'control at a distance', with the aid of electricity, was applied during his service in Munich when he developed techniques for exploding gunpowder mines under water as a means of countering the threat to Napoleon's invasion of Russia. He successfully exploded mines under the Neva river in 1812 and, as a captain in the 3rd Sumsk regiment, took part in the Russian invasion of France, and while in Paris he continued his experiments on exploding mines, this time under the Seine. Returning to Munich, he turned his attention to 'signalling at a distance'. He had previously helped Sömmerring to develop his electrochemical telegraph in Munich and assisted with demonstrations of his apparatus, and so was familiar with the requirements of telegraphy. He was also aware of the work of two German scientists, Carl Gauss and Wilhelm Weber, who in 1833 had operated an electromagnetic telegraph in Göttingen 'consisting of a double line of wires carried over the houses and steeples of Göttingen' [9] for their experiments in the transmission of electricity, but who could not afford the time to pursue their telegraphic researches any further [15].

Schilling's telegraph consisted of a Schweigger galvanometer carrying, in a suspension cord just above the needle, a disc painted black on one side and white on the other, which enabled the rotation of the needle to be seen clearly when current is passed though the coil (Figure 2.2). To render the instrument more sensitive the restraining effect of the Earth's magnetic field was nullified by means of an ingenious method devised by Ampère and known as 'astatic operation'. In the normal design of a galvanometer the needle turns under the influence of an applied magnetic force against the pull of the Earth's magnetic field. In the astatic galvanometer two needle magnets are mounted together on a common spindle with their poles opposed (Figure 2.3). The magnets are nearly, but not

Table 2.1 Electric telegraphs to the year 1850

Year	Inventor	Type	Features
1753	'C. M.'	Electrostatic	26 wires; pithball detector
1767	Joseph Bozolus	Electrostatic	Spark detector
1773	L. Odier	Electrostatic	Multi-wire
1774	Georges Le Sage	Electrostatic	24 wires; pithball detector
1777	Alessandro Volta	Electrostatic	Spark detector
1787	M. Lomond	Electrostatic	One wire, spark detector & code book
1790	Gustave Chappé	Electrostatic	Synchronous operation
1794	Reusser	Electrostatic	36 wires, spark detector
1795	Don F. Salvá	Electrostatic	26 wires; Leyden jar
1797	Tiberius Cavallo	Electrostatic	One wire, spark & code book
1800	Don F. Salvá	Electrochemical	Multi-wire; electrolysis detector
1802	W. Alexander	Electromagnetic	Multi-wire; electric relay detector
1809	S. T. Sömmerring	Electrochemical	Multi-wire; electrolysis detector
1810	J. R. Coxe	Electrochemical	Electrolysis detector
1811	J. S. Schweigger	Electromagnetic	Galvanometer detector
1813	J. R. Sharpe	Electrochemical	Electrolysis detector
1816	F. Ronalds	Electrostatic	26 wires; pithball detector
1820	A. M. Ampère	Magnetised needle	25 pairs of lines
1824	E. Smith	Electrostatic	24 wires; spark detector
1824	W. Sturgeon	Electromagnetic	6 wires; electric relay detector
1826	S. Porter	Magnetised needle	Multi-wires in glass tube
1827	Harrison Gray Dyer	Electrostatic	26 wires, spark and litmus paper
1828	St Amand Tde	Electrovoltaic	Electroscope detector
1830	A. Booth	Electromagnetic	Electric relay detector
1831	J. Henry	Electromagnetic	Bell detector
1832	P. L. Schilling	Magnetised needle	Six needles and coding
1832	S. F. B. Morse	Electromagnetic	Single wire and coding
1833	Gauss and Weber	Magnetised needle	Single wire and coding
			Electromagnetic generator
1836	K. Steinheil	Magnetised needle	Two needles and coding
		Electromagnetic	Acoutistic telegraph using two bells
1837	E. Davy	Magnetised needle	Single wire and coding
1837	Cooke & Wheatstone	Magnetised needle	Five needles used as pointers
1837	W. Alexander	Magnetised needle	30 wires; uncovers letter windows
1837	Du Jardin	Electrovoltaic	Single wire; electroscope detector
1837	M. Ponton	Magnetised needle	8 wires: display cards and code
1837	L. Magrini	Magnetised needle	Three needles used as pointers
1837	Stratingh	Electromagnetic	Relay detector operating 2 bells
1838	E. Davy	Electrochemical	Electrolysis detector
1838	Coooke and Wheatstone	Magnetised needle	Two needles and coding
1839	De Heer	Electric shock	Operator detecting through fingers
1840	Cooke and Wheatstone	Electromagnetic	ABC telegraph; ratchet mechanism
1843	R. Smith	Electrochemical	Electrolysis detector
1844	L. Bréguet	Electromagnetic	ABC telegraph
1846	A. Bain	Magnetised needle	Single needle; recording telegraph
1846	W. Siemens	Electromagnetic	ABC telegraph; ratchet mechanism
1846	R. E. House	Electromagnetic	Printing telegraph
1846	L. Bréguet	Magnetised needle	Two needles and Chappé code
1847	E. Stöhrer	Electromagnetic	ABC telegraph; ratchet mechanism
1848	W. Henley	Magnetised needle	Two needles used as pointers

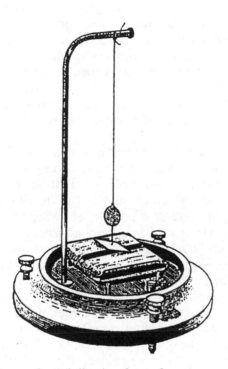

Figure 2.2   The indicator for Schilling's telegraph system

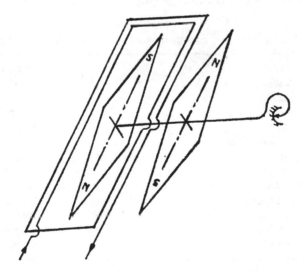

Figure 2.3   Principle of the astatic galvanometer

quite, equal in strength so that by applying the magnetic field to only one of them the rotation is made much easier, and the nullifying action of the Earth's magnetic field is almost cancelled out. To control the damping on the unrestrained movement of the astatic needle pair, Schilling introduced a small vane carried on the suspension cord beneath the needle (not shown in Figure 2.2), which rotated in a trough of mercury. The Schilling needle detector was an extremely sophisticated device, details of which were copied by many later experimenters, including Cooke with his railway telegraph.

Schilling used the sequence of movements of the disc produced by an incoming signal, in accordance with a pre-arranged binary code, to indicate the different letters of the alphabet. Letting 'b' equal a current in one direction, causing the black side of the disc to be shown, and 'w' equal a current in the reverse direction, showing the white side, he put A = bw, B = bbb, C = bww etc.[1] In a more advanced version he used a six-needle telegraph with a sending arrangement consisting of a keyboard, like that of a piano, having 16 keys in pairs of one black and one white key. Each key on being depressed made contact with a battery, as indicated in the circuit diagram shown in Figure 2.4. An additional galvanometer served as an alarm device so that when its needle was deflected it set in motion a clockwork mechanism which rang a bell, a method identical to that used in Sömmerring's system. The telegraph operator pressed a pair of keys of the same colour, one of which was connected to the common return wire shown in the diagram. The second key acted to control the direction of the current sent down the line, and so determined whether the black or the white face of the suspended disc was displayed.

Table 2.2 *Francis Bacon's* Exemplum alphabeti biliterarii *(1605)*

A	aaaaa	N	abbaa
B	aaaab	O	abbab
C	aaaba	P	abbba
D	aaabb	Q	abbbb
E	aabaa	R	baaaa
F	aabab	S	baaab
G	aabba	T	baaba
H	aabbb	U	*not used*
I	abaaa	V	baabb
K	*not used*	W	babaa
J	abaab	X	babab
L	ababa	Y	babba
M	ababb	Z	babbb

[1] The three-position, two-letter binary code developed by Schilling has a long history. It goes back to Francis Bacon's *Exemplum alphabeti bilitterarii* (1605), in which he proposed the five-position, two-letter code shown in Table 2.2, and used it for military applications.

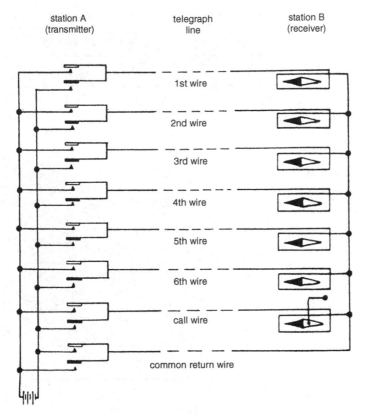

station A
(transmitter)

telegraph
line

station B
(receiver)

1st wire

2nd wire

3rd wire

4th wire

5th wire

6th wire

call wire

common return wire

Figure 2.4   Circuit diagram for Schilling's six-needle telegraph
Source: Yarotsky, Telecommunications Journal, 1982, reproduced with permission of the
International Telecommunications Union

Schilling designed this system for the Cyrillic alphabet; his code table is given in Table 2.3. For the Roman alphabet only five keys plus a common wire key would have been needed.

The six-needle telegraph was widely demonstrated throughout Europe in 1832 by Schilling and other university lecturers, and although it was never installed as a working commercial or military system, it implemented many of the techniques later to be incorporated in commercial telegraphs. At one of the lectures describing Schilling's apparatus, given by Professor G.W. Muncke at Heidelberg University in 1836, a young man from England, who happened to be in Heidelberg for an entirely different reason, sat in on the talk. He was William Fothergill Cooke, and his attendance at the talk and the demonstration that followed had a seminal effect on the progress of telegraphy in Britain.

Table 2.3 Schilling's six-needle code table (needle movement П = right, Л = left)

	Needle position							Needle position					
	1	2	3	4	5	6		1	2	3	4	5	6
А	П						Ф	Л		Л			
Б	Л						Х		П		П		
В		П					Ц		Л		Л		
Г		Л					Ч			П		П	
Д			П				Ш			Л		Л	
Е			Л				Щ				П		П
Ж				П			Ы				Л		Л
З				Л			Ю		П	П			
И					П		Я		Л	Л			
К					Л		1	П	П	П			
Л						П	2	Л	Л	Л			
М						Л	3		П	П	П		
Н	П	П					4		Л	Л	Л		
О	Л	Л					5			П	П	П	
П			П	П			6			Л	Л	Л	
Р			Л	Л			7				П	П	П
С					П	П	8				Л	Л	Л
Т					Л	Л	9	П		П		П	
У	П		П				0	Л		Л		Л	

2.4 Cooke and Wheatstone

The impression Cooke gained at the Heidelberg talk, with its promise of the transmission of intelligence by telegraph, caused him to abandon his erstwhile ambition (to follow in his father's footsteps in anatomical science) and to 'apply himself to the practical transmission of telegraphic intelligence' [16].[2] This he did with enthusiasm, and within a few weeks had succeeded in constructing his first electric telegraph, which consisted of three magnetised needles controlled by keys at the end of a set of six lines. By depressing the keys in a given order he was able to form 27 separate signals which could be transmitted and recognised at the receiving end as letters of the alphabet. He did not consider making use of a sequence of needle movements to extend this number of indications, as Schilling and others had. In the practical implementation of his invention he ran into difficulties (he had no technical experience in the

---

[2] It is unlikely that Dr Cooke senior raised any objections to this sudden change of course since he was well acquainted with Ronalds' telegraph, having on several occasions cooperated with Ronalds in experiments in his garden at Hammersmith.

business of prototype development, and next to no knowledge of electricity) and sought help, first from Michael Faraday, with whom he was acquainted and who encouraged his endeavours, and then from Dr Paul Roget (author of *Roget's Thesaurus*), who referred him to Charles Wheatstone, a professor at King's College, 'who was knowledgeable in these matters'. This was a most fortuitous suggestion: the two investigators pooled their resources and produced a series of inventions which firmly established commercial telegraphy in the Britain of the 1840s.

Charles Wheatstone had been appointed to the chair of experimental physics at King's College in 1834, and was to stay there until his retirement. His major researches were in sound and light, but by the time of his appointment he was already developing ideas for the transmission of information by telegraphic means. Wheatstone's entry into partnership with Cooke brought with it a much needed scientific competence. Cooke was essentially the business partner of the two and, although he had certain ideas about the use of the telegraph in the public domain, particularly on the railways, he did not understand the scientific basis upon which the telegraph worked. This Wheatstone provided. He was familiar with the relationship between current, movement of the needle and the length (resistance) of the controlling wire. His knowledge of Ohm's law and the discussions he had in his meeting with the American scientist Joseph Henry during the latter's visit to London in 1837 [17] enabled him to suggest methods of increasing the sensitivity of the process, principally by the use of many turns of fine wire in the coils surrounding the needles, and in an astatic operation for the signalling device which was eventually taken up by the railway. Cooke's contribution was, however, a vital one, and his energy and business acumen were responsible for the rapid development of their inventions during the following decade. This is reflected in the terms of the partnership – Cooke took 10 per cent of the profits as a business fee, and the remaining 90 per cent was divided equally between them.

The 1840s were a period of rapid growth for the railways, which Cooke quickly saw as a way in which his telegraph lines could gain the major advantage of a protected route as well as offering signal opportunities for the railway companies. It was through his father that he was introduced to Charles Vincent Walker of the Liverpool & Manchester Railway, who acquainted him with some of the company's communication problems. In particular, a need existed for reliable communication between a train and a distant stationary engine, used for rope-hauling a train through a tunnel or up an incline; and also to avoid the possibility of a head-on collision between trains travelling in opposite directions on the same section of single-line track. While this did not immediately lead to telegraph equipment being installed on the railways (the railway company was to continue to use its pneumatic warning whistles for some time to come), it did provide an opportunity for Cooke, with his partner Wheatstone, to perfect their equipment ready for its future adoption by the railways [18]. This occurred in 1838, when a line of telegraphs, the first completely developed telegraph system used for commercial purposes, was installed for the Great Western Railway following Brunel's recommendation, between Paddington and West Drayton, a distance of about 21 km. This was working as far as Hanwell by the following year and extended to Slough by 1843. Earlier, Cooke and Wheatstone had secured a patent, signed by William IV in June 1837, on 'Improvements in Giving Signals and Sounding Alarms in Distant places by means of Electric Currents transmitted through Metallic Circuits'. This patent was comprehensive enough to prevent a number of contemporaries, who were anxious to enter the telegraphy market, from competing in any effective way. They included Edward Davy, a surgeon, who had experimented with the electric telegraph and in 1836 had laid down a line of copper wires in Regents Park, successfully demonstrating his system to two railway companies, before taking out a patent in 1838 [19]. Davy emigrated to Australia before following this up, and thereby ceased to be a rival to the Cooke and Wheatstone concern. He did, however, leave behind him one important contribution: the telegraph relay, which he called the 'electrical renewer'. This invention allowed the lengthy lines to be broken up into short sections, each of which received a new signal through the relay of the previous section and in turn transmitted a copy of the received signal with its own relay and a local source of energy, thus 'renewing' the signal at each stage of the extended line. This device removed the necessity of using large currents to compensate for the leakage that inevitably accompanied long lines. Davy's 1837 patent for his relay was bought out by Cooke's company, the Electric Telegraph Company, a few years later for £600. Another telegraph invention, first shown at the Society of Arts Exhibition in Edinburgh in 1837, was William Alexander's electric telegraph. This was a multi-needle telegraph device consisting of 30 magnetic needles, each carrying a light-weight screen which was rotated out of view when affected by a current-carrying conductor surrounding the needle. The rotation of a screen revealed a printed letter, as shown in Figure 2.5. The 30 copper wires and a common return wire connected the transmitting keyboard with a panel of such indicators. Professor Alexander initially tested the device over 7 km of connecting wire in a chemistry lecture room at the University of Edinburgh. It was never used to provide a commercial service, and Alexander eventually withdrew his pending claim for patent rights, 'acknowledging the superiority of Cooke and Wheatstone's plans' [20].

A more serious contender to Cooke and Wheatstone's application of the magnetic needle to telegraphy was Professor Karl August Steinheil of Munich, who in 1836 became one of the first to use a magneto-electric generator as the energy source, following the discovery of induced

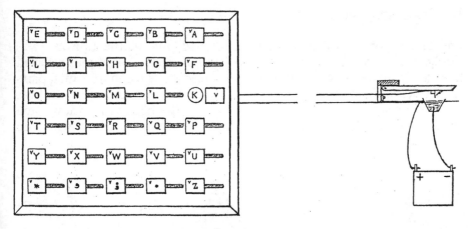

Figure 2.5   Alexander's electric telegraph
Source: Official Catalogue of the Great Exhibition, 1851

current by Faraday in 1831. In Steinheil's telegraph system two needles were actuated, one by positive and the other by negative currents, via only one connecting wire. Each needle carried a small reservoir of ink and a pen which, upon being depressed, marked a strip of paper drawn along by a clockwork mechanism. Steinheil used a coding system in which marks of equal length were drawn either above (positive current) or below (negative current) a line. By using two to four such marks, the system was capable of distinguishing $2^4 + 2^3 + 2^2 + 2 = 30$ letters. The full alphabetical code (Table 2.4) bears a striking resemblance to the later Morse code; the system was capable of printing just over six words per minute. In a second version of his equipment, Steinheil arranged for two bells of different notes to be struck, one having a low note corresponding to the arrival of a negative current and the other having a high note for positive current, again anticipating Morse, who was developing his sounder in the United States at that time. Steinheil's telegraph system was widely used in Germany and other countries of Continental Europe in the following decade. In 1849 he helped establish a railroad telegraph system between Munich and Augsburg, and in the same year was appointed to organise the Germano-Austrian Telegraph Union, for which he set up a 142 km line between Munich and Salzburg, together with several other lines [21]. He received a further call to organise a similar system for the Swiss government in 1852, but had however by this time adapted his equipment to Morse's code system which he advocated for use 'throughout all of Europe' [22].

2.5 Telegraphy on the railways

The first experimental telegraph to be tried out on the railways was Cooke and Wheatstone's five-needle device in 1837, sometimes referred to as the 'hatchment' telegraph. The five-needle telegraph, in which pairs of needles combined to point to a selected letter (Figure 2.6), was extremely easy to use. To enable a return to the neutral position, each needle was weighted at one end. The telegraph was a sophisticated

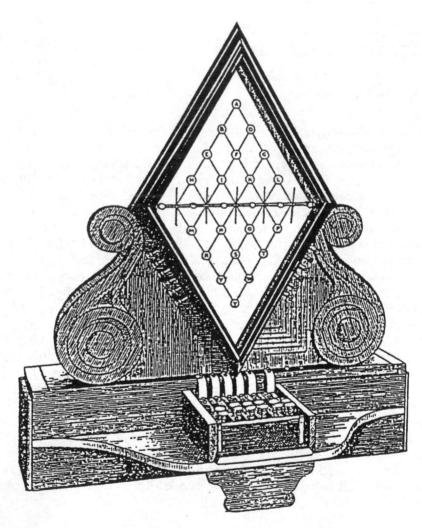

Figure 2.6   Wheatstone's five-needle 'hatchment telegraph', first used on the Euston-Camden railway line in 1837
Source: Official Catalogue of the Great Exhibition, 1851

arrangement containing two new features added by Wheatstone. The first was astatic operation, previously applied by Schilling, but further improved by winding the coil surrounding the magnetised needle on a ferromagnetic core instead of on a non-magnetic former. The second significant feature lay in the connections between the five controlling switches and the coils, which were based on a permutating principle whereby the telegraph wires were connected in various ways to form different circuits for the different signals, thus increasing the range of symbols which could be transmitted over a pair of wires. The telegraph ran from Euston station to Camden Town, a distance of about 2 km. As with Schilling's telegraph, the operator needed to press two keys to transmit a letter, and for each key had to select the correct positive or negative potential, a selection carried out automatically by the choice of black or white keys in the Schilling telegraph. It is interesting to compare the coding arrangements used by the two inventors: Schilling's code for his six-needle instrument is given in Table 2.3, and that for Cooke and Wheatstone's code for their hatchment telegraph in Table 2.5. The former was probably simpler to operate since the polarity selection would be easier to determine by always choosing pairs of keys of the same colour, and the system was flexible enough to allow further symbols to be added, such as the numerals indicated in Table 2.4. However, at the receiving end Cooke and Wheatstone's system provided an immediate identification for a given letter, as the needles were arranged vertically in a diamond pattern, so that the two needles could together point to the letter selected. A similar arrangement was devised by Magrini, who used three 'pointing needles' to achieve the same ends [23].

The hatchment telegraph was not long in operation, however. Within three years the wires were damaged and the apparatus ceased to work. The five-wire system proved to be too expensive and, before the next railway installation was carried out, a simpler two-needle instrument, requiring a coding book, was brought into use (Figure 2.7). This was installed on the Paddington to West Drayton line mentioned earlier, and fully opened by 1839, leading to widespread application in the next few years along the railways radiating out of London to many parts of the country. The first of these was the Blackwall Tunnel Railway, which opened in 1840. This was a rope-hauled railway in which an endless rope or cable between the Minories and Blackwall was driven by a stationary engine. A major problem in this particular short stretch of railway was that of safety. When the engine began to pull the rope along its 6.5 km length, all the carriages attached to it moved simultaneously. It only needed one carriage to become detached for the inevitable collision to take place and, since the carriages were continually being hooked onto and released from the cable to permit passengers to enter or alight, this remained an ever-present hazard. The installation of the Cooke and Wheatstone telegraph system along the

Table 2.4 Steinheil's two-needle code table (C, J, Q, U, X and Y not used)

A	‾‾‾ / – –	L	‾‾‾ / – –	0, (O)	– – –		
B	– – / – –	M	– – – /	1	– – – / –		
D	– / –	N	– – /	2	– – – / –		
E	‾‾‾ / –	O	‾‾‾ / – – –	3	– – – / –		
F	– – / –	P	– – / – –	4	– – – /		
G	– – / –	R	‾‾‾ / – –	5	– / – – –		
H	– – – – /	S	– – / – –	6	– / – – –		
Ch	‾‾‾ / – – – –	T	– / –	7	– / – – –		
Sch	– – / – –	V	– – / –	8	‾‾‾ / – – –		
I	– – /	W	– – / – –	9 (G)	– – / –		
K	– / – –	Z	– – / – –				

line to warn of accidents and stoppages proved highly successful, and on at least one occasion the press commented on its usefulness in transmitting news of an accident on the line, so preventing a fatal collision.

By 1845 the telegraph lines extended to railway stations on the Great Western Railway at Falmouth, Southampton, Liverpool, Edinburgh and Holyhead [24]. The east coast main line to Scotland was equipped with telegraph apparatus in 1846–47, and London was linked to Dover in 1846. This latter contract provided Cooke with the opportunity to form his own company to carry out the work, the Electric Telegraph Company, which acquired not only the Wheatstone patents but also

Table 2.5 Cooke and Wheatstone's code table for the 'hatchment telegraph' (C, J, Q, U, X and Z not used)

	Needle position						Needle position				
	1	2	3	4	5		1	2	3	4	5
A	R				L	M	L	R			
B	R			L		N		L	R		
D		R			L	O			L	R	
E	R		L			P				L	R
F		R		L		R	L	R			
G			R		L	S		L	R		
H	R	L				T			L		R
I		R	L			V	L			R	
K			R	L		W	L				R
L				R	L	Y	L				R

those of Alexander Bain and other inventors. This Cooke was able to do when, in 1843, he bought out Wheatstone's share for a fixed sum, leaving Wheatstone to develop his scientific ideas outside the company, one of which was the enormously successful printing telegraph (discussed later). By 1852 the company had become responsible for establishing over 6000 km of telegraph lines in Britain, much of which was laid down for the railway companies. Of the seventeen metropolitan offices opened for the company, eight were at railway termini and were open 'day and night' [25]. Although some earlier lines had been buried (the West Drayton wires were encased in a hollow iron tube below ground), this practice was discontinued in later construction. The problems of insulating the long telegraph lines had been solved by suspending them alongside the track on porcelain cups spaced at frequent intervals.

Some of these lines were used for a specialised railway telegraph which made an enormous contribution to railway safety. This was a single-needle, three-position instrument capable of indicating 'line clear', 'train on line' and 'line blocked', and came into use in about 1867 [26]. It was still in use in the 1930s, and made possible the somewhat haphazard time-interval method of regulating traffic on the line, whereby a train was allotted a specific time to travel a given section. This was replaced by a more positive space-interval system, known as 'block working', in which only when there was a clear section of the line would the signalman concerned allow a train to proceed. A similar block system was installed in the German railways, commencing with the Austrian–German railway in 1870, designed by Siemens and Frischen.

In America some use was made of an English patent for electric

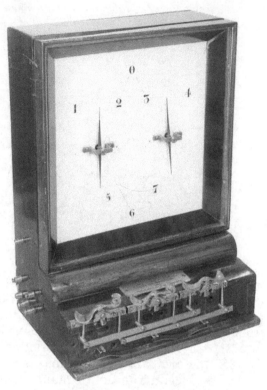

Figure 2.7   Cooke's two-needle telegraph
Source: With acknowledgements to British Telecom Archives

telegraph communication between trains on the same line, using the railway lines themselves for the electrical connection. The patent was registered by a William Bull of London in 1860; rail sections were to be used 'to indicate at a station the progress of a train'. The idea was taken up in 1867 by William Robinson, inventor and member of the American Railway Association, who accompanied his installation with an intensive advertising campaign. Robinson's 'Wireless Electric Signals' (Figure 2.8) was a closed-circuit system in which the wheels of the train provided an electrical connection between the pair of rails when the train entered a control section, automatically setting the visual signals at 'danger' (the red light M in Figure 2.8), and effecting section block working for the train system. To isolate each section of the line the separate sections were insulated from one another by wooden splices so that the progress of a train could be indicated in terms of 'location, direction, rapidity, and length; thus all necessary information regarding moving trains is automatically announced every few minutes at the stations' [27]. The system was initially installed on the Philadelphia and Erie Railway in 1872, and

Figure 2.8   Robinson's 'Wireless Electric Signals'
Source: Brignano, 'History of Railway Signals', 1981

later copied by a number of other railroads until superseded by the Western Union Morse code system in the 1880s, which had by then became the standard communication and warning system for all the American railways.

While telegraph systems were being applied to the problems of railway signalling in Britain, America and the mainland of Europe, in Britain they were not widely used by the business community or the general public for non-railway business, and made comparatively little impression at first on the public despite an intensive advertising campaign (Figure 2.9). The rash claim that the signal travels at a rate of 280 000 miles in one minute/second (the copywriters were a little unsure on this point!) was clearly at variance with the actual speed of message transmission, which of course included delivery by messenger boy at the receiving end and probably several sets of reception, decoding, recoding and retransmission as the message passed through different telegraph stations. One incident which did help to make the availability of the service more widely known was the well-advertised arrest of a suspected murderer by the name of Tawell in Slough, made possible by the telegraph network. He was seen to board the Paddington train at Slough, and a full description was telegraphed through to the main line station. One of the coding peculiarities of the telegraph system at that time was its limitation of transmitting only 20 letters of the alphabet, (C, J, Q, U, X and Z were not available), thus presenting a problem to the telegraph operator responsible for transmitting Tawell's description as 'a man in the garb of a Quaker'. With commendable resourcefulness the operator sent the word 'Kwaker', which was duly recognised at the receiving end, and the suspect was arrested as he left the train [28].

2.6 Dial telegraphs

Skill in operating the needle telegraphs was generally acquired by a period of work with the railways, and unless a well-trained operator was available the needle systems were difficult and expensive to use. Consequently a number of telegraph companies developed simpler systems that did not need highly trained operators. They were, however, much slower in use and, although valuable for light traffic, where the line was a busy one and a trained operator could be employed, the two-needle telegraph was preferred.

Early in their business cooperation, Wheatstone improved an early invention of Cooke's into just such a simple system: the first commercial dial telegraph, introduced in 1840, which he named the ABC Telegraph (Figure 2.10). This contained a dial inscribed with the letters of the alphabet around its periphery. It revolved by clockwork, and could be stopped at any part of its revolution when the selected letter shown on the dial was opposite an aperture in a fixed plate. This was similar in principal to Ronalds' design described earlier, but whereas Ronalds used static electricity Wheatstone used electrical impulses from the sending station acting on an electromagnet controlling a ratchet movement. The sender first set the dial to the letter required, as seen through the aperture, either by setting a crank handle or depressing a given telegraph key. This actuated the clockwork mechanism, which, through the ratchet, rotated the dial in one direction and sent out a series of electric pulses, energised by a battery, corresponding in number to the set position of the dial. The pulses passed along the transmission line to operate a relay

THE WONDER of the AGE!!

INSTANTANEOUS COMMUNICATION.

Under the special Patronage of Her Majesty & H.R.H. Prince Albert.

THE GALVANIC AND ELECTRO-MAGNETIC

TELEGRAPHS,

ON THE

GT. WESTERN RAILWAY.

May be seen in constant operation, daily, (Sundays excepted) from 9 till 8, at the

TELEGRAPH OFFICE, LONDON TERMINUS, PADDINGTON AND TELEGRAPH COTTAGE, SLOUGH STATION.

An Exhibition admitted by its numerous Visitors to be the most interesting and ATTRACTIVE of any in this great Metropolis. In the list of visitors are the illustrious names of several of the Crowned Heads of Europe, and nearly the whole of the Nobility of England.

"This Exhibition, which has so much excited Public attention of late, is well worthy a visit from all who love to see the wonders of science."—MORNING POST.

'The Electric Telegraph is unlimited in the nature and extent of its communication; by its extraordinary agency a person in London could converse with another at New York, or at any other place however distant, as easily and nearly as rapidly as if both parties were in the same room. Questions proposed by Visitors will be asked by means of this Apparatus, and answers thereto will instantaneously be returned by a person 20 Miles off, who will also, at their request, ring a bell or fire a cannon, in an incredibly short space of time, after the signal for his doing so has been given.

The Electric Fluid travels at the rate of 280,000 Miles per Second.

By its powerful agency Murderers have been apprehended, (as in the late case of Tawell,)—Thieves detected; and lastly, which is of no little importance, the timely assistance of Medical aid has been procured in cases which otherwise would have proved fatal.

The great national importance of this wonderful invention is so well known that any further allusion here to its merits would be superfluous.

N.B. Despatches sent to and fro with the most confiding secrecy. Messengers in constant attendance, so that communications received by Telegraph, would be forwarded, if required, to any part of London, Windsor, Eton, &c.

ADMISSION ONE SHILLING.

T. HOME, *Licensee.*

Nurton, Printer, 48, Church St. Portman Market.

Under the Special Patronage of Her Majesty

And H. R. H. Prince Albert

GALVANIC AND MAGNETO

ELECTRIC TELEGRAPH,

GT. WESTERN RAILWAY.

The Public are respectfully informed that this interesting & most extraordinary Apparatus, by which upwards of 50 SIGNALS can be transmitted to a Distance of 280,000 MILES in ONE MINUTE,

May be seen in operation, daily, (Sundays excepted,) from 9 till 8, at the

Telegraph Office, Paddington, AND TELEGRAPH COTTAGE, SLOUGH.

ADMISSION 1s.

" This Exhibition is well worthy a visit from all who love to see the wonders of science."—MORNING POST.

Despatches instantaneously sent to and fro with the most confiding secrecy. Post Horses and Conveyances of every description may be ordered by the ELECTRIC TELEGRAPH, to be in readiness on the arrival of a Train, at either Paddington or Slough Station.

The Terms for sending a Despatch, ordering Post Horses, &c, only One Shilling.

THOMAS HOME, *Licensee.*

N.B. Messengers in constant attendance, so that communications received by Telegraph, would be forwarded, if required, to any part of London, Windsor, Eton, &c.

G. NURTON, Printer, 48, Church Street, Portman Market.

Figure 2.9   An early telegraph advertisement, 1845
Source: Marlaad, 'Early Electrical Communications', 1964

Figure 2.10   Wheatstone and Cooke's ABC Telegraph
Source: Fleming, '50 Years of Electricity', 1921

which actuated the ratchet in the receiving device to rotate an indicating dial to the required letter; this dial was returned to its initial state also by a clockwork mechanism. Care was needed by the operators at both ends to ensure that the dials of the instruments pointed to the correct letter. Elaborate rules were devised to ensure that this and other conflicts did not occur; a typical set of such rules for Post Office operators is reproduced in Figure 2.11. In a later instrument Wheatstone used an alternating current generator, operated by a crank handle and located in the base of the instrument. Its effect was to send a series of positive and negative pulses down the line as the handle is turned to rotate the dial, as described above, with the action terminating when the depressed key is reached.

This was one of the last cooperative ventures of the two pioneers – in 1840 they quarrelled on the prior invention of the telegraph. Although their differences were somewhat mollified by the intervention of Sir Marc

POST OFFICE TELEGRAPHS

INSTRUCTIONS FOR WORKING

PROFESSOR WHEATSTONE'S INSTRUMENTS

TO WORK THE TELEGRAPH

(An ABC Instrument consists of the Communicator, the Indicator & the Bell)

1. The handle of the Instrument attached to the Communicator, which causes the armature of the magnet to rotate, must be kept in continuous motion by one hand, while the fingers of the other are employed to manipulate the stops or keys. Care must be taken not to stop the motion until the end of the message and indeed slightly in advance; or the Index at the other end fails to come up to +.

2. A key need not be continuously pressed down, it will suffice merely to touch it: but another key must not be pressed down until the index, or pointer, has arrived at the letter previously indicated.

3. The same key cannot be pressed down twice in succession: to repeat a letter touch the preceding key and without waiting for the arrival of the Index at that letter, touch again the proper key.

4. Before commencing to send a message the Index of the Communicator and Indicator of all the Instruments must point to the cross +.

5. If by inadvertence the Index of the Communicator has been left at a letter, it must be brought to the cross + before the Telegraph is adjusted.

6. The brass switch at the back of the Indicator must always, when the Instrument is not in use, be turned to the letter "A".

7. To call attention for the purpose of sending a message, turn the switch of the Indicator from "A" to "T"; then turn the handle and let the needle pass from + to +, this will ring the bell at the other end of the line. Wait an interval of time sufficient to allow a reply to be sent; if no reply be given, continue to call.

8. The Receiver will notify his attention by repeating the signal — + to +.

9. If the Operator at the other end cannot attend at once, he will send 'W' for 'Wait' to show that he has received the message.

10. A short time must be allowed the Receiver before sending the message, to enable him to put his indicator in accord with his Communicator, if it be found by him to be wrong.

11. At the end of each word the needle must be brought to +.

12. If by accident the needle of the Indicator becomes misplaced, so as to be unintelligible, the Receiver must break in by causing his needle to rotate. The Sender will immediately stop sending. Both Sender and Receiver will then set the needles at + and time must be given for this, and the Receiver will give "R" for "Repeat", and the last word he understood. The Sender will then repeat the message from that word.

13. Every word and every initial letter or part of a word used for abbreviation must be followed by the +.

14. At the end of the message, the needle must be turned round from + to + twice, except when intending to send a second message, then once or once in reply to once round.

15. The Receiver must repeat this double revolution.

16. To signify figures, use the semi-colon and then the + after sending them; the Receiver then knows that the signals mean Figures and not Letters.

Figure 2.11 Post Office rules for operating a Wheatstone ABC instrument
Source: From Post Office Instructions, 1900

Brunel and Professor Daniell as adjudicators, who diplomatically summed up the situation by stating, 'it is to the united labours of the two gentlemen, as well qualified for mutual assistance, that we must attribute the rapid progress which this invention has made since they were associated', in 1846 they went their own way. Cooke became increasingly concerned with managing the expanding networks, while Wheatstone continued to make contributions to the science and technology of telegraph transmission [29].

Dial telegraphs were also being installed on the railways in Germany in 1843. The first, on a short section between Aachen and Ronheide, used an imported Cooke and Wheatstone apparatus and included two additional wires to operate an alarm bell. The system was built to support a rope-hauled railway up a slope towards Ronheide, using a stationary engine – a repeat of the arrangement made by Cooke for the London–Blackwall railway. Main-line telegraphs followed during the next year, using equipment manufactured by Fardeln, Kramer and Siemens & Halske. In France, François Clement Bréguet, a clock and instrument maker, developed his own version for the public telegraph networks, while Bouvillon devised a version which was used on the railway between Paris and Rouen.

The first main-line telegraph connection to be completed in Germany was designed by William Fardeln and installed on the Taunusbahn railway linking Kastel, Beibrich and Wiesbaden in 1844 [30]. Fardeln travelled to London in 1840 and stayed for two years, taking every opportunity to study the telegraph techniques then being developed for the British railways. After his return to his home town of Mannheim, he was invited to construct a system for the small Taunusbahn railway. He based his design on that of Cooke and Wheatstone, and installed it in a charming Black Forest clock casing (Figure 2.12). Unfortunately, although his design for the telegraph worked well, the same could not be said for his line construction. Whereas Cooke hung his wires between insulators mounted on iron poles, the Fardeln line consisted simply of wooden poles containing a saw-cut at the top into which the wire was wedged and then covered with a lead sheet. As a result the current became so weak through leakage to the ground that, before it had travelled 100 km, it would no longer operate the receiving instrument [31].

The Siemens device (Figure 2.13) appeared in 1846, and was a simplification to the extent that the sending and receiving instruments were identical, both worked by the same current. This ensured perfect synchronisation between them, lack of which sometimes occurred with its competing Cooke and Wheatstone design, which never achieved popularity on the Continent. Instead of relying on clockwork, Werner Siemens used a step motor to rotate an indicating pointer (Figure 2.14). When the operator at the sending station closed the circuit, the pointer of his instrument began to rotate, and since the similar device at the

Figure 2.12   Fardeln's dial telegraph
Source: Deutsches Museum Archives, Munich

Figure 2.13   Siemens dial telegraph
Source: Siemens Museum, Munich

receiving station was included in the same circuit it also rotated in synchronism. The dial of the instrument consisted of a series of 30 finger keys, each marked with a letter of the alphabet. When one of these keys was depressed, the pointer of the sending instrument was mechanically arrested and the circuit interrupted, causing the pointer at the receiving instrument to stop at the same point. In later instruments an electric bell was included, and a magneto-electric generator, operated by a handle, supplied power to the step motor mechanism. The Siemens dial telegraph

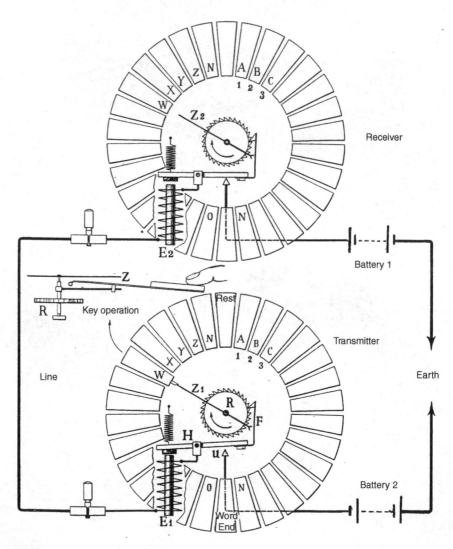

Figure 2.14   Operation of the Siemens dial telegraph
Source: Siemens Museum, Munich

was widely used in Germany on the commercial networks, and in the Russian telegraph network [32], but met its end in 1848 when a speech by the King of Prussia at the opening of the Berlin parliament took 7½ hours to transmit by dial telegraph. A competing operation based on Morse equipment, which had just arrived from Britain, took only 1½ hours to transmit the same speech. The inevitable result was the curtailment in the manufacture of the dial telegraph by Siemens in favour of Morse equipment, produced by the company in 1849, and the subsequent adoption of Siemens' Morse equipment throughout the telegraph network in Germany and elsewhere in Europe. Dial telegraphs continued to be used on the railways in Germany, even long after reliable Morse code equipment began to be available. This was largely due to a new design of a magneto-electric dial telegraph by Siemens in 1856, which eliminated the need for a battery and was rugged and reliable in operation [33]. It was used for railway work in Russia, Sweden and Turkey, together with a similar device, constructed somewhat more cheaply, and patented by August Kramer of Nordhauser. In France, where an alphabetical telegraph designed by Bréguet had been developed in 1844, the system was in use for many years. This was more complex than any of the earlier devices, requiring two cranks to simulate the indications required by the Chappé semaphore system, already in use throughout France in 1844. The Chappé code, with its 64 code variations, was applied to Bréguet's alphabetical system and to other French electrical telegraph systems 'in order to avoid retraining of their semaphore operators', and in the process retarded the development of telegraphy in France for several years [34].

2.7 Codes and ciphers

Despite the widespread use of cheap galvanised iron wire for the early telegraphs, the cost of linking stations together was by far the major expense of setting up a telegraph system, the terminal equipment accounting for only a small part of the installation cost. Financiers looked askance at multi-wire systems, and systems with a single connection wire, coupled with an earth return, were much preferred. However, speed was also important and single-wire systems such as the dial telegraphs, although simple and cheap to use, were painfully slow in operation. In order to identify the 30 or 40 different symbols required in written information transmission, a process of coding the current impulses travelling down the line was seen as essential, and the code used needed to be one easily learnt by the operators.

We need here to distinguish between codes and ciphers. In a *code* the letters of the alphabet are replaced by symbols. An important group of codes used in telegraphy are the two-level or binary codes, of which the

Morse code is the best-known example. In a cipher the letters containing the message are replaced by other letters, either singly on a one-to-one basis, in which case the message will not be shortened, but still affords some secrecy, or with a group of letters represented by a single letter or number. The latter scheme was introduced into the operation of the mechanical semaphore towards the end of its period of use. It requires a cipher-book (often incorrectly referred to as a codebook) and a high order of accuracy in transmission.

The coding system often used in electric telegraphy was a binary code in which recognition was by either the duration or the polarity of the transmitted impulse. The most successful of these relied initially upon polarity, probably because this was simple to apply with the needle instruments then in use. Steinheil and Schilling's codes were of this kind, although their systems were both multi-needle devices. The choice of coding representation varied with the implementer. Steinheil allocated the simplest codes to the letters that occurred most frequently in the German language, and Schilling did the same with the Russian alphabet. Morse is said to have weighed the quantity of lead type found in each compartment of a compositor's tray for each letter, allocating the simpler code combinations to the heavier groups of type. When commercial telegraphy was much advanced and business users could rely on the accuracy of transmission, their thoughts turned to the use of time-saving ciphers which could reduce the actual number of signals transmitted and hence cut the cost of a telegram. Such ciphers were developed in America during the 1880s, particularly by the railroad companies. Each company had its own set of ciphers and a cipher-book in which a single transmitted word or set of numbers could relate to quantity, type and destination of goods. Towards the end of the century, after many railroads (and other private companies [35]) had been using their own (different) telegram ciphers for some time, the American Railway Association promulgated a Standard Cipher Code, compiled from a 'Phillips' code already in fairly general use [36], and this was adopted fairly universally. The cipher-book they produced had almost 750 pages, printed on fine India paper with a thumb index. It covered all contingencies from Accident and Arrival to Wrecks and Worsted, and was 'intended as an open code and its object is brevity, not secrecy; but it may be made secret by previous arrangement between correspondents in the ordinary manner'. The use of such time-saving ciphers became essential for the even more expensive transmission of 'cablegrams' over the submarine cables, and were widely used by the news transmission services (discussed later).

Until Morse began his experiments in 1832, the idea of using a duration-related code for the transmitted electric current had not been considered. When it was, the whole industry began to change. Whereas the needle systems of Cooke and Wheatstone, with their polarity-related indication, ushered in the first effective electromagnetic telegraph systems in the 1840s (a period corresponding, incidentally, to massive expansion of the railway network in Britain), across the Atlantic other systems were on their way. Morse, uninfluenced by the success of needle systems, was developing his single-wire system using an efficient duration-related coding technique which was to dominate telegraph communications for the remainder of the century and into the first half of the next.

References

1   APPLEYARD, R.: 'Pioneers of electrical communication' (Macmillan, London, 1930)
2   FAHIE, J. J.: 'A history of electric telegraphy to 1837' (F.N. Spon, London, 1884)
3   GARRATT, G.R.M.: 'The Electric Telegraph', *in* SINGER, HOLMYARD, HALL and WILLIAMS (Eds): 'A history of technology', vol. 5 (Oxford University Press, Oxford, 1958), pp. 650–51
4   RONALDS, Sir F.: 'Descriptions of an electrical telegraph and of some other electrical apparatus' (1823), Library of the Institution of Electrical Engineers, London
5   PRESCOTT, G.B.: 'History, theory and practice of the electric telegraph' (D. Appleton, New York, 1880), pp. 427–30
6   'Old time telegraph history', *Electrical World*, 1903, **41**, pp. 113–16
7   DIBNER, BERN: 'Aldini, Giovanni', *in* Gillispie, C.C. (Ed.): 'Dictionary of scientific biography' (Charles Scribner, New York, 1970), vol. 1, pp. 107–8
8   ALDINI, G.: 'Essai théorique et expérimentale sur le galvanisme' (Paris, 1804), p. 218
9   HAMEL, J.: 'Historical account of the introduction of the galvanic and Electro-magnetic Telegraph', *Journal of the Royal Society of Arts*, 1859, **7**, 22nd July and 29th July
10 COXE, J.R.: *Thomson's Annals of Philosophy*, February 1816, **7**, p. 162
11 SMITH, R.: *The Practical Mechanic and Engineer's Magazine*, 1846, **1**, p. 239
12 Correspondence held in the Ronald's Collection at the Institution of Electrical Engineers, London.
13 AMPÈRE, A.M.: *Journal für Chemie und Physik*, 1820, **29**, pp. 275–81
14 SCHWEIGGER, J.S.C.: 'Fortsetzung dieser Vorlesung in der Versammlung in der Gesellschaft zu Halle', *Journal für Chemie und Physik*, 1821, **31**, pp. 7–17
15 FEYERABEND, E. Von: 'Das hundertjährige Jubiläum der elektrischen Telegraphie', *Elektrotechnische Zeitschrift*, 1933, **46**, pp. 1109–11
16 COOKE, W.F.: 'The electric telegraph; was it invented by Professor Wheatstone?', vol. 2 (W.H. Smith, London, 1866), p. 14
17 HENRY, J.: 'Journal of European trip 1837' (1837), manuscript held in Archives at the Smithsonian Institution, Washington D.C.

18   COOKE, W.F.: Letter, 1st April 1837, *in* WFC papers, vol. 12, Archives of the Institution of Electrical Engineers, London.

19   'Davy, Edward', *in* 'Dictionary of national biography', (1888), vol. 14, p. 185

20   COOKE, W.F.: Letter, 2nd December 1837. *in* WFC papers, vol. 1, Archives of the Institution of Electrical Engineers, London.

21   SCHORMAIER, M.: 'Telegraph und Telephon in Bayern' (1886)

22   Letter from Steinheil to Schaffner, 9th July 1854, *in* SCHAFFNER, T.P.: 'The telegraph manual' (Pudney & Russell, New York, 1859)

23   DAWSON, K.: 'Electromagnetic telegraphy; early ideas, proposals and apparatus', *in* 'History of technology', vol. 1 (Mansell Publications, London, 1976), pp 113–41

24   COOKE, W.F.: 16th January 1845, *in* WFC papers, vol. 2, Archives of the Institution of Electrical Engineers, London

25   LARDNER, D.: 'The electric telegraph popularized' (Lockwood, London, 1855), p. 273

26   LEACH, M.E.: 'The growth of the electric telegraph on the railways of Britain', IEE Professional Group S7 Summer meeting, Twickenham, 1st–3rd July 1988

27   BRIGNANO, M.: 'The search for safety: a history of railway signals' (Union Switch & Signal Division, American Standard Inc., Pittsburgh, 1981)

28   TEGG, W.: 'Posts, telegraphs, telephones' (William Tegg, London, 1878), pp. 182–6

29   BOWERS, B.: 'Sir Charles Wheatstone FRS 1802–1875' (HMSO, London, 1975)

30   FEYERABEND, E. 'Der Telegraph von Gauss und Weber im Werden der elektrischen Telegraphie' (Reichspostministerum, Berlin, 1933)

31   WEBER, M.M. von: 'Das Telegraphen und Signalwese der Eisenbahn' (Weimar, 1867)

32   SIEMENS, C.W.: 'Electrical instruments and telegraphic apparatus', *in* 'Record of the International 1862 Exhibition' (R. Mallet, London, 1862)

33   HEINTZENBERG, F.: 'Zur Geschichte des Doppel-T Ankers', *Siemens Zeitschrift*, 1940, **1**, pp. 1–5

34   DAUMAS, M. (Ed.): 'A history of technology and invention, vol. III, 1725–1860' (John Murray, London, 1979), p. 383

35   MORGAN & WRIGHT: 'Cable code' (Morgan & Wright, Chicago, 1894)

36   AMERICAN RAILWAY ASSOCIATION: 'The standard cipher code of the American Railway Association' (American Railway Association, New York, 1906)

Chapter 3

Commercial telegraphy

Only a few years after Oersted's discovery that an electric current pro-
duces a magnetic field, another practical demonstration of the effects of
a unidirectional electric current was being shown at the Royal Society of
Arts in London. This was William Sturgeon's soft-iron electromagnet,
invented in 1825, and used in the next major advance in the development
of telegraphy [1]. The implementation of a device whereby the flow of
current in the coils of an electromagnet could effect mechanical movement by the attraction of an iron pole piece must have been well known
to the telegraphy experimenters in the 1830s. Yet it remained largely
unused while efforts were being made by Cooke, Wheatstone, Steinheil
and others to establish a working system based on magnetic needles.
Perhaps it was because these efforts were so successful that the greater
potential of the electromagnet was not realised until well into the 1840s.
It was not, however, an electrical experimenter but the Professor of
Literature of the Arts of Design at the University of New York who was
to make this advance.

3.1 Morse and single-line working

Professor Samuel Finley Breese Morse, already a well-known painter
(two of his portraits may be seen today in the White House), was returning from a trip to Europe on the packet-ship *Sully* when he fell into
conversation with a Professor Jackson, a celebrated geologist, on the
subject of electricity and magnetism. The year was 1832. It was a long
voyage home from Le Havre to New York (it took almost a month on the
packet-ship that year). Apart from some knowledge of chemistry in the
composition of electric power cells, gained from lectures he had attended
under Professor Silliman in his college days, the subject was new to
Morse and exercised his imagination vividly. Once home, he occupied

himself with a series of experiments at his university with the help of a colleague, Professor Gale, a chemist, which resulted in a crude but working telegraphy apparatus capable of communicating over a distance of about a kilometre. Morse's initial idea was to arrange coded lead type along a wooden stick to transmit words. Each 'coding stick' would thus possess a unique series of spaces and 'teeth', representing a coded version of the letters, and telegraph operators would draw these sticks past a pair of electrical contacts, connected between battery and line, thereby transmitting a series of pulses to a receiver at the other end. He quickly abandoned this doubtful mechanism, substituted a simple spring-loaded switch – referred to from the very beginning as the 'Morse key' – and directed his efforts towards perfecting the equally unique receiving device, shown in Figure 3.1. It was a device constructed out of materials that were easily to hand in an artist's workshop. The framework of the receiving device was an artist's canvas-stretching frame, at the top centre of which was suspended a freely moving 'marking lever', which in Morse's first attempt was simply a pencil making contact with a strip of moving paper. The lever holding the pencil was free to move from side to side under the influence of an electric current flowing in the coils of an electromagnet. To move the strip of paper uniformly past the recording pencil, he made use of the works of a clock, shown to the right of the photograph. The transmitter consisted of a source of electric current to be turned on and off with the Morse key, and sent down the line to flow through the coils of the magnet in the receiver.

Initially the effect of the received current on the magnet's coils was weak, and the system would operate only over less than 12 m of wire. Here Professor Leonard Gale, his colleague at the university, was able to help. Gale was familiar with the work of Joseph Henry, the American experimenter, on the efficient construction of electromagnets. Following Henry's techniques, Morse rewound his coils with many turns of fine wire and was then able to extend the range of the instrument up to 16 km [2]. The transmission lines also presented a minor problem to Morse. He thought at first that it was necessary to cover the wires with insulating material and even enclosed them in a lead tube, not having heard of the experiments by Gauss, Weber, Steinheil and others who used suspended bare wires successfully for their telegraphs. In transmitting signals down his wires he made use of a binary code of his own devising, consisting of a series of short and long pulses of current, each set representing a particular letter or number, the set of combinations chosen being particularly efficient when transmitting an English text. Figure 3.2 is a reproduction taken from Morse's notebook, made on the voyage home, which shows clearly his original intention to use the code for numbers only. He considered several ciphers using these numbers, such as 252 representing 'England' and 56 for 'Holland', and visualised them being used to transmit governmental business 'so that secrecy could thereby be achieved'.

Figure 3.1 Morse's recording instrument, 1837
Source: *Deutsches Museum Archives, Munich*

With his prototype equipment he was soon in a position to place on exhibition at his university a model of his 'recording electric telegraph', and the public were invited to see it. Before he was able to produce an instrument for commercial exploitation, however, he needed two things; skill in practical design and finance. He found them both in Alfred Vail, an experienced graduate from the university having substantial mechanical expertise, and the helpful background of his father's Speedwell ironworks in Morristown, New Jersey, which latter enabled him to solicit the

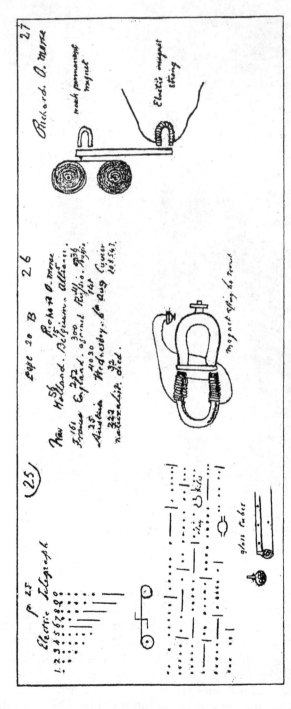

Figure 3.2   Reproduction from Morse's notebook, showing his original code
Source: Morse, 'Samuel Morse: letters and papers', 1914

financial support needed to manufacture a prototype equipment.[1] In 1837 Morse, together with his now accredited partners, Alfred Vail and Professor Leonard Gale, filed a caveat for his invention at the Patent Office in Washington in the same year as Cooke and Wheatstone filed their patent in Britain. But Morse, who worked in ignorance of their discoveries, had defined a telegraphy system which was quite different. He described its principle features himself in the wording of his caveat as

1   a 'marking instrument', consisting of a pencil, pen or print-wheel,
2   use of an electromagnet to impress the instrument on a moving strip of paper,
3   a 'system of signs' (Morse code) identifying the information transmitted,
4   a 'single circuit of conductors' [3].

This combination was indeed new, and it enabled Morse and his partners to initiate the installation of electric telegraph stations across America and influence the role of telegraphy on a worldwide stage. As mentioned above, Morse intended originally to use his code in the form of a cipher in which all the words of the English language would be given a unique number, and only the number transmitted. After Vail joined him he extended his code to include all the letters of the alphabet, together with various punctuation signs and some other symbols. The complete code, shown in Figure 3.3a, is now referred to as the American Morse Code. When this reached Europe a number of changes were made, yielding the International Morse Code shown in Figure 3.3b, which soon became the standard for almost a century, replacing the American Code on 1st July 1913 for commercial stations in the United States by international agreement.

To his original list of features Morse added a fifth, sometime between 1835, the completion of his first instrument, and 1837, the date of its public announcement. This was a transmitting relay which extended the range of his telegraph signal by recreating it for further transmission after about 32 km. In the biography of his father, E.L. Morse states that, 'It is only fair to note that the discovery of the principle of the relay was made independently by other scientists, notably by Davy, Wheatstone, and Henry,' but claims that 'Morse antedated them by a year or two and could not possibly have been indebted to any of them for the idea.' He was nevertheless refused an English patent for his complete system during a visit to Europe in 1838 on the grounds that it had already been published. He had more success on the European mainland, and was awarded a French patent in the same year. The relay had no difficulty in

[1] The workshops at Speedwell were fully equipped, having constructed most of the machinery for the SS *Savannah*, the first steamship to cross the Atlantic, and in 1838 Vail was able to set up working telegraph instruments which demonstrated the Morse system at Speedwell and later in New York, Philadelphia and Washington.

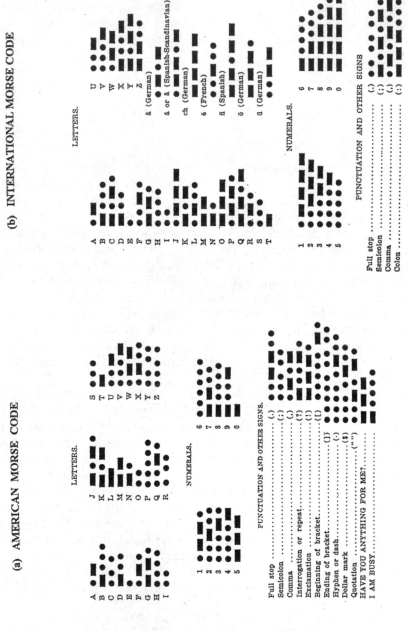

Figure 3.3   (a) The original (American) Morse code. (b) The international Morse code

being accepted in his own country for patent rights, and the 'Morse Relay' patent later created problems for his competitors, who found that long-distance telegraphy using the Morse (or any other) system was unworkable without a repeating relay.

Morse's demonstration and explanation of his telegraph in Washington in 1837 was well received, but it was not until 1842 that Congress finally awarded a grant of $30000 to establish the first experimental telegraph line in America, between Washington and Baltimore, a distance of 64 km. Morse preceded his work on the Baltimore line with a number of experiments, one of which concerned the choice of a suitable power supply to operate his lines [4]. For this he used a Grove battery, recently invented by William Robert Grove in 1842. It was by far the most powerful battery available at the time, but it used platinum as one of its plates, which made it rather expensive. The power and reliability that Morse obtained was, however, politically very desirable – as became apparent in the events that followed. It was fortunate for Morse and his partners that the year his telegraph first became operational coincided with a Democratic Convention in Baltimore, where the votes for the presidential nomination required no less than nine successive ballots, the results of each to be telegraphed to Washington as they occurred. The excitement generated among the senators crowding into Morse's receiving room in the Supreme Court augured well for future approbations from Congress, and his telegraph system became firmly established as a vital American facility. Other lines quickly followed, and by 1846 telegraph lines were installed between New York and Boston, Buffalo and Philadelphia, and Baltimore and Washington. All were achieved with the aid of private finance from several companies created by Morse and his partners, who now included F.O.J. Smith, a former Congressman.

3.2 Telegraph companies in the United States

It was at this early stage that Congress was approached to persuade the government to buy out the existing telegraph patents from Morse and his partners, with a view to establishing a central government telegraph service, to be run by the Postmaster General, then Clive Johnson. Congress was warned unequivocally by Johnson that 'irreparable damage might result if this private enterprise were allowed to proceed unchecked'. The warning went unheeded, and the only line in government hands, the original line from Washington to Baltimore, was leased out to one of the recently formed private telegraph organisations, the Magnetic Telegraph Company. This ushered in a period, described by R.L. Thompson as 'an era of methodless enthusiasm' [5], in which a veritable plethora of new companies joined the stampede for communication profits. A list of the most important of them, with the names of their principal developers, up to about 1850, is given in Table 3.1. They competed with one another for telegraph poles, wires, equipment and operating staff, quarrelled over routes and patent rights, and in the process established over 1 million km of telegraph wires throughout all of America.[2]

Table 3.1 The main North American telegraph companies in 1852, and their principal developers

Company	Year established	Developer
Magnetic Telegraph Co.	1845	A. Kendall
Washington & New Orleans Co.	1846	A. Kendall
Western Telegraph Co., Baltimore	1848	F. O. J. Smith
H. O'Reilly Contract Co., Philadelphia	1845	H. O'Rielly
Atlantic & Ohio Telegraph Co.	1848	H. O'Rielly
Pittsburgh, Cincinnati & Louisville Co.	1847	H. O'Rielly
New Orleans & Ohio Telegraph Co.	1847	A. Kendall
Ohio & Mississippi Telegraph Co.	1848	H. O'Rielly
Illinois & Mississippi Telegraph Co.	1849	H. O'Rielly
Ohio, Indiana & Illinois Telegraph Co.	1847	H. O'Rielly
Lake Erie Telegraph Co.	1847	H. B. Ely
Erie & Michigan Telegraph Co.	1849	F. O. J. Smith
North-Western Telegraph Co.	1846	H. O'Rielly
New York & Erie Telegraph Co.	1847	F. O. J. Smith
New York, Albany & Buffalo Telegraph Co.	1852	A. Kendall
New York & Mississippi Valley Printing Co.	1851	F. O. J. Smith
Montreal Telegraph Co.	1847	O. S. Wood
New Brunswick Electric Telegraph Co.	1848	L. R. Darrow
Nova Scotia Electric Telegraph Co.	1849	S. Cunard
New York & Boston Magnetic Telegraph Co.	1845	F. O. J. Smith
New York & New England Telegraph Co.	1849	H. O'Rielly
Maine Telegraph Co.	1847	J. Eddy

The companies not controlled by Morse and his partners shared a common problem caused by the legal strength of the Morse patents in the United States. They had to pay the necessary royalties to use the different items of telegraph equipment, reach an agreement with the companies, or develop their own unique equipment. An early pioneer of the telegraph in the United States was Henry O'Rielly, one-time postmaster of Rochester, who, impressed by the activity of Morse and his contemporaries in establishing telegraph lines out of Washington, decided to join this burgeoning industry and develop his own set of lines, seeing this as an ideal business venture. He was aware of the patent difficulties, but was able to circumvent them in a unique manner. Together with Amos Kendall, who had been his

[2] Rather less in fact than the lines established at this time in Europe, Asia and Africa, which together totalled about 1.64 million km.

former superior as Postmaster General, he agreed to raise capital for the construction of certain lines (initially from Philadelphia to Pittsburgh), for operation by a Morse company, in exchange for receiving immunity from the Morse patent rights [6]. The 'O'Rielly Contract' saw the beginning of a massive construction of telegraph lines expanding westwards across the United States, which prospered initially due to the low tariff rate charged to newspaper companies, who made full use of them. However, the contract was full of ambiguities and loopholes, and O'Rielly himself was never very scrupulous in observing its terms.

It was the power of litigation, exploiting these discrepancies, that finally broke O'Rielly's power over the expanding telegraph networks in 1848. This period has been regarded by Harlow as 'the era of litigation' [7], and these litigations certainly engendered an unprecedented enmity among scientists and others concerned with telegraphy. Probably no other invention in history has brought about as much 'legal embroilment, bitterness, venom, backbiting, slander, perjury, and other chicanery' [8]. Morse himself was obsessed with real or imagined transgressions against his patent rights, and spent much in legal fees, rendering him at times nearly bankrupt. There were, however, other problems for O'Rielly and the telegraph industry. The appalling winter weather experienced as the lines pushed westwards, and the vandalism of lines by native Americans, settlers, and the activities of rival companies intent on pushing their own lines into new areas, all contributed to a diminution of O'Rielly's influence on further expansion. Despite these difficulties, other entrepreneurs were active, and several secured the assistance of inventors who had original ideas for sale which avoided, or partially avoided, the Morse patents. In looking for areas of exploitation the telegraph pioneers also considered a number of large public concerns which they felt could make extensive use of telegraph facilities.

Unlike the railways in Britain, the US railroads were slow to make use of this new form of swift communication, and it was not until 1852 that a main line – the Erie – began to use telegraphy in its operation. Other railroad companies soon followed in the booming expansion of the network that took place mainly between 1852 and 1860, using the telegraph system to promote safe travel and at the same time providing the telegraph companies with protected routes for their wires. The method of railroad control in the United States at the time was abysmal. On the Erie railroad a 'leading train' had right of way along the track for an hour, and then had to wait for a further hour at a specified stopping point before it could proceed. The only alternative was to have a man preceding the train carrying a red flag; after waiting for twenty minutes the train could proceed until it caught up with the flag, and the process was repeated after another interval. This system was not changed until block working, made possible by telegraph signalling, was put into place. As soon as the railroads accepted the need for telegraph signalling, progress was rapid, and during the period of expansion some 37000 km of telegraph wires were constructed by the railroad companies alone, adding considerably to the number of commercial lines already in existence. Once familiar with the idea of using telegraphy for railway signalling, consideration was given to communicating with the moving train itself. Edison and Phelps in 1881 independently suggested ways of doing this, using almost identical technology. The principle was to construct a large induction coil with the wire coiled longitudinally around the carriage, as shown in Figure 3.4, and to include this in a circuit containing a buzzer, telephone, Morse key and a battery. An insulated conducting wire was laid alongside the track and connected with a similar arrangement in a trackside signalling office. Two-way communication was achieved by inductive coupling between the loop carried by the carriage and the trackside office, and information was exchanged by Morse code. The system was in use on a limited number of railroads for several years and exhibited at the Edison display at the Paris 1889 International Exhibition [9].

The railroads and the press were now major users of the commercial telegraph network. To a large extent the telegraph companies were in the beginning news agencies, and the operator-managers were expected to gather news items about their own localities and, when requested, to telegraph these on for a fee. Use of the telegraph by journalists brought with it a particular language of abbreviations, used to save both money and time in telegraphing news through from the reporting agents. This was widely applied in the United States, where the use of such abbreviations almost became an art form [10]. Typical abbreviations were *potus* for 'President of the United States', *yam* for 'yesterday morning', *gx* for 'great excitement', *ogt* for 'on the ground that', and *scotus* for 'Supreme Court of the United States'. The telegraph operators themselves also formulated their own codes to facilitate rapid exchange of information. These varied between different companies and groups of operators, common abbreviations being *ii* for 'I am ready', *ga* for 'go ahead' and the imperative *sfd* for 'stop for dinner'.

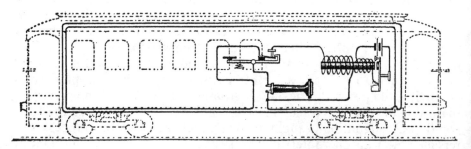

Figure 3.4   Inductive communication with a moving train
Source: Mayer, 'American Telegraph Systems', 1898

The enterprise of the telegraph companies, as well as providing a vast network of wires linking towns and railroads, also promoted the development of new methods of transmission and reception of the now universal Morse code. To secure freedom from their dependence on Morse patents, a number of the larger telegraph companies adopted two alternative printing telegraphs: the House printing telegraph, and the one developed by Bain. The House printing telegraph, shown in Figure 3.5, was designed by Royal Earl House of Vermont, a self-taught inventor, and patented by him in 1846 [10]. In principle it was very similar to Wheatstone's ABC Telegraph, which had been patented in 1838, and also employed a print-wheel containing the letters of the alphabet, each formed of lead printing type. The wheel was rotated step by step by successive impulses of current operating a ratchet action for an electromagnet relay. By controlling the number of impulses transmitted, the wheel was caused to stop at the required arc corresponding to the letter transmitted, and a hammer action then pressed the letter type onto a strip of paper, thus printing out the message. The transmission of impulses corresponding to letters of the alphabet was effected by a

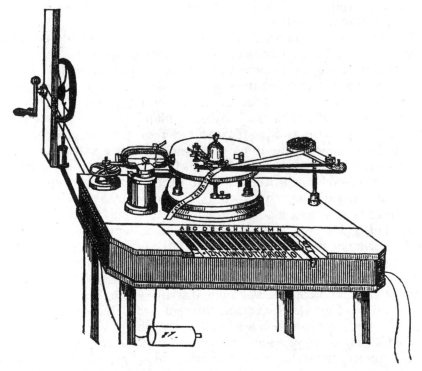

Figure 3.5   The House printing telegraph
Source: Turnbull, 'The Electromagnetic Telegraph', 1852

piano-style keyboard, which can be seen in Figure 3.5. A revolving cylinder, acted upon by the keyboard, controlled a rotary switch which passed a continuous stream of impulses until arrested by the operator pressing down the appropriate letter key. Thus the number of pulses transmitted down the line depended on the location of the current letter key with respect to the previous letter being transmitted. It was therefore essential that the transmitter cylinder and the receiver print-wheel be kept in exact synchronism, and a considerable amount of design effort was applied by House to achieve this end. The printer was eventually phased out of use as new printers were developed which were less compromising in their adjustment and performed at a faster rate. In its heyday the House printer was a popular system, capable of printing messages at a rate of more than 50 wpm. An extensive range of telegraph lines equipped with his printing telegraph were laid down between New York, Boston, Washington and Philadelphia, by the New York and Mississippi Valley Printing Telegraph Company (later to become the Western Union Company) [11] and by other companies extending their lines westwards to Cleveland and Cincinnati between 1847 and 1855. One of these lines, using House's printer, extended from Poughkeepsie through Rochester and Buffalo to reach St Louis in January 1852 and formed the longest line then operated by one company, almost 2500 km in length.

The Bain system, which was used by several of the United States telegraph lines, included an automatic punched paper tape sender and a new electrochemical telegraph receiver, both brought into the country in 1848 by their inventor, Alexander Bain, a Scotsman. Bain was born near the town of Thurso in 1811. His education was minimal (at one stage he was employed as a herdsboy in the Scottish highlands), but by perseverance, aided by the opportunity of employment as a journeyman clockmaker in London, he was eventually to construct and patent a variety of significant mechanical and electrical inventions [12]. In his printing telegraph, patented in 1846, he substituted for the Morse key a pre-prepared and perforated paper ribbon about 14 mm wide, drawn by a clockwork mechanism so as to pass over a conducting cylinder. A single hole in the tape represented a dot, and a series of three holes represented a dash. A metallic spring or stylus made contact with the drum through these holes and caused an electric current to be transmitted, forming the dots and dashes of the Morse code. Bain's receiver was conceived in an earlier patent of 1843, and consisted of a metal stylus which was brought down by the action of an electromagnet, through which the received electric current passed, to bear upon a second paper ribbon which was impregnated with potassium prussiate. As the paper was drawn by a clockwork mechanism over the stylus, a series of bright blue lines was printed on the ribbon. It was reputed to be twice as fast as the House instrument and three times as fast as Morse's printer [13].

In the mid-1850s, two further inventions were added to the United States telegraph networks. The telegraph line operators at that time were noted for their skill in 'reading' the Morse printing receiver by the sounds it made as the relays closed, which differed slightly for a dot and a dash. This led Vail and Morse to produce the Morse sounder, which became the universal receiving instrument in all stations whether or not it was accompanied with a printing receiver. In his commercial dealings concerning the sounder, Morse insisted that it was implicit in his first patent and was 'just the pen-lever deprived of a pen' [11]. The sounder, shown in Figure 3.6, consisted simply of an electromagnet acting on a pole piece, mounted on a pivoted sounding lever, and working between two stops. When the magnet was energised and then released, the sounder emitted a click which could be recognised as resulting from either a short or long period of excitation. In a later version, and the one employed by the British Post Office as standard equipment, a double-plate sounder was used. This became known as 'Bright's bells' a British adaptation of the sounder, designed by Charles Bright in 1855. By means of a polarized relay, the received dot and dash signals were made to activate one of two relay strikers acting on bells of different tones or, in a later variant, sounding plates of brass or steel.[3]

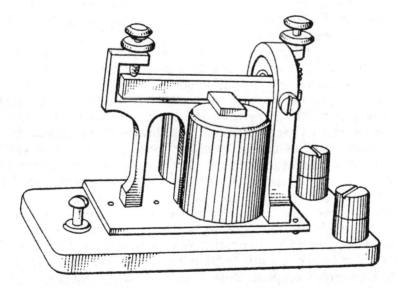

Figure 3.6   A Morse sounder
Source: McNichol, 'American Telegraphy Practise', 1913

[3] Charles Tilson Bright (later Sir Charles Bright) was to become one of Britain's foremost telegraph engineers in the nineteenth century, responsible for major advances in submarine cable technology, to be described in Chapter 5.

A second device was a new recording instrument, designed by David Hughes of Kentucky, and used by the American Telegraph Company in 1855. Hughes was born in London but spent much of his working life in the United States. He was a brilliant inventor, employed by the American Commercial Printing Company at the time when he devised his printing instruments. In 1856 he sold his invention to the company upon securing a US patent (No. 14,917) for it. The Royal House patents for a printing machine were also held by the company, and the joint ownership of the patents for these two advanced machines later enabled them to produce a third machine, designed by Phelps of the company, which incorporated the best features of both the House and Hughes systems, and became widely used by the United States telegraph companies. It was, however, the original Hughes machine that was first offered to the American Telegraph Company in 1855 for use on its extensive network. The main feature of the Hughes machine was its free-running mode of operation (see later), which was considered a vast improvement on the House system developed seven years earlier. It also used a piano-type keyboard and printed at 60 wpm, successfully circumventing most of the Morse patents. However, its development and patent costs needed to be met, and the American company concerned was anxious to share its purchase cost for the Hughes patent rights with other companies, before it would permit the Hughes machine to be used by them. This was to lead to much wider cooperation between a number of American telegraph companies in the coming years.

The American Telegraph Company was one of the most successful firms at the time, and was voracious in creating and absorbing further telegraph 'territories'. Many considered that it was now time to limit the unbridled competition between companies, and particularly the aggressive behaviour of the American Telegraph Company, since this was beginning to restrict the availability of an efficient public service. In 1855 a moratorium on this continued rivalry between companies was proposed jointly by Morse and Amos Kendall, his business partner and later owner of several important telegraph networks. They invited leaders of some of the largest private companies to a series of meetings to 'devise a plan for harmonizing all interests and protect existing lines' [14]. The final outcome was the 'Treaty of Six Nations', an agreement between the American Telegraph Company; the New York, Albany & Buffalo Telegraph Company; the Atlantic & Ohio Telegraph Company; the Western Union Telegraph Company; New Orleans & Ohio Telegraph Lessees; and the Illinois & Mississippi Telegraph Company. This was largely a territorial agreement which divided most of the eastern United States between the six companies, but they also divided the cost of the Hughes patent purchase amicably between them and arranged to share some lines and equipment for their mutual benefit.

The agreement was short-lived. By 1861 the nation was at war with itself, and the various telegraph companies saw their lines broken and their stability impaired, in many cases beyond recovery. After the civil war (discussed in the next chapter) the composition of the telegraph companies changed completely, and even more new companies were formed. Reid, in his *The Telegraph in America*, lists some 217 companies for the year 1886, many holding only a few tens of kilometres of line.[4] At about this time Congress was actively considering some form of nationalisation for the telegraph companies. It had an option to do so as a result of an 1866 Act of Congress which stated:

That the United States may at any time after the expiration of five years from the date of the passage of this Act for postal, military, and other purposes, purchase all the telegraph lines, property, and effects of any or all of said companies at an appraised value, to be ascertained by five competent disinterested persons, two of whom shall be selected by the Postmaster General of the United States, two by the company interested, and one by the four so previously selected.

Senator C.C. Washburn, in his immensely detailed speech to Congress on 22nd December 1869, made an overwhelming case for so doing, citing the success of the British Post Office following the Telegraph Act of 1868, together with similar governmental intervention in telegraph services in Belgium, Switzerland, Prussia and other European countries. He quoted figures for the installation and maintenance of a nationwide telegram service at a uniform (and lower) rate than existed in any area of the United States at the time [15]. Despite his eloquence, the facts and figures he provided, and the support of much of Congress and the Press, the telegraph service was never centralised as were similar services set up by the posts, telephone and telegraph services in Europe. Opposition came from within Congress and outside it, mainly activated by Western Union interests. Not surprisingly, a monopoly gradually emerged during the next decade in the shape of the powerful Western Union Company [16], which cooperated with the United States Post Office to provide a public service paralleling the well-established letter post in America.

The Western Union Telegraph Company was established in 1851 as the New York and Mississippi Valley Printing Telegraph Company, its first task to build a line from Buffalo to St Louis, Missouri. It became Western Union in 1856 and was one of the first telegraph companies in the United States to make use of the railroads to establish its lines. This was arranged by agreeing to a contract giving free transmission of railroad messages, all having priority over non-railroad messages, which did much to promote a safe and efficient railroad network in the United States. In

---

[4] The same thing happened thirty years later in connection with the submarine telegraph, but then the capital investment required was bigger, whereas with line telegraphy it could be profitable to set up a comparatively short line with very little capital, attracting a large number of developers within a very short space of time.

return, the railroads laid telegraph lines along their protected routes and provided a chain of Western Union offices to house the telegraph equipment at all its major stations (Figure 3.7). These offices were eventually linked to Western Union central offices in nearby towns, thus providing a nationwide telegraph service.

The man responsible for the consolidation of the telegraph companies into a single enterprise in this way was Hiram Sibley, a self-made businessman and promoter. After practising as a shoemaker at an early age, he entered a cotton factory and started up a machine shop at the age of twenty-one. He was already a rich man, engaged in banking and real estate, when he entered the telegraph industry and organised the New York and Mississippi Valley Printing Telegraph Company. In 1854 he formed an association with Ezra Cornell, and between them they formed the Western Union Telegraph Company with Sibley as managing director [17]. During the 1850s he arranged mergers and takeovers of minor telegraph companies, a process that continued relentlessly as the telegraph network pushed westwards and continued throughout the American Civil War. In 1860 Sibley was awarded a contract by Congress, under the Pacific Telegraph Act, to build a line across America from east to west. The line was completed by Western Union in 1861, making this company the first to establish telegraphic communication across the land

Figure 3.7   A Western Union station
Source: The Shelburne Museum, Vermont

from the Atlantic to the Pacific, despite the Civil War.[5] The line was connected with local networks already operated by Western Union. After the transcontinental railroad was completed in 1869, the telegraph lines were moved to follow its route [12]. Western Union found itself in a particularly good position during the Civil War. Its lines generally ran from east to west and were almost unaffected by the war, unlike their main rival, the American Telegraph Company, whose lines ran from north to south and were frequently damaged or destroyed by the competing armies. After the war Western Union continued to grow, absorbing more than five hundred smaller companies, and increasing its number of telegraph offices from 132 in 1856 to over four thousand in 1866. By 1902 it controlled over 1.5 million km of telegraph lines and two international submarine cables [18], and became the first industrial monopoly in the United States.

Some part of Senator Washburn's hopes was realised in the operational working of the company offices. Messages, or mailgrams as they were then termed, could be handed in at a government post office, which then arranged to forward them to Western Union for onward transmission. When received at the Western Union office nearest to the addressee they were delivered by the nearest government post office as a local letter. Thus the government post office took responsibility for delivery, subcontracting the Western Union monopoly as the executive carrier – a process quite different to the national service common in almost all other countries. In 1859 Western Union set up its own cipher code to improve the efficiency of its nationwide service, now handling millions of words per year for transmission over the telegraph networks. This was known as the Standard 92 Code, in which common sentences and phrases were replaced with selected numbers in the range between 1 and 92. Several of these abbreviations have survived to be used by the amateur wireless fraternity and until quite recently, by the Morse coastguard stations (see Chapter 11), including the well-known 73, 'best regards', and 88, 'love and kisses'.

The tremendous expansion of the telegraph network, with its offices and supporting organisations, in the last few decades of the nineteenth century was an important source of employment across America. By 1900, Morse operating had reached a peak in Western Union, with the relay office in Chicago, for example, employing 880 operators on a single shift. The operators became highly skilled and were referred to colloquially as 'brass-pounders', in acknowledgement of the repetitive tapping, hour after hour, on an unyielding brass Morse key (Figure 3.8a). Many operators developed muscular pain in their sending arm and sometimes

---

[5] Although fictional, the magnitude of the problems encountered in the task of establishing almost 5000 km of line across the entire continent are described realistically in Zane Grey's novel *Western Union*, published in 1937.

actual paralysis (which we would now describe as repetitive strain injury), a serious problem with all telegraph operators [19, 20]. In an effort to lighten this work the American companies adopted a new form of Morse key, which they termed the 'bug key', more properly known as the Vibroplex key, patented by Horace G. Martin in 1904 (Figure 3.8b). It employed a vibrating contact point which made a series of dots automatically, the operator exercising control to limit the number of dots required for a given character, thus reducing the work to transmitting dashes and arresting the vibrating contact as required. Although widely used for commercial and military purposes in America, it was never taken up officially in Britain, except by some amateur wireless operators.

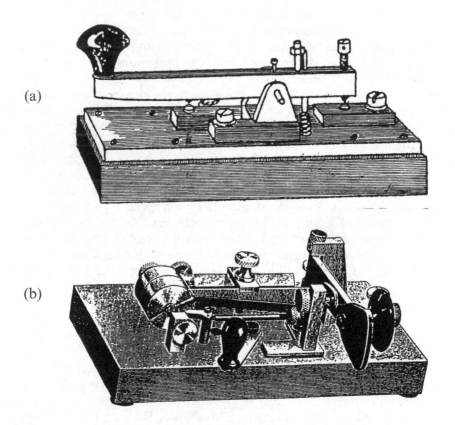

Figure 3.8   Two variations on the Morse key: (a) an international Morse key and (b) an American 'bug' key

3.3 Development in Britain

In Britain and Continental Europe the struggle to establish telegraph lines was less frenetic and, although competition and inter-company rivalry were just as keen, fewer companies were involved. The telegraphs followed the railways more often than in the United States, and the telegraph offices for the public were often at the stations – not too convenient for customers to reach, as railway stations were often situated at the edge of small towns, or sometimes a little way outside. As the telegraph network expanded, however, offices were established in town centres, linked to the railway station by extension lines, where they joined the main telegraph network. The full importance of telegraph operations in Britain was realised after 1851 when links were established with the continent of Europe, so that 'a London wine merchant could telegraph his suppliers in the middle of France and ask about the grape harvest and speculate accordingly' [21].

The coding feature of Morse's invention was quickly appreciated and taken over by the railway and other commercial lines in Britain. A single-needle telegraph was made to respond to the Morse code by keying in a positive or a negative current, corresponding to the dots and dashes of the Morse code, which made the needle swing to the left or right of its neutral central position; mechanical damping had to be taken into account to ensure a precise indication. This movement could be followed by a skilled operator, although later the needle instruments were replaced completely by the Morse sounder, described earlier, which gave faster operation and was less fatiguing for the operator.

For a number of years the two devices were operated together at the central telegraph stations established in the centre of large towns. Sir Francis Bond Head, an engineer and director of the London & North-Western Railway [22], has left us with a vivid description of activities in a central telegraph station situated at Lothbury in the City of London in 1849 (Figure 3.9) [23]. This was a large building having three galleries. On the ground floor, over a long counter in the reception hall, messages to be transmitted were handed in to a clerk.

These messages were written on a half-sheet of paper, nearly one half of which is pre-occupied by a printed form, to be filled up by the name and address of the writer, as also of the person to whom his communication is addressed; the charge for the message, answer, porterage, or cab-hire; the date and hour at which the message is received etc., etc. . . . As fast as these messages are written, they are, one after the other, passed through a glass window to a small compartment on the ground floor, termed 'the Booking Office', where, after having been briefly noted and marked . . . they are put into a small box, a bell is then rung, and at the same instant up they fly, through a sort of wooden chimney, to the attic regions of the building, to 'The Instrument Department' . . . we found therein four or five intelligent-looking boys, from fourteen to

Figure 3.9   The Central Telegraph Station, London, in 1849
Source: Head, 'Stokers and Pokers', London, 1849

fifteen years of age, and eight little instruments . . . On every instrument there is a dial, on which is inscribed the names of the six or eight stations with which it usually communicates . . . As fast as the various messages for delivery, flying one after another from the ground floor up the chimney, reach the level of the instruments, they are brought by the superintendent to the particular one by which they are to be communicated, and its boy, with the quickness characteristic of his age, then instantly sets to work. His first process is, by means of the electric current, to sound a little bell, which simultaneously alarms all the stations on his line . . . In return a corresponding signal announces to the London boy that he is ready to receive it. By means of a brass handle affixed to the dial, which the boy grasps in each hand, he now begins rapidly to spell off his information by certain twists of his wrists, each of which imparts to the needles on his dials, as well as to those on the dials of his distant correspondent, a convulsive movement designated the particular letter of the telegraphic alphabet required . . . While a boy at one instrument is thus occupied in transmitting a message written by its London author in ink which is scarcely dry, another boy at the adjoining instrument is, by the reverse of the process, attentively reading the quivering movements of his dial, which by a sort of St. Vitus's dance are rapidly spelling to him a message via the wires of the South-Western Railway, which word by word he repeats aloud to an assistant, who, seated by his side writes it down on a sheet of paper, which as soon as the message is concluded descends to the 'Booking Office', where, inscribed in due form, it is without delay despatched to its destination by messenger, cab, or express, according to order . . . In a corner of one of the attics in which the eight electric instruments are placed there stands a small very ordinary-looking piece of cheap machinery composed of a few wheels, giving revolution to a small cylinder, upon which there has been wound a strip of bluish paper half an inch wide and about 60 yards [55m] in length. As this insignificant thread of paper slowly unrolls itself . . . it receives from a little piece of steel wire about a quarter of an inch [6mm] long, and about the size of a large needle, a series of minute black marks, composed of 'dot and go one,' – two dots, – two dots and a line, – two lines and a dot, – and so on. But who makes the dots? The answer in a few words explains the greatest mechanical wonder upon earth. The little dots and lines marked upon the narrow roll of paper revolving in a garret of the London Central Telegraph Station, are made by a man sitting in Manchester, who, by galvanic electricity, and by the movement of a little brass finger-pedal, is not only communicating to, but is himself actually printing in London information which requires nothing but a knowledge of the dotted alphabet he uses to be read by any one to whom it may either publicly or confidentially be addressed!!

Within a very short period of time the equipment at the London Central Telegraph Station was enhanced to contain many more instruments. On the first and second floors 176 instruments of five different kinds were to be seen, some of which belonged to the Anglo-American and Indo-European companies, concerned with submarine cable systems, and who originally occupied part of the building before it was taken over by the nationalised Post Office (see later). These were:

Morse printers	84
Single-needle receivers	76
Bright's bells	11
ABC telegraphs	4
Double-needle receivers	1

and were used to communicate with all the offices in the metropolitan districts, extending as far as Windsor, and to instruments placed in certain private or government establishments, such as Buckingham Palace, various hotels, the War Office, the Treasury, the Horse Guards and the House of Commons. On the third floor, or Provincial Gallery, there were 131 instruments:

Morse printers	66
Single-needle receivers	21
Bright's bells	13
Wheatstone automatic receivers	20
Hughes automatic printers	11

They were installed to communicate with all the telegraph stations in Great Britain (190 stations in 1871). By 1874 the number of instruments was increased to almost 600, and the operating staff to 1240 instrument clerks, of whom 740 were women, responsible for more than 26000 messages per day [24]. To assist in communication with local offices in the London metropolitan area, an elaborate system of pneumatic tubes was also used [25]. This followed a proposal by Josiah Latimer Clark, a telegraph engineer working for the Electric Telegraph Company, and was installed into place between the Stock Exchange and the Central Telegraph Office. By 1865 the increase in telegraph traffic led the company to extend its London pneumatic tube network and to install similar systems in Liverpool, Birmingham and Manchester.

The Morse recorder was introduced into Great Britain in about 1850, a few years after its first telegraph company was formed. The Electric Telegraph Company, established in 1846, controlled all of Cooke and Wheatstone's patents and had the largest share of telegraph wires in the country, some 80000 km in 1868 (excluding its share in Atlantic and Continental submarine cables). By that time two other types of telegraph recording instrument were in use, tested by the Electric Telegraph Company on its lines connecting London, Manchester, Newcastle, Glasgow and Edinburgh. These were the printing system of David Hughes, already in use in the United States, and the Wheatstone automatic apparatus.

Wheatstone's automatic system was patented in 1858 and is remarkable for its method of preparing the information to be sent – 'off line', as we would now say – using a separate keying machine for this purpose. The paper tape produced contained a series of punched holes which were recognised by the transmitting and recording instruments as representing the Morse code characters constituting the message. By separating out the message preparation from the actual process of transmitting and recording, these latter operations could be speeded up, with consequent saving of time and money. The Wheatstone system was not brought into wide use in Britain until 1867, when it operated at a speed of 70 wpm; later versions ran at up to 600 wpm. Its operation is discussed later in this chapter.

3.4 The Telegraph Acts of 1868–69

By 1868, a significant date in the history of the British telegraph industry, the public telegraph network consisted of almost 150 000 km of wire and over 3000 stations, plus another 1000 stations provided by the railway companies [26]. There were then five major telegraph companies in Britain, the largest (and oldest) of which was the Electric Telegraph Company. None were operating without criticism. The reputation of the companies was at a low ebb, with frequent errors, delays and high tariffs. Their rates compared unfavourably with those of their counterparts on the Continent, particularly with the Paris district network, which offered rates half those of Britain, and furthermore guaranteed delivery of telegrams within half an hour.[6] The general malaise in Britain was seen as resulting from inter-company rivalry and the costly duplication of services. The Continental telegraph services had been state-owned since their inception in the 1840s, and their generally improved service was considered to stem from a central unified operation. In Britain, the case for similar reform was being pursued by advocates both within and outside parliament. By the late 1860s all these telegraph companies had become involved in preliminary discussions with the Postmaster General, at that time Frank Ives Scudamore [27], who had been campaigning for the telegraphs to become an arm of the postal services for some time.

Scudamore was an experienced Post Office man, having joined the company at the age of eighteen in 1841 on completion of his schooling at Christ's Hospital, where his uncle, Sir Charles Scudamore, held a post. His rise through the Post Office hierarchy was rapid. He was the first to establish a scheme for government savings banks, which is in use today, and in 1856 he advised on the desirability of the state acquiring (nationalising) the British telegraph companies and assuming a monopoly of use for their network of lines. These discussions eventually led to decisions in parliament, incorporated in the Telegraph Acts of 1868–69, the brief for which is set out in the opening paragraphs of the 1869 Act:

[6] This was perhaps a little unfair, since the Paris network had an extensive pneumatic tube system which was classed with the electric system for statistical purposes. The cost of telegraph transmission over the tube remained fixed no matter how long the message, and often appeared more cost-effective than the electric telegraph, in which a cost-per-word count was in operation.

Whereas by The Telegraph Act 1868, her Majesty's Postmaster General is empowered to purchase, in the manner therein mentioned, the whole or any part of the undertaking of any telegraph or other company authorised to transmit telegraphic messages in the United Kingdom, except the undertakings of the Atlantic Telegraph Company and the Anglo-American Telegraph Co. Ltd . . . is required to make one uniform charge for the transmission of telegraphic messages throughout the United Kingdom . . . the Postmaster General is required . . . to pay the several companies therein mentioned all reasonable costs and expenses incurred by them . . . and . . . shall have the exclusive privilege of transmitting telegrams within the United Kingdom [28].

When it is appreciated that the term 'telegraph' was later to embrace both line and wireless telegraphy, it will be seen that the Acts were major pieces of government legislation. The telegraph companies referred to are listed in Table 3.2 and their positions in 1868 are discussed below. They included the five major British telegraph companies plus Reuter's Telegraph Company, acting for the press. The railways were initially opposed to nationalisation but eventually consented to its extension to their telegraph systems also, subject to a separate agreement with the government.

3.4.1 The Electric & International Telegraph Company

The Electric & International Telegraph Company was created through a merger of the Electric Telegraph Company with the newly formed International Telegraph Company in 1855, the latter being concerned with submarine telegraph communication to the Netherlands and telegraph lines within that country as well as Belgium. The parent company was established in 1846 and had purchased patents for the various systems devised by Cooke and Wheatstone and was, at least initially, in a strong position to compete with the other telegraph companies. In addition it possessed a near monopoly of way-leave rights to construct telegraph lines along many of the railways, and by converting the contracts from way-leaves to exclusive

Table 3.2 Companies named in the Telegraph Acts (Great Britain) of 1868–69

Company	Date registered	Length of line (km)	Payment made (£1000)
The Electric & International Telegraph Co.	1846	16 000	2939
The British & Irish Magnetic Telegraph Co.	1849	7600	1244
The London & District Telegraph Co.*	1859	555	60
The United Kingdom Electric Telegraph Co.	1860	2700	562
The Universal Private Telegraph Co.	1861	223	184
Reuter's Telegram Co.	1851	—	726

* Became the London & Provincial Telegraph Co. for a short time from 1867.

rights it was successful in compelling other companies to use roads or canals for their telegraph routes. By the time of the Telegraph Act of 1868 it was the largest telegraph company in the country, owning 16000 km of line, 1300 telegraph stations in Britain and Ireland, and employing 3000 skilled operators. It also had control of three Continental cables [29].

A major use for this network was to transmit news directly from London to other major cities and thence to local newspaper proprietors and others. A description of this operation is found in a copy of Manchester's *Freelance* in 1868, billed as 'an entertaining journal of humour and interest'. In a piece describing the 'Telegraph Office in 1868' the writer comments on the quantity of telegraph information sent from London early every morning, containing

extracts from the early papers e.g. *The Times, Daily News,* City Articles, lists of the New Cabinet, panegyrics upon Mr. Gladstone by the *Daily News* and *Daily Telegraph,* and diatribes against him by the *Herald, Standard,* etc. From this supply the newspapers in large towns draw their pabulum for their second edition and ... supply their subscribers with the latest intelligence.

The writer claimed a speed of transmission of 30 wpm using a two-key Bright's bell sounder for this news transmission with the operator, 'receiving and writing ten copies at once, preparing eight folio sheets with an ivory stile pressing on alternative layers of black carbonised and white tissue paper' [30].

Some of the technical equipment used by 'the Electric', as it was popularly known, is shown in Figure 3.10, reproduced from the *Illustrated London News* of 1868, and shows the variety of equipment in the London Central Telegraph Office at that time, including the Hughes Printer described earlier. In 1852 an interesting innovation was agreed between the company, the Astronomer Royal Sir George Airy, and Charles Walker, Telegraph Superintendent of the South-Eastern Railway company, to transmit Greenwich Mean Time (GMT) to London Bridge Station and, via several different railway companies, to Edinburgh and other cities. Until the maintenance of GMT became a legal requirement in 1880, different communities throughout Britain kept their own local time, related to the local meridian. The 1852 agreement was gradually adopted throughout the country, becoming a standard by the end of the year, for railway timetable compilation and train running. A public method of announcing correct GMT was the 'time ball', one of which can still be seen today above the Strand in London. This was a large ball, held magnetically at the top of a vertical pole and released to slide down the pole in response to a time signal at precisely 13:00 hours, telegraphed from Greenwich.[7] Time signals announced in this way were vital to the

[7] The time ball at Deal was installed on top of the 1816 Semaphore Building at the entrance to the naval dockyard. The building is now open as a museum.

*Figure 3.10   Some of the equipment in use at the Central Telegraph Office in
1868*
Source: *Illustrated London News, 1886*

public, shipping and business organisations before the introduction of radio time signals in the 1920s [31].

3.4.2 The British & Irish Magnetic Telegraph Company

The closest rival to the Electric was the British Electric Telegraph Company, which in 1857 amalgamated with the English & Irish Magnetic Telegraph Company to become the British & Irish Magnetic Telegraph Company, with its headquarters in Liverpool. It operated extensive lines along those railways not exclusively secured by the Electric, mainly in the north of England, Scotland and Ireland. To connect with London the company laid a series of ten wires in troughs along trunk roads via Birmingham, Manchester, Carlisle and Glasgow, using the new insulation material, gutta-percha. This was one of the first companies to use a machine-generated electricity supply in place of batteries, hence the word 'Magnetic' in its title. Many of its lines were underground, in contrast to the Electric and the London companies, which made extensive use of lines carried on poles or suspended from the roofs of houses. In its operational practices the Magnetic was one of the first to make use of female telegraph operators. The role of 'telegraphiste' was popular with unmarried women since, as noted at the time, 'to girls of respectable parents of moderate means . . . unless they married, only two occupations were open viz., either to become nursery governesses or milliners – positions anything but agreeable or profitable'. They received no pay during training, and subsequent emolument depended on the speed they achieved, with a ceiling for speeds of 10 wpm or above when ten shillings (50p) per week was earned [32].

3.4.3 The London & District Telegraph Company

The London & District Telegraph Company, established in 1859, was formed to provide cheap telegraph communication for the metropolis, initially within 6.4 km of its central telegraph office at Charing Cross, later extended to 32 km. The initiative for this enterprise came from the Waterlow brothers, Sydney and Alfred, who had separate establishments in the City, half a kilometre apart, which they wished to link by private telegraph. The crowded built-up area separating them made the laying of underground cables an expensive operation. The alternative, proposed by Sydney Waterlow, was to make use of roof-to-roof wiring across the tops of the congested buildings of central London. Galvanised iron wires were supported by telescopic iron masts mounted on roofs or fixed to chimneys, and although this was cheaper than burying the cables, it did necessitate obtaining permission from householders and making wayleave payments to them. The process of obtaining permission from a multitude of householders was long and tedious, requiring, it was said,

4000 interviews with 1900 people [33]. The system was far from perfect, and carelessly installed wires posed some danger to pedestrians. However, rooftop wiring was a common method at the time, and could also be found in Paris, Brussels, New York and other large cities.

Like the Magnetic Telegraph Company, 'the District' was a pioneer in the use of female operators. Soon after its establishment, 45 'novices' were recruited to be trained under a matron in the intricacies of transmission and reception using Morse signals, operating single-needle instruments. Incumbent male clerks in the telegraph offices were strictly instructed not to speak to the telegraphistes ' under pain of instant dismissal'. Despite this and other constraints in their employment, the position of telegraphiste was popular and became the norm for the new nationalised telegraph offices established after 1870.

The company originally intended to complete the coverage of central London with 100 stations, at which messages could be handed in for onward transmission to the station nearest to the addressee, final delivery of the decoded message to be completed by messenger boy. However, it ran into severe difficulties with its expansion plans. In the mid-1860s the company was able to announce the opening of telegraph offices on the new Metropolitan Railway, a useful route for their surface cables to western districts of London. Some changes in equipment were made, with the introduction of a dial instrument developed by the W.T. Henley company, and a similar instrument from Siemens & Halske produced at their new manufacturing branch in England. An important private network was installed linking the London Fire Brigade stations and a number of private businesses. These businesses were sufficiently confident of the service to announce that they would now be able to accept orders 'placed by telegraph'. However, although the system eventually achieved profitable operation, after several years of loss, it completed only about 80 of its 100 planned stations since, apart from the problems arising from roof-to-roof wiring, its performance was marred by errors and slow delivery of messages. A major cause of both was the continual coding, transmission and decoding, which might be necessary several times during the transmission of a single message – a potential source of error on each occasion. The idea of a 'through switchboard' at the stations, to permit the messages to flow through in coded form to the next station and so avoid the errors inherent in the process of decoding, coding and retransmission, was not to occur until after the nationalisation of the telegraph companies in 1870.

3.4.4 The United Kingdom Electric Telegraph Company

The United Kingdom Electric Telegraph Company was formed to develop telegraph communication in Britain over public highways. It was proposed by Thomas Allan in 1850 but did not become operational until

1860, once capital was raised. The intention was to introduce a system having a fixed uniform rate over major trunk routes, initially over a route linking London with Birmingham, Manchester and Liverpool. The transmission of messages at a standard rate was popular with business users and was used extensively by banking interests in the City. Agreement was reached with Grand Junction and other canal companies to use their towpaths for the telegraph lines, and these became the main routes for the lines when the company became fully operational in 1863. By this time the Hughes fully automatic printing telegraph was available. This instrument removed the necessity for decoding the Morse transmission, so that recipients could be given a plain-language printed version of their message as soon as it was received at the central telegraph office. Further trunk lines were opened, and by 1865 some 15000 km of wire was in place linking the main cities in Britain. Some of these routes were carried underground in 2-inch (50 mm) cast iron pipes containing several wires wrapped in tarred yarn. After nationalisation by the Post Office a similar size of pipe was used on many routes in London, several of which contained as many as 120 wires in each pipe [21].

3.4.5 The Universal Private Telegraph Company

The Universal Private Telegraph Company was established by Alderman David Salomons, later to become Lord Mayor of London, who included on his governing board two well-known engineers, William Fairbairn of Manchester and Charles Wheatstone. The latter's cooperation gave the company access to the full range of Wheatstone's inventions. 'The Universal' based its entry into the telegraph market in 1860 on a promise to offer uniform (and cheap) telegram costs over the major trunk lines, as 'the United' had done, initially linking London with Birmingham, Manchester, Newcastle, Leeds and Liverpool. It also offered to provide direct telegraph facilities between particular points, such as business establishments or private premises, rather than provide a general telegram service to the public of the kind offered by the District company. Although it was compelled to establish its lines along canal towpaths because of the stranglehold of the larger companies on railway routes, it mainly succeeded in its aim. Within the cities it also made use of roof-to-roof wiring, but made less obtrusive by bringing together as many as a hundred separate copper wires into a single cable, from which two subscribers could be linked by individual wires leading away from the main cable at two appropriate points on its route. Since the company's policy was to link specific business premises to one another by private lines, no public offices were needed. Also in keeping with its policy of employing a minimum of staff, the company offered the use of Wheatstone's ABC telegraph instruments, described earlier, which enabled a message to be transmitted, letter by letter, without the need for a Morse-trained operator. A remotely operated electric bell accompanied the ABC telegraphs and, since these incorporated a magneto-electric generator, the subscriber was able to turn a handle to generate the electricity to initiate the alarm – known discreetly as the 'annunciator' so as not to frighten the user!

3.4.6   The Reuter's Telegram Company

The final company named in the Telegraph Acts of 1868–69 was not a public telegraph company, but the news-gathering Reuters company. This was established in 1851 and had its headquarters in London, serving banks, brokerage houses and leading business firms. In 1859 Baron Julius Reuter contracted with the Electric & Magnetic Companies for the exclusive right to supply foreign news telegraphed to all the towns in Britain. Its first newspaper subscriber was the London *Morning Advertiser*. Other publishers followed, and Reuter's agents became widely established in most towns on the Continent. This service became firmly established as a result of a news coup in 1859 when Reuters was able to transmit to London a speech by Napoleon III which foreshadowed the Austro-French Piedmontese war in Italy. When the conflict between the French and Austrian armies commenced, Reuters had representatives with both armies, giving it almost a monopoly on war news.

By 1865 the Turkish cable route to India was operational and available for public use. Julius Reuter quickly saw that if he had his own private cable across the North Sea to connect with his offices in Germany, he would be able to make use of the India cable directly and would virtually own the quickest telegraph route between Great Britain and the East for news purposes. He began negotiations with the Hanoverian government and obtained a concession from their king to land a cable on the island of Norderney off the coast of north-west Germany. The Hanoverian government would then connect this to new overland wires to be built to Hanover, extended to Hamburg, Bremen and Cassel, to be used solely for the Norderney cable messages. This offered an enormous commercial advantage and after the Norderney cable was laid by Fleeming Jenkin in 1866 it established Reuters as the premier news service of the day. The cable itself passed into British government hands in 1870, along with other assets of the Reuters company, but not before Julius Reuter had amassed one of the many fortunes made out of the telegraph and its cables in the latter days of the nineteenth century.

3.4.7   The Exchange Telegraph Company

One of the companies which remained outside the Post Office nationalisation arrangement was the Exchange Telegraph Company. This was founded by Sir James Anderson and George Baker Field, and was also a private news distribution service [34]. It was incorporated in 1862, although it did not begin to do substantial business until 1872, when a licence was granted by the Postmaster General authorising the company to set up a system of telegraphy in London 'within nine hundred yards [820 m] of the Stock Exchange'. Its purpose was to permit direct working between any 'Stock or Commercial Exchange and the Offices of any Broker, Share-dealer, Banker or Merchant'. Similar arrangements were granted later in other towns for the Stock Exchange Company, linking the local stock exchanges in Liverpool, Manchester, Leeds, Birmingham, Edinburgh, Glasgow and Dublin with a central office. An additional service was provided in these towns for a 'telegraphic call system' by which subscribers were enabled 'at any hour of day or night to call a Messenger, Cab or Policeman and also give the alarm of fire'. In its second decade of operation the company became a serious rival to Reuters in providing news, particularly business news, gathered from its own agencies in Britain and abroad, to exchanges, clubs and the evening newspapers, and was particularly active during the war in South Africa when it produced a popular paper entitled *The Telegraph News of the Exchange Telegraph Company*.

3.5 Faster, cheaper telegrams

After the transfer of most of the British telegraph companies into the control of the Post Office, a number of new developments began to appear on the telegraph network. A policy of clearing away all of London's rooftop lines was put into effect, and they were instead placed 'safely and quietly underground' [32]. At the same time exchange switchboards were installed in all the main distribution centres. From February 1870 it became possible to send a telegram for one shilling [5p] from one end of the British Isles to the other. The popularity of the 'shilling telegram' did much to endear the nationalised service to the general public, as did the provision of more telegraph offices in the larger towns in Britain and elsewhere, and resulted in a vast increase in the public use of the telegraph service. The press was particularly enthusiastic about the reduced rates (which were even cheaper for these privileged users), and by the end of the century the number of words transmitted by the telegraph in Britain had increased from 4.2 million in 1874 to 15.7 million in 1899 [35]. Some administrative changes were also carried out to improve the efficiency of the service. A two- or three-letter code was used to identify the towns of origin and destination. Important locations were given two-letter codes, for example AB (Aberdeen), YZ (York), GW (Glasgow) and EH (Edinburgh), and their use materially reduced transmission time for domestic telegrams, as did the provision of registered (and short) 'telegraphic names' for commercial companies and individuals [36].

Technical developments that began to be implemented in the 1870s originated mainly from the American networks. The first of these enhanced the capacity of the telegraph lines by permitting two messages to be sent simultaneously down the same line in opposite directions, a technique known as 'duplex telegraphy'. Two methods were used; the differential relay invented by J.G. Gintle of Vienna in 1853, and developed in practical form by J.B. Stearns in 1872; and a bridge technique, proposed by Alexander Muirhead a few years later in 1875, in response to the particular problems of submarine transmission. The differential relay is shown in Figure 3.11. It was based on an electromagnet with its coil split into two halves, half the current going through each half of the coil. At the transmitter the magnetic effects of the current in the two halves cancelled so that the relay did not close, and the current was simply sent down the line without affecting its own receiving relay. At the receiver the current arriving from the line passed through the two halves of the coil windings so as to reinforce each other, and the relay was operated. A similar arrangement worked in the opposite direction along the same pair of wires. The Post Office began to adopt duplex working extensively in about 1873, and brought quadruplex circuits into use a few years later. Quadruplex working, first put into practical form by Edison in 1874, enabled four messages to be sent simultaneously, two in each direction along the same wire. The system involved the use of two differentially wound relays at each end of the line, with a pair of keys controlling two relays at each station. One of these was a polarised relay which responded only to a reversal of current through it, while the other relay was non-polarised and responded only to the strength of current through it, independent of direction. Given these two different responses to the transmitted current, only two keys were necessary, one to reverse the direction of the current and the other to increase the current when depressed, the two actions being recognised by the appropriate relay at

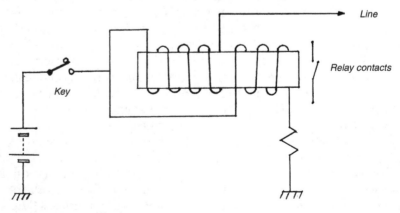

Figure 3.11   The differential relay

the other end of the line. The output current was then further duplexed by the action of a differential relay at each end of the line as described above.

In the Bridge method, shown in Figure 3.12, an artificial line formed part of a Wheatstone bridge circuit, together with two equal resistors, R_1 and R_2, and the transmission line itself. With the bridge balanced, the transmitted current through the key passed directly down the line to a similar bridge device at the other end, causing no response by the local detector, the galvanometer G. A current arriving from the line, however, passed through this detector via the artificial line to the ground return, causing the galvanometer to respond. A similar bridge arrangement at the distant end enabled duplex operation to be achieved. When used for submarine cable operation, the artificial line included a substantial line capacitance and inductance, as well as resistive elements and careful balancing of these components to accommodate differing line characteristics (see Chapter 5) [37, 38].

As well as increasing the density of the telegraph traffic, attention was paid to increasing the rate of transmission. By 1870, just after the Post Office took over responsibility for telegraphy in Britain, a speed of 60–80 wpm was the highest obtainable, even with off-line preparation of the message to be transmitted. In the following thirty years this was increased to 400–600 wpm, some ten times the speed for the most expert manual transmission, and was achieved mainly by wide use of the Wheatstone high-speed recorder system.

At the larger telegraph offices new central switchboards were installed for through connection of telegraph messages and, at the same time, two associated innovations were put into effect to improve the efficiency of

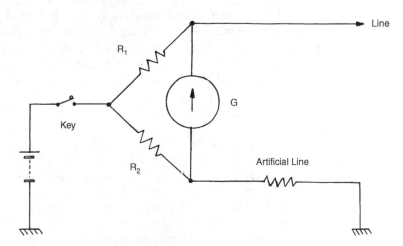

Figure 3.12   The bridge duplex method

the office. The first of these was the use of lamp calling and clearing procedures. This has been quoted as saving 4000 transactions a day at the central telegraph office in London in 1871, reducing the average cross-London delay from 18 to 5 min, and was quickly adopted by other exchanges throughout Britain. The second was the gradual conversion to central-battery working which, in the London area alone, resulted in the abolition of over 1000 scattered groups of primary batteries in telegraph installations. Technical developments in primary cell design also allowed the replacement of expensive Daniell and bichromate cells, still in use, with an improved Leclanché cell. This was coupled with the availability of electric power mains, which made charging easier and considerably reduced the cost of battery maintenance. The techniques of central battery working were extended to heavy users of telegraph services which had their own exchanges, so that, for example, the many newspaper offices in Fleet Street were able to replace several thousand cells by power supplied by lines from the central telegraph office in London [39]. Towards the end of the century the Van Rysselberghe system of combined telephone and telegraph transmission lines was put into effect on the Continent and in America, although it was never widely used in Britain [40]. This was an arrangement much favoured by the telephone companies of the time, and the addition of an earth return enabled the twin metallic telephone lines to be used as one line of a telegraphic system, thus allowing both Morse transmissions and speech to be carried without additional lines. An LC filter was included in the telegraph line to prevent the rapid changes in current that occurred during telegraphy from affecting the telephone lines by mutual induction [41].

The telegraph equipment used in Continental Europe in the second half of the nineteenth century differed little from that available in Britain. Extensive use of dial telegraphs, mentioned earlier and manufactured by Siemens, extended into the German-speaking and Russian countries, while the French made use of a complex Bréguet version, as did the Belgian telegraph companies, which also applied the Rysselberghe shared line system in their networks after the use of telephones became widespread. The improved performance of the Wheatstone automatic printer in the 1890s led to it gradually supplanting Siemens and imported American printers in Continental Europe, and it became widely used until it in turn was challenged, by the Creed printer, a decade later. Although many developments in telegraph design were subject to patent law (English and US patent rights were favoured as the German patents of this period offered less protection to inventors), much technical information was exchanged between European countries, many of which supported the International Telegraph Union, set up in 1865 with the aim of encouraging technical and economic cooperation among its 20 member countries. This cooperation smoothed the problems that arose in linking telegraph networks across national boundaries, and by 1900 the availability of telegraphic communication became almost universal, for every town and hamlet and at fairly uniform rates.

The arrival of the thermionic valve in the first decade of the twentieth century gave telegraphy a new tool for improving the telegram service. One of the ways in which this was used was in the development of new frequency multiplex systems in Britain, Germany and the United States in the 1920s. The principle was to encode a Morse signal of dots and dashes into a train of a.c. impulses transmitted within the audio frequency band. Several of these encoded signals, each imposed on a different audio frequency carrier, could then be summed and passed down a single pair of wires as a frequency division multiplex (FDM) system. At the receiving end a set of bandpass filters, each tuned to the relevant audio frequency, was arranged so as to separate out the individual encoded Morse signals from the multiplexed transmission [42]. Siemens & Halske provided one of the first such systems seen in Britain. This was installed between London and Manchester, and provided six sending channels on one pair of wires and six return channels on another pair. At the sending end six triode oscillators were keyed on and off through relays activated by the Morse signals to be sent, combined and fed through a common amplifier valve to a telephone line pair. At the receiving end the complex waveform was amplified and fed to six bandpass filters connected in parallel. Each filter passed its own frequency, corresponding to the transmitted oscillator frequency, and the result was passed to a separate rectifying valve in the receiving apparatus, at that time connected to a Wheatstone recorder. Later a GEC system of similar design was widely installed throughout the British trunk network. Other FDM systems, designed by AT&T and carrying twelve channels in each wire pair, were installed in the 1920s over a wide area in the eastern United States, while in Germany the Siemens & Halske system was applied to the underground telegraph network to cover most of the principal towns, with links to Switzerland and Austria. Between 1930 and 1940 the voice frequency multiplex scheme was extended to the growing telephone network and became a uniform FDM system carrying both telephone and teleprinter transmission throughout Britain (see Chapter 11).

3.6 Recording and printing

Coding, recording of coded messages, and decoding were operations that required a certain amount of operator skill, and since they were often repeated several times and at different telegraph stations during the transmission of a single message, they also introduced delays. This was not welcome in a commercial environment. Quite early in the development of practical systems for telegraphy transmission, efforts were being made to produce a recording or printing telegraph which would receive and print out the message at the receiving station. A recording telegraph would provide a visual record of the coded transmitted information, whereas a printing telegraph would directly print the decoded information. Both were examples of automatic telegraph machines that could send or receive messages without the intervention of a skilled operator – a long desired development in long-distance telegraphy since the invention of the early (and slow) ABC dial telegraphs.

Before the general adoption of Morse code, several printing devices were tried to simplify the process of information transmission and allow it to be carried out by unskilled operators. The ABC telegraph of Wheatstone has been mentioned earlier; in 1841 Wheatstone devised an ingenious modification that would print out, albeit rather slowly, the actual letter transmitted. It was driven by a clockwork motor and used an electromagnetically operated escapement. The rotating letter dial of the ABC instrument was replaced by a print-wheel looking very much like the daisy-wheel once common in computer printers. This had 24 flexible radial arms carrying the printing type. A drum carrying a sheet of paper covered with carbon paper was located close to the printing arm. In operation the print-wheel was rotated until the desired type was at the right position over the paper, when a mechanical action, driven by a clockwork device, released a weighted hammer which struck the type and impressed a carbon copy of the letter onto the paper. The drum revolved by one letter space, and the hammer mechanism was reset ready for the next letter. A similar system devised by B.S. Yakobi in 1850 has been described by Yarotsky [43]. Neither was used in an operational telegraph network. A special-purpose printing telegraph used widely in the United States was the 'Stock Ticker', invented by Edward Calahan from Boston. This employed a type-wheel limited to printing stock market quotations on a moving paper tape so that pairs of figures could be transmitted quickly between the stock exchanges and directly to brokerage houses. Edison simplified Calahan's machine by redesigning the type-wheel and printing hammer mechanism – and his commercial intervention with this popular invention was instrumental in launching Edison on his electrical engineering career. Calahan is also noted for his invention of the central battery system which came into wide use later in telephone exchanges. Apparently his stock ticker, though popular with finance offices, was held responsible for damage to carpets and furniture due to leakage of sulphuric acid from the batteries it used! The notion of maintaining a battery system remote from the customers and able to provide power for a number of stock tickers located in separate offices was Calahan's solution to this problem.

Steinheil's recording telegraph, described in Chapter 2, printed his 'dot' code onto paper by deflecting magnets, which were connected by capillary tubes to cups of ink. This was widely used in Germany, Austria and Russia in the 1830s and was probably the first recording telegraph to be part of a working network. Morse's own design of telegraph recorder, the Morse recorder, was the first of a series of practical recording devices which began to appear from the 1840s. The paper tape produced had the code for the characters inscribed as zigzag shapes which needed to be read by a skilled operator. An improved version, the 'Morse inker', replaced the side-to-side motion of the recording lever or stylus with a vertical motion or the use of an inked wheel, so as to impress a series of short and long lines on the moving paper strip. Other recording tele-graphs prepared an inked record of the transmitted code in a similar manner to the Morse inker, with slight differences in the mechanism for controlling the inking process using a lever, wheel, inking pad or liquid ink reservoir. Typical were those designed by Bréguet and Siemens & Halske, used mainly in Continental Europe, including Russia [44]. Others, such as the Bain printer described earlier, relied on an electro-chemical printing process and was used predominantly in the North American networks. Several of these later recording devices, including the Bain printer, received coded information, not directly from the incoming signals after generation by a relay, but via a paper tape punched with a series of holes through which a pair of contacts met to initiate the originally transmitted code. This paper tape could be pre-pared off line, using a simple current-controlled mechanical punch, with the advantage that it could be fed into the recorder at a greater speed than originally prepared and so speed up the process of transmission. These potentially faster recording devices were generally referred to as 'automatic recorders'.

In Britain, and later in Continental Europe, universal acclaim went to Wheatstone's automatic high-speed recorder, an instrument he con-tinued to develop for most of his working life [45]. It was patented in 1858 but not brought into service until 1867. The method of reading the data entered into Wheatstone's recorder was unique at the time. Whereas in the Bain and other printers the tape reader simply pulled the tape across a metal surface, and metal springs pressing on the tape made contact through the holes, in the Wheatstone system two metal rods made contact through a pair of holes, producing a sequence of positive and negative pulses accordingly. When a dot was read the two pulses occurred in quick succession; when a dash was read the gap between the pulses was longer thus producing a current, first in one direction and then in the other (as with a dot transmitted but with a longer interval in the case of a dash). These were interpreted by polarised relays which responded to the direction of the current such that dots and dashes could be printed on the moving tape. The Wheatstone recorder consisted of three distinct parts: a perforator upon which an operator prepared a message off line by punching holes in a paper tape; a transmitter which sent the recorded Morse code message over the line, controlled by the presence of holes in the tape at a much faster speed than its preparation entailed; and finally a recorder which could be a Morse inker to record the coded message onto another paper tape for visual inspection. The perforator consisted of three punches in line across the paper and was worked by hand, by electromagnets or with pneumatic power. One row of equidistant small sprocket guidance holes were pre-punched along the centre of the paper tape. On either side were punched larger holes; a telegraphic dot was represented by two holes in line with each other across the tape and on either side of the sprocket holes; a telegraphic dash was represented by two holes, one ahead of the other. The transmitter recognised the different spatial separations of these pairs of holes, controlling the start and stop of a current pulse, which was then transmitted down the line and on to the recorder. Note that the code was impressed on the tape not by the operation of a single connected Morse key, but by a pair of keys on the instrument itself, each controlling the flow of a positive or negative current and demanding a different form of manual skill from that of the conventional Morse key.

Such labour-saving printing devices did not receive universal acceptance. A rearguard action was fought in 1856 by David McCallum with his 'Globotype telegraph', in which coloured balls were released electrically from a series of storage containers, each corresponding to a pre-arranged code, to line up in a glass tube where the numbers or letters represented could be read. He decried printing telegraphs:

They are all very ingeniously contrived . . . but why attach such cumbersome expensive machinery for the purpose of printing the letters? If a message must be printed, why not have a man in the office to do it? He will do it better than it can be done by galvanic power attached to a telegraph wire, etc. etc. [46].

The Wheatstone transmitter was one of the first telegraphic devices which made use of 'double-current' working, which considerably improved the accuracy of Morse transmission, especially over long lines such as submarine cables (discussed in a later chapter). The technique was first applied by C.F. Varley in 1854, then employed by the Electric & International Telegraph Company [47]. The principle was to transmit a unidirectional current down the line during the spaces between the Morse dot and dash signals, and to reverse the direction of this current when either a dot or a dash is transmitted. This caused a very positive change in the position of a relay-controlled switch, and so reduced the risk of error [48]. According to Varley, its major advantage was to overcome the retarding effect of the higher line capacity that became apparent when gutta-percha insulation replaced air as the dielectric medium, particularly in submerged circuits. To 'key in' the requirements of the double-current relay, as well as the differential and polarised relays used for duplex working, the simple Morse key, consisting of a single circuit closure, was replaced by a changeover switch where the line connection was to the rocking arm of the key, while the two ends of the key bar made alternative polarity connections to the line.

By making use of double-current working and ensuring the smooth passage of the tape (oiled paper was used), the Wheatstone transmitter could be made to send pulses along the line some ten times faster than a manual Morse code operator could achieve and, in its final form, was capable of continuous operation at 600 wpm. The Wheatstone equipment, in substantially the form described above, was operated by the Post Office in Britain up to about 1930, the longest operational period for any of the telegraph devices developed in the previous century.

The quest for simplification and avoidance of the need to identify the recorded Morse characters and write out the plain language equivalent led to a number of direct-reading printing telegraphs in which the message could be printed directly in readable type. One of these was that of R.E. House, described earlier, and was used in several of O'Rielly's American networks. In operation the users found that the synchronisation between the transmitter and receiver controlling motors was not too reliable, and before long the system was replaced by the superior Hughes printing telegraph in 1854. All previous printing telegraphs, including the House system, had been based on a slow step-by-step motion, but in the Hughes printer the inventor replaced this by a synchronised, free-running wheel system in both transmitting and receiving motors. When one of his controlling keys was pressed a corresponding pin, chosen from a set of 28 pins arranged in a circle, was raised, electrical contact was made with a horizontal continuously rotating arm at a precise time in the printing cycle, and in doing so sent a synchronising pulse down the line. This controlled the receiving motor which, at certain instants, pressed a gummed paper slip against the type-wheel, kept inked with printer's ink from a roller, and so recorded the selected character from those engraved on the periphery of the wheel. The type-wheel was constrained by a governor to run at a uniform rate, and the wheels at the sending and receiving stations were arranged to run in step. A major reason for its success was the free-running isochronous type-wheel it employed, which printed out each letter without stopping. It also had the great advantage of simultaneously printing out the message at both the sending and receiving ends. This enabled the type-wheel to be rapidly reset if this proved necessary during transmission. Various modifications of this widely successful machine were brought out during the next three decades. The best-known of these was Phelp's modification, used extensively within the United States. This combined the successful features of the Hughes and House instruments but, though somewhat faster than either of them, it was difficult to drive, requiring more motive power – often supplied by a steam engine!

With the passing of the nineteenth century, new methods of direct-reading telegraph printers made their appearance, particularly for long-distance lines. The principal contestants were the Creed system and that of Baudot, the latter having a modified form developed by Murray in the United States. These all used methods of coding and keying in the message quite different from those used by Morse, and represent a major departure from the basic techniques of binary coding that had been in use for so long. Together with the later development of the teletypewriter, they signalled the end of hand-keying operations, and their consideration is deferred to Chapter 11.

3.7 Overhead or underground?

Since the early days of the electric telegraph, a major decision was whether to use overhead wires, supported on insulators, or bury the wires to protect them from the hazards of weather or mechanical interference. The problem was one of cost: underground wires needed to be covered in adequate insulation which would not deteriorate with the passage of time, and the labour of excavating a suitable trench could be excessive, especially in city built-up areas. On the other hand, making use of a simpler construction of suspended wires required suitable wayleave arrangements. This was not a difficulty where the lines followed the route of a private enterprise, such as a railway or canal, but it was a major problem when the wires had to cross the rooftops of a crowded city. Further, the hazards of the wires becoming detached from their fixtures through bad weather or accident could (and did) present a danger to the passing public. For these and other reasons underground installation was the preferred option, and while most countries operated a mixed system incorporating some above-ground wiring and some underground, wires were buried underground whenever it was feasible or economic to do so.

In Great Britain the decision to 'go underground' was not taken lightly. Beresford notes that 'In many towns the Companies had been obliged to go underground and frequently to do so for short sections along country roads' and comments that 'the extravagant outlay on the great underground mains of Germany has always been a mystery, and can only be accounted for by supposing that practical telegraph engineers had nothing to do with the decision' [49]. Major Beresford was concerned since his Royal Engineers were at that time providing considerable assistance to the Post Office with this civilian work, providing extensions to underground lines, and replacing unsightly and hazardous overhead lines, particularly in large towns (see Chapter 4). Nevertheless, the pace of replacement and transfer to underground working in Britain quickened in 1886, after a heavy snowstorm brought out the danger of unanchored overhead wires, and curtailed services over a wide area of

central London. The Electric Telegraph Company was a major propon-
ent of underground telegraph cables and, with the installation of their
new central telegraph station at Founder's Court, Lothbury, enclosed
cables in 3-inch (75 mm) cast iron pipes, laid chiefly under the pavements,
to link their various substations and clients' premises. These pipes carried a number of cables, mainly insulated with gutta-percha, the number
gradually increasing to as many as 20 four-wire cables (quads) insulated
with gutta-percha and wrapped in cotton tape. The work of placing
London's telegraph (and later telephone) network underground was continued by the Post Office after nationalisation in 1871, again using a
standard 3-inch diameter pipe, with test boxes at intervals providing
access from street level, to assist the maintenance of the now extensive
network.

In Germany an extensive network of underground cables was constructed in the 1870s. As early as 1839, when the first Prussian telegraph
lines were proposed, Major O'Etzels, later responsible for installing the
Prussian networks, stated that 'over long distances the lines should not
be carried through the air in order to protect them from accident, wanton
destruction and breakage' [50]. The first underground telegraph cables
from Berlin to Schöneberg were buried 0.9 m deep by the side of the
railway and insulated with rubber or wool impregnated with wax and
turpentine. Several cables were laid underground (mainly by Siemens &
Halske) from Berlin to Frankfurt, Köln, Oderberg, Stettin (now
Szczecin, Poland) and Hamburg, initially to meet the communication
requirements of the newly formed German National Assembly. The
work was hurried; Siemens' advice on the construction of the buried
lines was not heeded and a poor performance was achieved. However, the
cables were afterwards replaced with gutta-percha insulation and the
network extended to cover much of Prussia and Saxony [51]. The work
of installing this vast network of underground cables was shared by
the two firms of Siemens & Halske and Felten & Guilleaume in about
equal proportions. They both made use of gutta-percha exclusively for
their lines, despite its high import cost from South-East Asia (Great
Britain, with its settlements in Malaya, had almost a monopoly of this
valuable insulation material). By 1890 Germany had the longest underground telegraph cable system in the whole of Europe, all of it insulated
with gutta-percha, protected with a covering of sand and plaster and
enclosed in lead or iron tubes. Figure 3.13 shows the final extent of
the German underground network. The numbers by the side of the
lines indicate the year of installation. The network extended into what
was then known as Pomerania, most of which was ceded to Poland
and Russia after the Second World War. A comparison of the extent
of underground telegraph lines completed in European countries a
little earlier, in 1881, is given in Table 3.3 (compiled from telegraph statistics provided by the International Telegraph Union [52], Berne, for

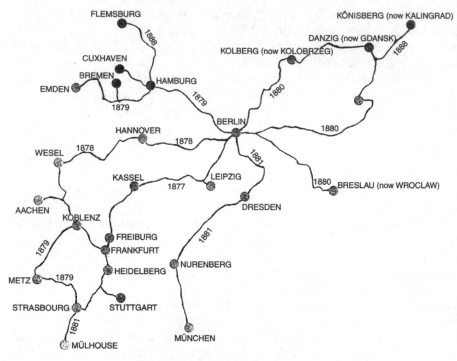

Figure 3.13   The German underground telegraph network in 1890
Source: Siemens Archive Document, Berlin

Table 3.3 *Underground telegraph lines in Europe in 1881*

Country	Cable length (km)*
Germany	5500
France	851
Britain	771
Russia	202
Holland	96
Switzerland	46
Austria	30
Belgium	11
Romania	11
Denmark	3
Total	7251

* The use of multicore cables gave a total wire length of 69 232 km.

1881[8]). Some 200 km of underground cable was also established in Russia, and consisted of cables radiating out from St Petersburg and part of the route from there to Warsaw. These were all laid and maintained by Siemens & Halske using gutta-percha insulation and enclosed, as were the Prussian and Pomeranian cables.

In the United States overhead wires still continued to be used in cities, where a combination of rooftop fixings and supporting poles was often prominent (Figure 3.14). The process of transferring telegraph circuits underground in the United States was not without its difficulties. The need for high insulation and the avoidance of leakages were technical problems which were quickly solved; more insidious effects were the induced currents arising from adjacent telegraph wires

Figure 3.14   Street telegraph communication in the United States
Source: *La Nature* **44**, *1886*

[8] The Bureau of the International Telegraph Union was established in 1868, initially to act as a clearing house for international communications concerned with line telegraphy. It assumed the same role for international wireless communication in 1906.

or from electricity supply cables. This caused some problems in the adjustment of polarised relays, and was one of the reasons for the abandonment of the sounder with its heavy current requirements as the new century advanced. Political objections were also advanced by commercial interests to the wide use of underground installation of telegraph cables. It was reported in New York that 'physicians were found to certify that the opening up of the pavements would very likely give rise to an epidemic of cholera' [53]. A British journal, *The Electrician,* noted that, as London streets are always in a chronic state of dissolution, either Londoners are impervious to cholera or there is no cholera in the composition of the London pavement!

3.8 Telegraphy in British India

Outside Europe and America the largest operating telegraph network in the second half of the nineteenth century was to be found in the Indian subcontinent. Its development can be attributed to one man, William Brooke O'Shaughnessy, a doctor employed as a surgeon by the East India Company in 1833 who also taught as a professor of chemistry in Calcutta [54]. His interests ranged widely, and in 1839 he constructed an experimental telegraph line of almost 50 km in the Botanical Gardens of Calcutta. For signal detection he used the deflection of a galvanometer and, arising from his chemical interests, the decomposition of paper impregnated with potassium iodide, which changed its colour to reveal the presence of a signal when a current passed down the line. He used a synchronisation scheme similar to that of Ronalds, but found it difficult to achieve any useful accuracy with this arrangement. In 1847 sufficient official funding was secured to build a local electric telegraph to assist navigation for the port of Calcutta. This was a substantial system, including over 130 km of line and four telegraph stations at Calcutta, Mayapur, Bishnupur and Diamond Harbour, the end of the line. For the most part the line ran along the banks of the River Hoogli (Hugli), with short diversions to the stations, to which were added Kedgeri and Kukrahutty on the opposite side of the river in February 1852.

Unlike his counterparts in Europe and America, O'Shaughnessy used thick iron rod about 10 mm in diameter instead of iron or copper wire as his conductor. For this heavy rod he claimed the following advantages:

1   immunity from damage by storms;
2   if thrown down, it would still remain undamaged;
3   it could not be broken or bent maliciously without the use of tools;
4   the conductivity was so great that no insulation was required;
5   no straining up in tension was necessary;
6   it would tolerate rusting to an extent fatal to wire;
7   it could not be damaged by birds or monkeys.

He was rather too optimistic about his claims about insulation, particularly for the underground sections of the line, and found it impracticable to retain this heavy construction for the very extensive lines he later erected [55]. Part of this first line was overhead; the remainder was buried, with cloth tape wound round the wire and saturated with pitch, and the insulated lines placed within a row of roofing tiles half-filled with a mixture of sand and resin. This construction was found to be completely effective after 40 years, when a section of the line was excavated. O'Shaughnessy was much concerned with the problems of crossing rivers with his line and experimented with bare iron rod under the waters of the Hoogli, connecting the two banks over a distance of about 1200 m, and succeeding in transmitting weak signals across the river. He also tried using the water alone as a conductor with plates immersed in the water at either end and connected to his apparatus. Some success was achieved, but again the signals were weak and the battery power required was excessive [56]. Insulated underwater cables were effective but it proved difficult to maintain an effective insulation and protect the cables from damage caused by ships dragging their anchors and other underwater hazards. The problem was not finally solved until the 1870s, when a structure of strong steel wire spanning the river and supported by tall masts on either bank was introduced. The longest span achieved with this technique was constructed across the Krishna river north of Madras: it was 1500 m long, and was still in operation by the turn of the century.

In 1852 O'Shaughnessy's activities and the communication needs of the time finally convinced Lord Dalhousie, then Governor-General of Bengal, of the necessity for a nationwide telegraph network. He proposed in a minute to the Government of India that telegraph lines be constructed widely over the continent, commencing with lines from Calcutta (then the capital of British India) to Bombay, Agra, Peshawar and Madras. O'Shaughnessy was appointed Superintendent General of Telegraphs, and sent to London to give evidence to the Court of Directors on this proposal and arrange for the despatch of relevant telegraph material to India. It was during this visit that he became convinced of the value of the Morse system of telegraphy, and he sent out to India sets of every instrument then in use or proposed. A period of experimentation followed during which several thousand kilometres of lines were constructed across India, mainly using needle instruments. A Telegraph Act of India was passed in 1854 defining the provision and use of a national telegraph network. It stipulated penalties for establishing or maintaining unauthorised electric telegraphs, and prescribed punishments for cutting and stealing line and authorised the Telegraph Department to loan lines to the railways [57].

Two years later, in 1856, O'Shaughnessy made a second visit to Europe, this time concerned with the implementation of a full Morse system on his network, and to receive a knighthood for his work in India. He became impressed with the Morse implementations on the Continental European telegraph networks described by Werner Siemens, whose acquaintance he made in Germany, and ordered a supply of instruments from the firm of Siemens & Halske, a branch of which was by now located in Britain. From this date, Siemens became the main supplier to the Indian telegraph network, both for railway and commercial use. To assist in the introduction of Morse code into India, a 'class of young gentlemen' was formed at Gresham House in London, to be trained and sent out to India to introduce the Morse system. In 1857 the first detachment of 74 of these 'Morse assistants' went out to Calcutta, Bombay and Madras, and the change from the needle to the Morse system was begun. While O'Shaughnessy had been away, the installation of new telegraph lines had proceeded rapidly and over 1500 km of additional lines were in place by May 1857, including 500 km along the railway set up by the East Indian Railway Company. However, by the time he returned to Calcutta in December 1857, he did so in the middle of the Indian Mutiny and found many of his telegraph lines destroyed in much of Bengal, the Punjab and central India. In the next chapter the sequence of events leading to this is considered in some detail, as are its effects on telegraph operations for commercial and military life in India. Despite the equipment losses, murders of telegraph operators and low morale among the staff, O'Shaughnessy effected a remarkable recovery from an almost untenable situation. The old lines that had been destroyed were rapidly reconstructed, and new ones erected, and by the end of March 1858 they extended to a total length of 16000 km over the entire continent. During this time the network had been extended to Ceylon (Sri Lanka), including a cable across the Gulf of Mannar to link with the mainland. The laying of this cable was a difficult operation, carried out with native sailing vessels in bad weather. This was carried out by Mr Wickham of the Indian Telegraph Department, as described by O'Shaughnessy:

The operation was as difficult, the line as long, the navigation at least as dangerous, as that of placing the cable across the Straits of Dover, for which a squadron of steamers and costly machinery were employed. Mr Wickham performed his task under sail, and with no other apparatus than the rude windlass of a native vessel.

The cable, insulated with gutta-percha and protected with iron guards, continued in working order until 1867, when it was replaced.

On the reorganisation of the Indian telegraphs after the Mutiny, a number of improvements were introduced. These included the use of the Morse sounder in place of the single-needle system, several years earlier than its general introduction in Britain, and a lightning protector, which he called a 'discharger', connected to the line. This latter was important in a tropical region subject to many storms, and it proved extremely successful. It consisted of a bundle of iron wires or a tray of iron filings, connected to earth, through which the circuit wires passed, insulated only by a thin bituminous paper. While the insulation was adequate for weak telegraph currents, a lightning discharge would be expected to puncture the paper and pass harmlessly to earth. This technique, patented by Edward Highton as a 'lightning strainer', had been exhibited for the first time at the Great Exhibition of 1851, and O'Shaughnessy may have learnt of its properties during his visit to London [58]. The year 1858 also saw the introduction of 'telegraph stamps', which enabled telegrams to be accepted at places where no telegraph station existed. India was the first to introduce this useful administrative idea, which was copied in Europe and America. O'Shaughnessy returned to England for good in 1860, leaving behind him well-constructed lines extending to 18 000 km, 150 telegraph offices and a network which had passed over 200 000 messages in India and Ceylon during the previous year. Shortly after he left, the Indian lines were connected to a smaller telegraph network which had been developed in Burma (Myanmar), thus completing British telegraph communications in what was known as Further India.

After the Mutiny, a close liaison between the military and the Post & Telegraphs organisation in India resulted in the inauguration of a system for civilian training of British soldiers in electric telegraphy, a procedure that was later introduced in Britain following the formation of the Royal Engineers Telegraph Battalion (see Section 4.7). Unlike the British arrangement, however, the men were kept in training by employing them as paid operators in civil offices, which must have been helpful to the army telegraphist after he left the force. It certainly proved invaluable to the administration during emergencies, such as famine or flooding, when the civilian telegraph organisation was hard pressed.

O'Shaughnessy and his successors took considerable interest in training and education for telegraph operators and supervisors to work the Indian lines. Supervisors were expected to pass a preliminary competitive examination and attend the Royal College of Indian Engineering in England for further education before being sent out to India. Local operators began to receive all their training in India in the 1890s, and work in the telegraph offices became a sought-after position for the educated Indian. A textbook was written and published in India for training operators on the Indian railways and used in training operatives, previously recruited and trained in England. One of the training diagrams is reproduced in Figure 3.15, and shows the central place of the Morse sounder in the railway network of the 1890s, together with the use of the lightning discharger, discussed above [59]. Note that single current working used for the 'up train' and 'down train' circuits makes use of a two-way Morse key switch so that, in its normal receive connection, the line is connected to the sounder and is transferred to battery connection via the key only

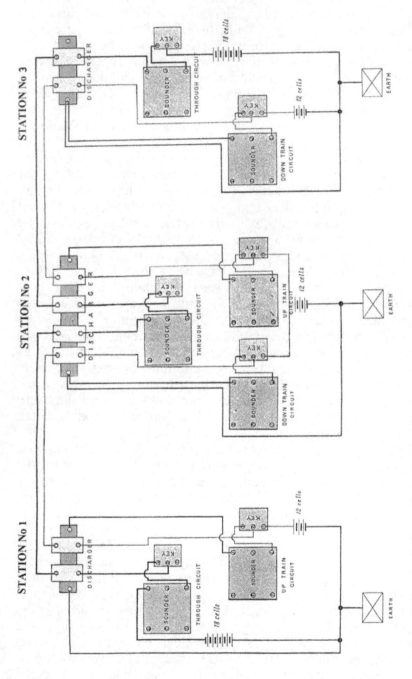

*Figure 3.15　An Indian railway training diagram
Source: James, 'Morse Signaller's Companion', 1894*

to send current down the line during transmission. The through circuits are thus maintained in the default operating condition, providing a constant current through the line connections of all three stations which may be interrupted by a keying operation at any station.

References

1 *Transactions of the Royal Society of Arts*, 1825, **63**, p. 37
2 TAYLOR, W.B.: 'An historical sketch of Joseph Henry's contribution to the electro-magnetic telegraph' (Smithsonian Institution, Washington, 1879)
3 REID, J.D.: 'The telegraph in America' (Derby Bros., New York, 1879)
4 MORSE, S.F.B.: 'Experiments made with 100 pairs of Grove's battery', *Sillimans American Journal of Science*, 1843, **45**, pp. 390–94
5 THOMPSON, R.L.: 'Wiring a continent 1832–1866' (Princeton University Press, Princeton, N.J., 1947)
6 MORSE, E.L.: 'Samuel F.B. Morse, his letters and journals', vol. 1 (Houghton Mifflin, New York, 1914)
7 HARLOW, A.F.: 'Old wires and new waves' (D. Appleton-Century Co., New York, 1936)
8 MAVER, W.: 'American telegraph system: apparatus and operation' (Wm. Maver & Co., New Jersey, 1898)
9 HARNETT, R.: 'Wirespeed: codes and jargons in the news business' (Shorebird Press, San Mateo, Cal., 1997)
10 POPE, F.L.: 'Royal E House and the early telegraph', *Electrical Engineering* (NY), 6th March 1895
11 BREWER, A.R.: 'Western Union Telegraph Company, 1851–1901' (The Western Union Co., New York, 1901)
12 BURNS, R.: 'Alexander Bain, a most ingenious and meritorious inventor', *Engineering Science & Education Journal*, (1993) Vol 2, No 2, pp 85–93
13 PRESCOTT, G.B.: 'History, theory and practice of the electric telegraph' (D. Appleton, New York, 1866), pp. 127–34
14 Letter from Kendal to Morse, 27th August 1857, *in* Morse papers, vol. 38, Library of Congress
15 WASHBURN, C.C.: 'Speech of Hon. C.C. Washburn of Wisconsin in the House of Representatives, December 22, 1869' (F.J. Rives, Printer, Washington, D.C., 1869)
16 'War of the Western Union on the postal telegraph' (1873), Library of Congress Shelf No. HE7781.W25, Washington, D.C.
17 'Hiram Sibley', *in* 'Dictionary of American Biography' (Oxford University Press, Oxford, 1935), vol. 17, p. 145
18 'Statistical abstracts of the United States: 1902' (Superintendent of Documents, Washington, DC, 1903), pp. 418–20
19 'Telegraphists' cramp', *The Electrician*, 1908, **62**, p. 282
20 SMITH, MAY: 'A study of telegraphists' cramp' (HMSO, London, 1927)
21 CELORIA, F.: 'Early Victorian telegraphs in London' (Middlesex Archaeology Society, London, 1978), pp. 415–35

22 'Head, Sir Francis Bond', in 'Dictionary of national biography', vol. 25, pp. 324–5
23 HEAD, Sir FRANCIS: 'Stokers and pokers on the London & North-Western Railway, The electric telegraph and the Railway Clearing House' (John Murray, London, 1849)
24 'The Central Telegraph Office', *Illustrated London News*, 28th November 1874
25 'London telegraph station', *Journal of the Telegraph* (NY), 15th May 1871, **4**, (12), p 137
26 KIEVE, J.: 'The electric telegraph — a social and economic history' (David & Charles, Newton Abbot, 1973)
27 'Scudamore, Frank Ives', *in* 'Dictionary of national biography', Suppl. II, p. 153
28 'An Act to alter and amend "The Telegraph Act, 1868"', 31 & 32 Vict., c. 110, (Eyre & Spottiswoode, London, 1869), Chap. 73
29 WILSON, E.: 'Government and the telegraphs' (Macmillan, London, 1868)
30 'A telegraph office', *Freelance*, 12th December 1868, **3**, p. 399
31 BERESFORD, C.F.C.: 'The Deal time ball', *The Military and Telegraph Bulletin*, 15th September 1884, pp. 51–2
32 DURHAM, J.: 'Telegraphs in Victorian London' (The Golden Head Press, Cambridge, 1959)
33 'House top telegraphs', *All the Year Round*, November 1859
34 SCOTT, J.M.: 'Extel 100, the centenary history of the Exchange Telegraph Company' (Exchange Telegraph Company, London, 1972)
35 BROWN, L.: 'Victorian news and newspapers' (Oxford University Press, Oxford, 1985)
36 MACKAY, J.A.: 'Telegraphic codes of the British Isles 1870–1924' (Mackay, Dumfries, 1981)
37 STRANGE, P.: 'Duplex telegraphy and the artificial line', *Proceedings of the Institution of Electrical Engineers A*, 1985, **132**, (8), pp. 543–52
38 McNICOL, D.: 'American telegraphy practice' (McGraw-Hill, New York, 1913)
39 ANON: 'Telegraphy', *Post Office Electrical Engineers' Journal*, 1956, **49**, pp. 166–72
40 BAKER, E.C.: 'Sir William Preece, Victorian engineer extraordinary' (Hutchinson, London, 1976), p. 319
41 MOURLON, M.C.: 'The Van Rysselberghe system of simultaneous telegraphy and telephony', *The Electrician*, 1885, **14**, pp. 459–60, 476–7
42 CRUICKSHANK, W.: 'Voice frequency telegraph', Journal of the Institution of Electrical Engineers, 1929, **67**, pp. 813–31
43 YAROTSKY, A.V.: '50th anniversary of the electromagnetic telegraph', *Telecom Journal*, October 1982, **49**, pp. 709–15
44 SIEMENS, C.W.: 'The electrical exhibition in Paris — telegraph apparatus', *in* 'The record of the international exhibition 1862' (R. Mallet, Paris, 1862), pp. 529–44
45 BOWERS, B.: 'Sir Charles Wheatstone, 1802–1875' (HMSO, London, 1975)
46 McCALLUM, DAVID: 'The Globotype telegraph' (Longman Brown & Green, London, 1856)

47 VARLEY, S.F.: 'Varley's double-current system', *The Military and Civil Service Telegraph Bulletin*, 15th August 1887, pp. 329–31

48 POST OFFICE ENGINEERING DEPT: 'The Wheatstone telegraph system' (HMSO, London, 1924)

49 BERESFORD, C.F.C.: 'Records of the postal telegraph companies', *The Military Telegraph Bulletin*, 15th April 1884, (6), p. 19

50 CRAEMER, P. 'Geschichte der Deutschen Fernmeldekabel' (E.S. Mittler, Berlin, 1940), p. 6

51 SIEMENS, W. Von: 'Kurze Darstellung der an den Preussischen Telegraphlinien mit unterirdischen Leitungen gemachten Erfahrungen' (1851), Siemens Archive Document, Berlin

52 'The Berne Bureau', *in* 'Yearbook of wireless telegraphy and telephony' (Wireless Press, 1920), pp. 954–6

53 'Electrical progress', *The Electrician*, 9th May 1885, **14**, p. 532

54 BRIDGE, J.A.: 'Sir William Brooke O'Shaughnessy MD FRS FRCS FSA', *Royal Society Notes and Topics*, 1998, **52**, pp. 103–20

55 LUKE, P.V.: 'Early history of the telegraph in India', *Journal of the Institution of Electrical Engineers*, 1891, **20**, pp. 102–22

56 FAHIE, J.J. 'A history of wireless telegraphy 1838–1899' (Blackwood, Edinburgh, 1899)

57 'The Telegraph Act of India', *Bengal Almanac*, 1854, (34)

58 BEAUCHAMP, K.G.: 'Exhibiting electricity' (Peter Peregrinus and the Institution of Electrical Engineers, London, 1996), p. 83

59 JAMES, C.S.: 'The Morse signaller's companion' (1894), IEE Library document, London

Chapter 4

Military operations

From its beginnings, telegraphy was considered by governments to be a powerful adjunct to military operations. This was particularly so in France: following the successes obtained with Chappé's semaphore system by Napoleon's army, the first electric telegraph line was constructed by the government for its own use in 1845. Indeed, in July 1847 the French minister of the interior was to declare in the Chamber of Deputies, 'La télégraphie doit être un instrument politique, et non un instrument commerciale' (The telegraph must be a political instrument, not a commercial one). It was five years after the construction of France's first electric telegraph line before any commercial use was allowed, and even then governmental use had priority. In Prussia military use was dominant when the first government telegraph line was established in 1847, and the public were not permitted to use the state system until 1849. Similar arrangements were in force in Austria, Holland and Russia, while in Britain, although the mechanical semaphore system was initially a military venture, the electric telegraph was shared fairly amicably by commercial and military interests. In the United States commercial considerations prevailed from the first, even during the Civil War, which took place in the midst of a major expansion of the commercial telegraph service.

In this chapter the use made by the military of electric telegraphy up to the end of the nineteenth century is considered. To a major extent this became the province of governments and the armies, with navies relying on other methods for distant communication. The breadth and scope of these applications began to intensify towards the end of the century, when the first tentative steps in wireless telegraphy began to affect military communication in all the services, but this development will form the context of later chapters. Here the gradual acceptance of line telegraphy by the military is assessed in terms of its value in field service to the commanders and participants in the many conflicts that filled the second half of the nineteenth century.

In Britain the Admiralty was the first to use telegraphy, its semaphore system connecting it with naval bases at Portsmouth and Plymouth. Wellington enlisted the help of the Royal Navy in the Peninsular War and a chain of visual telegraph stations was established between his headquarters, near Lisbon, and Badajoz, 200 km distant. To communicate with the fleet, a system of a mast and yardarm support was used from which hoists of up to five hanging balls, one flag and one pendant were hung, indicating the numbers 1 to 999 according to their position, according to a numerical cipher system. Wellington's commanders used a portable semaphore telegraph designed by Gamble (shown in Figure 1.7), and a number of these became available for military use. The Gamble semaphore was used to augment the defences of the military lines at Torres Vedras which stretched to the Tagus river, said to include a hundred forts and entrenchments. Perhaps this encouraged the formation of the first army telegraph section, a small group attached to divisional headquarters, set up by Wellington in 1814, equipped with poles, flags and balls, and using a codebook evolved from the naval flag dictionary. Use of the electric telegraph was slow in being assimilated into military planning, and had to await the urgent requirements of the Crimean campaign in 1854–56, at a time when its commercial use was already well established in Britain.

4.1 War in the Crimea

A major aim of the war was to check the expansion of Russia towards Constantinople, and prevent the disintegrating Ottoman Empire from falling within the Russian sphere of influence. To this end, capturing the Russian naval base at Sevastopol was seen as an essential first step. Even before the start of the campaign the Russians held the advantage in communications since a working semaphore system between their headquarters in Moscow and Sevastopol was in place, based on the Chappé system. This enabled short messages to be conveyed in about two days. In early 1854 the Russians placed an order with Siemens & Halske, which had already established several electric telegraph lines in their country, to 'construct as quickly as possible an overhead telegraph line from Warsaw to St Petersburg'. This was followed by additional lines in the north to Helsinfors, Kronstadt and Revel, and south to Odessa and Sevastopol on the Black Sea, a telegraph network which covered a total distance of 10 000 km, extending from present-day Poland and Finland down to the Crimean Peninsula (Figure 4.1). These lines, completed by 1855, were a considerable help to the Russian authorities in controlling the movement of troops and war matériel, and, not least, in enabling direct communication with Berlin to arrange for the shipment of heavy war equipment from Germany [1]. The Morse telegraph equipment developed by the

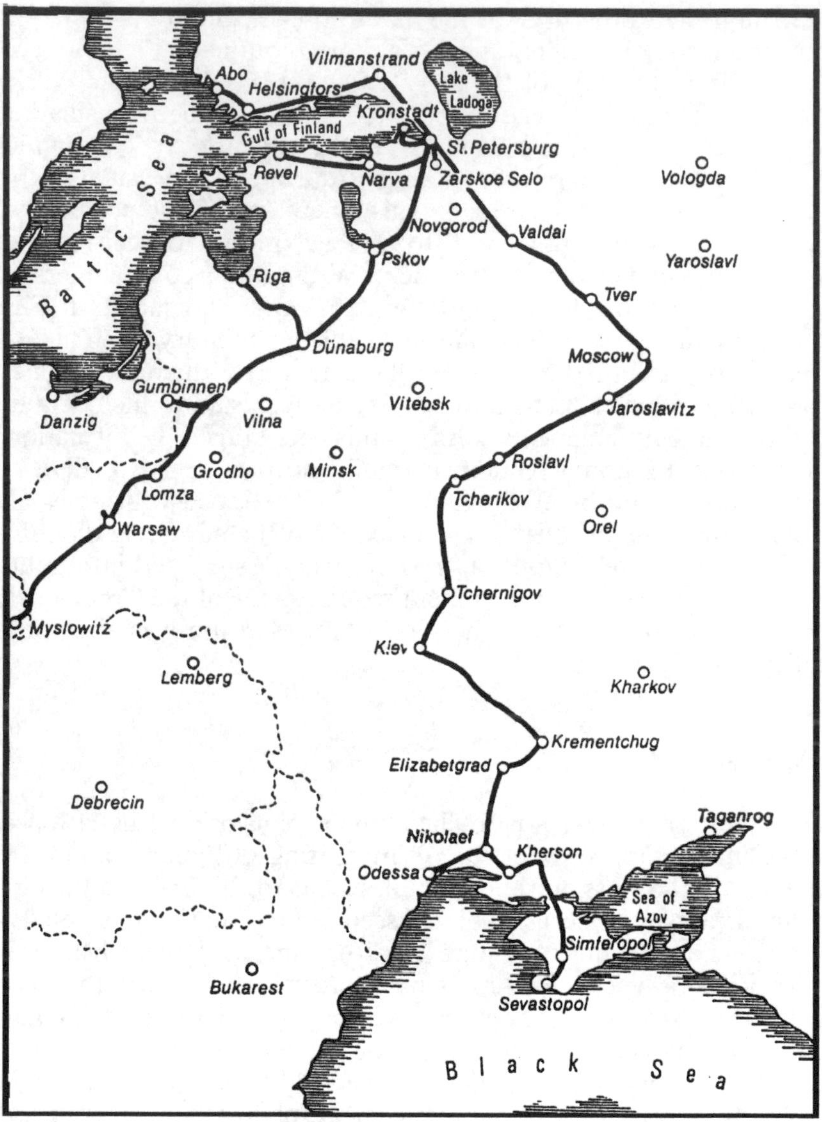

Figure 4.1   The Russian telegraph network to the Crimea in 1855
Source: Siemens, 'Recollections', 1889

German company for this network included for the first time in 1854 the
Siemens automatic writing system, shown in Figure 4.2, in which a
punched paper tape was prepared off line for data transmission by a
three-key punch, similar to the Wheatstone automatic system described
in the previous chapter. Indeed, Werner Siemens noted that, 'Wheatstone

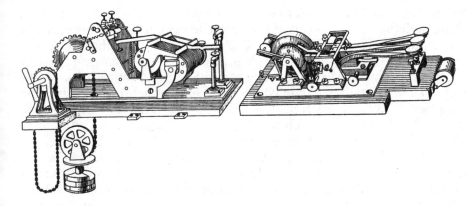

*Figure 4.2   Siemens' Crimea telegraph equipment, consisting of an automatic
            writing system (left) and a three-key perforator (right)*
Source: Siemens, 'Recollections', 1889

made good use of my three-key punch in 1858 for his electromagnetic express writer, without however naming the source whence he derived it' [2].

The Allies (Britain, France and Turkey) made use of a mixed system. Messages would be carried by a steamer from the Crimean Peninsula to Varna in Bulgaria, and then by mounted couriers, galloping from Varna to Bucharest, and finally via ordinary postal routes to London or Paris, taking from 12 days to 3 weeks to arrive at their destination. In 1855 France and Britain paid for an electric telegraph line to be laid between Varna and Bucharest, a distance of 200 km, passing through Shumla, Silistria and Rustchuk, the expense of its construction shared equally between them, and the line constructed by the French 95th Regiment. At Bucharest the line joined the trans-European telegraph network to Paris, which was linked via the cross-channel submarine cable, in place since 1851, to the War Office in London.

To facilitate communication between the armies in the Crimea, the Electric Telegraph Company had provided two wagons, each fitted with 20 km of cable, batteries and instruments, which enabled the armies to set up field lines, specifically linking the British headquarters in Balaclava with the French headquarters in Kamiesch. Some 34 km of line was eventually laid, and eight telegraph stations established. The wire was intended to be buried, and was insulated heavily with gutta-percha for this purpose, but this was not always achieved, leaving the wire exposed to frequent damage. The work was carried out by Royal Engineer sappers who had been sent to the Electric Telegraph Company in London to receive preliminary training in laying and working telegraph lines. The telegraph instruments used were a special field-pattern Cooke and

Wheatstone single-needle device, mounted on a tripod with a battery box slung underneath, modelled on the needle telegraph used so successfully on the railway network in Britain. The company also supplied an apparatus, designed by Latimer Clark, their chief engineer, intended to lay the cable just below the surface of the ground at a rapid rate. This took the form of a converted plough in which the cable was fed from a drum on a hauling wagon through a hole in the attached plough to fall into a trench created by the plough, which was filled automatically with earth as the plough moved forwards (Figure 4.3). However, the ground in the Crimea was too hard for the device to be used, and a narrow trench had to be dug manually and later filled in; but the idea was used later in other theatres of war. Burying the cables only 100 mm or so below the ground did not afford much protection, and breaks in the lines were frequent. The sappers, whose task it was to repair these breaks, were not as yet particularly skilful in the job. A contemporary account by a civilian volunteer described the repairs:

The wire was very frequently broken from being so near the surface; and when this was the case, the sappers who worked the telegraph went after the fault and, having found it, they twisted the wires together, wrapped them with gutta-percha, and melted this with a lucifer. In consequence of the numerous bad joints from this mode of making them, the insulation was very defective. When we arrived, I of course saw that they were made properly, the linesman doing it instead of the sappers. [3]

Lack of sufficient trained personnel from the sappers' companies led to a decision by the Royal Engineers to supplement them with civilian staff, recruited directly from the Electric Telegraph Company. In 1855, Alfred Varley, an electrician with the company, was sent out with 15 civilians. They were deployed in various locations for a while until the rigours of military life caused them to move to Constantinople, or to Varna to work

Figure 4.3   Latimer Clark's cable plough
Source: *The Military Telegraph Bulletin, by courtesy of the Royal Corps of Signals*

the lines. The telegraph facilities in the peninsula and the connections from Varna to the European telegraph network were gradually put in order, but what was missing in these communication arrangements was a link across the Black Sea from Varna, from where the extended European telegraph network could be reached, and Balaclava at the southern tip of the peninsula (Figure 4.4). This would have to be a submarine cable, but the required length of 550 km was longer than had yet been attempted by any of the cable companies.

A British company, R.S. Newall, was charged with the task, which had to be completed early in 1855 if it was to have a useful effect on the conduct of the campaign [4]. Newall was well equipped to provide cable-laying equipment on board the ships. In addition to its expertise in producing cables for several submarine telegraph companies, including the cables used in the cross-channel routes from England to France in 1851–52, they were manufacturers of winding drums with friction-band brakes, much used in colliery winding equipment, which were essential for sea-laying operations. The starting point for the Crimean cable was Kaliakra Castle on the shores of Kaverna Bay, near Varna, where the ships to lay the cable were moored. These were the steamer *Argus*, chartered by R.S. Newall, and two navy ships, HMS *Terrible* and HMS *Spitfire*, which also carried the whole of the telegraph apparatus and the military personnel to operate it, consisting of men of the Royal Artillery

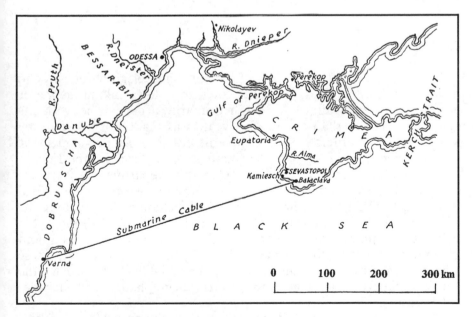

Figure 4.4   The Crimean Peninsula, showing the position of the submarine cable
Source: Nalder, 'History of the Royal Corps of Signals', 1958, reproduced by permission of the Royal Signals Institute

and Royal Engineers. The cable itself was unarmoured, except at the ends where it reached the shore, and was insulated throughout with gutta-percha. The frail nature of the cable made this a delicate operation, carried out often at a speed of only 3 knots.

During the operation the cable itself was used to communicate with Kaliakra to check that it remained intact. It was laid in just eighteen days in April 1855, and was to last until December, shortly after the fall of Sevastopol, which formed the last piece of news transmitted through the cable before the unprotected cable broke and could not be repaired [5]. It had served its military purpose, however, and enabled news of the campaign to reach both the military headquarters and the public. The speed of communication, now reduced to about 24 h, was considered a mixed blessing by the commanders in the field. Whereas previously, under the slower methods of communication used, they were left to exercise their own tactical judgement, with the more efficient methods the armies were commanded nearly as much from London and Paris as from the seat of war itself. The British Commander-in-Chief, General Simpson, was inundated with enquiries from the War Department on pettifogging matters of minor administrative detail, and was reputed to have remarked that 'the telegraph has upset everything!' [6]. The facility was also much used by Louis Napoleon from his headquarters in Paris and was to result in the resignation of his General Canrobert, who complained of over-zealous meddling in the conduct of the campaign [7].

4.2 The Indian Mutiny

By 1857, the date of the Mutiny, the commercial telegraph network in India was well established, with over 6000 km of lines linking the major centres of communication. Although primarily a civilian network, its rapid expansion in the previous five years was made possible only by assistance, under orders from the government to every department of state, in which the military played a significant role. From March 1856 to May 1857 alone, when the Indian Telegraph Department was super-intended by Lieutenant Chauncey, and Captain Stewart of the Royal Engineers, 1500 km of new line was installed including 500 km along the railway line from Burdwan (Barddhamar) to Patna [8]. In addition to the civilian telegraph network, the army made wide use of the heliostat and heliograph, the latter being worked at night by the use of limelight. These techniques of visual communication became highly developed by the army in India, worked alongside the electric telegraph, in the last decades of the century.

In May 1857 a large part of the sepoy army stationed at Meerut mutinied, shot their officers and marched on to Delhi. The revolt spread to many areas of British India (and is well documented: see for example

Reference [9]). The mutineers, a number of whom had received training in telegraphy while attached to civilian offices, were well aware of the value of the telegraph in military operations, and quickly destroyed the lines in the areas that they held. The first section of the line destroyed was between Meerut and Delhi, but not before the Delhi office had sent out a message informing stations on the network of the events taking place [10].[1] This enabled the Punjabi authorities to disarm their native regiments before they became aware of the success of the rebel sepoys at Meerut, which considerably reduced the effects of the Mutiny and probably saved British India from complete collapse. In the Delhi area the destruction of telegraph lines continued with lines cut between Agra and Indore, Cawnpore (Kanpur) and Delhi, and those between Allahabad, Cawnpore and Lucknow wrecked.[2] However, the line between Delhi and the Punjab was kept open during the whole time of the siege of Delhi, despite the massacre of Charles Todd, in charge of the station, and several of his staff. The morale in the telegraph offices during the Mutiny was high, and this helped to keep the part of the network not in the mutineers' hands operating, thus materially assisting the British Army's efforts to restore peace and re-establish the network after the Mutiny was over. One superintendent's report noted that

Signaller Higgins, at Beowra, buried his instruments before he left. The office was soon in flames, but the instruments were uninjured; and when General Mitchell, on the 15th September, attacked and dispersed the forces [he was able] to dig up his instruments and re-open communication on the same day' [10].

After the suppression of the Mutiny in 1858 the British Government assumed direct control over affairs in India in place of the now disbanded East India Company, and the military arrangements for control and defence of the subcontinent were brought under direct control of the British crown. This included the establishment of signal units within various sapper and miner regiments. Several of these units took part in the second Afghan War in 1878, when two of the companies became classified as telegraph units. One accompanied the column that marched

---

[1] The telegram announcing the rise of the mutineers sent to Rawalpindi from Brigadier Frazer, the officer commanding in Delhi, reads:

Message from Delhi 11th May
From Brigadier Frazer,   To Amballa
    Comdg at Delhi   To Brigadier Comdg at Amballa
Number of Words 78 Service

Cantonment in a state of siege. Mutineers from Meerut. 3rd Light Cavalry number not known said to be one hundred and fifty men cut off. Communication with Meerut taken possession of the Bridge of Boats. 54th N.I sent against them but would not act. Several officers killed and wounded. City in a state of considerable excitement. Troops sent down but nothing certain yet. Further information will be forwarded. Copy to be sent to Brigadier in Command Rawelpindle.

[2] At the end of the Mutiny it was calculated that the mutineers had destroyed 1500 km of telegraph wires causing 'a damage of five lakhs [500 000] of rupees'.

from Quetta to Kabul, and the other the column operating on the Khyber route, extending the civilian telegraph lines as they went by laying field cable up to the frontier. To assist recruitment to these and other telegraph units, a School of Army Signalling was opened in Bangalore in 1881, reducing the need to rely solely on military signal training in England.

4.3 The American Civil War

The next important military conflict in which the telegraph played a major role was the American Civil War. A feature of this war which made the telegraph so vital to both sides was the geographical spread of the network. In previous wars in which the telegraph had been used, it was often necessary to establish lines of communication with the forward troops. Control of the front via telegraph lines had not been essential, although it had been exercised in a desultory way in the Crimea. Certainly the need for frequent communication over thousands of kilometres with several large and widely distributed armies did not arise, and neither had there been a need to manage the moving and supplying of very large numbers of troops over a wide area – and both became features of the American Civil War. The well-established railroad network across the continent was used extensively by both sides, but particularly by the Union side, which had longer lines of communication to contend with.

Before 1861 there was no use of, or provision for, a separate military electric telegraph network. The United States Signal Corps would have been expected, at the very least, to experiment with the new technology, but relied on 'flags by day and torches by night [and] stood inadequate to the vast terrain and the huge dispersed forces engaged in the struggle' [11, 12].[3] Yet in 1861, at the commencement of the war, the civilian telegraph services were well advanced and covered almost all the country, except for the far west. The three major telegraph companies at the start of the conflict were the American Telegraph Company, the Western Union Company and the Southwestern Telegraph Company. The first two of these connected all the cities and a great number of the towns supporting the Union, whereas the Southwestern Telegraph lines were situated mainly in the Confederate States. A number of smaller companies were in operation, filling the gaps left in the coverage of the three larger companies. As a consequence the telegraph facilities needed by both sides were met initially by making use of this civilian network – a situation that was far from satisfactory to either military or commercial

---

[3] The situation in the Signal Corps was not to change until Augustus Greely, its first commander, pushed through changes during the Spanish–American war in 1898, when its most successful operation, the Santiago de Cuba campaign, was launched on the basis of an intercepted message by a technical officer in the Corps [13].

interests. The Union Government under Abraham Lincoln, established in Washington, and the newly fledged Confederate Government under Jefferson Davis in Richmond, Virginia, both thought that the commercial companies would be able to supply all their military needs, even to the extent of extending lines to the battle fronts and accompanying the troops on the march. The scale and needs of the conflict soon disabused the military of this notion, and gradually telegraph corps were established, albeit with equipment and men from the civilian organisations.

Lincoln and the Union side were the first to appreciate this requirement. Lincoln himself had spent an early part of his life as a civilian telegraph operator, and recognised the need for an organised military telegraph system long before Congress and many of his generals did. In February 1862, Lincoln took control of most of the nation's telegraph lines. The following year the United States Military Telegraph Department was created, staffed by civilians and separated from the United States Signal Corps, which had never successfully incorporated telegraph operations, and indeed there was some friction between the corps and the new Telegraph Department. The pre-war commercial network had connected cities, but armies often fought and moved outside cities. Portable, front-line electric communication for field operations was required. General Ulysses S. Grant recognised this and was generous in his praise for the work of the telegraph service in the field. 'The signal service was used on the march,', he said, 'and the moment the troops went into the camp the telegraph service, never needing orders, would set to work to set up its wires.' Telegrams from the field were essential to Lincoln in his role as Commander-in-Chief, enabling him to follow and supervise military action, and he would often read the translation of the Morse code over the operator's shoulder. In periods of stress he was known to send ten or twelve dispatches to various generals in a single day. Some of these dispatches were in cipher, and Lincoln is said to have invented the first of these for the Union side, a simple affair in which decoding consisted of reading backwards, stressing phonetics instead of spelling. Ciphers were used by both sides. More than a dozen new ciphers were devised during the war and discarded as soon as the enemy learned how to read them. The Confederate ciphers were simpler than those of the Union, and were frequently cracked by the northern experts [13].

During the first two years of the war the Confederates did not develop a government military telegraph service as did the north. Instead, the southern part of the American Telegraph Company was reorganised as the Confederate Telegraph Company, an independent unit which quickly dominated the industry in the south and was in effect a nationalisation by the Confederate Postmaster General, J.H. Reagan of the American Telegraph Company. A few generals carried telegraph instruments with them, but for the most part the Confederate troops relied on commercial lines.

At the beginning, however, for both sides the civilian telegraph network played an important role in mobilising the armies. In the north, immediately following the Confederate attack on Fort Sumter on 12th April 1861, which signalled the start of the war, Lincoln telegraphed all the states of the Union for 75 000 troops. Within three days he received replies which indicated that 100 000 volunteers were prepared to enter the Union army. The telegraph's first vital campaign success was in support of a Union effort to ward off an attack by Confederate forces directed against Washington. A telegram to the governor of Massachusetts for twenty companies to muster on Boston Common, and a similar telegraphic request for five companies from Pennsylvania to proceed to the capital, were answered promptly, as were requests for the railroads to transport the troops. Twenty-four hours' delay, and Washington would have been in the hands of the Confederates. Major Plum, in his impressive *The Military Telegraph During the Civil War*, claims that 'The telegraph, railroad and troops unitedly saved the Capital. Without either, Washington was lost' [14].

The telegraph was also to convey the news of the first Union defeat, but only with the aid of couriers to cover the last 16 km to Washington from the Union Army of the Potomac (General McDowell apparently found it impractical to complete the line – or perhaps he had not carried enough telegraph wire). This was the first battle of Manassas at the Bull Run river, where the 22 000 Confederate troops attacked 30 000 Union troops defending Washington. In the War Department telegraph office, a complacent party sat ready to listen to what they expected would be the concluding scenes of the rebellion. After a somewhat ominous and lengthy silence the telegraph finally brought news of a Union defeat and rout, the troops fleeing back to Washington. This battle made it clear that changes would be needed not only in the military command but in the way the telegraph was to be used. Fortunately the two changes came together. One of Lincoln's generals, supportive of the need for advanced military telegraphs, was made Commander-in-Chief. This was General George B. McClellan, who had studied the military telegraphs of the Crimean War. He gave charge of all lines within his department to Anson Stager, the general superintendent of Western Union, who organised a field telegraph system that moved with McClellan's advance into Western Virginia early in the war. Within five months Stager had strung 1830 km of wire for military use, much of it on the march at a rate of 12 to 20 km per day. Never again would the Union side be let down by inadequate performance of the field telegraph system.

The equipment used in the field was generally the same as was used in the commercial network, with much reliance placed on the Morse sounder for reception. Some attempts were made to equip regiments with mobile equipment, and cable and battery wagons were effective in carrying lines from the nearest civilian telegraph office to the front. A proposal to equip a 'movable telegraph train' by Major A.J. Myer proved very much a liability to the Union forces. Major Myer had originally served as the Union Signal Officer and had been responsible for setting up Lincoln's Military Telegraph Department, the forerunner of the Signal Corps in 1864. His telegraph train was equipped with instruments invented by a Mr G.W. Beardslee of New York. 'They work', he said, 'without batteries, and can be used by anyone who can read and write, after a day's practice.' They were in fact a form of dial telegraph with a controlling handle which not only pointed to the symbol it was required to transmit, but also operated an electromagnetic generator to produce the current to be transmitted down the line. The procedure for transmitting a message was extremely slow, and the current produced so variable, that it barely reached more than a few kilometres down the line. Its only merit was that Morse training was not required, so that it could be used by unskilled personnel. To combat the slow operation of the device Beardslee invented an alphabet of abbreviations for common words likely to be used by the military. Thirty wagons, complete with a Beardslee instrument, cable, poles, and so on, were purchased and distributed over the various Union armies and training establishments. While not successful for their original purpose, the trains proved effective for the military telegraphs once they were equipped with Morse keys and sounders, and were employed continuously until the end of the conflict [15]. A much more successful improvisation was the small portable keyboard instrument, shown in Figure 4.5, one of which was carried by every operator (of both sides), together with a short length of wire. Originally devised for line testing, they proved invaluable for line tapping to read the enemy's transmissions. In his book *Old Wires and New Waves*, Harlow recounts several incidents in which the portable line tapping instruments were used extremely effectively by young telegraph operators of both sides to determine the plans of the enemy [13].

Figure 4.5   Portable keyboard Morse instrument
Source: Coe, 'The Telegraph', 1911

As was the usual practice in commercial telegraphy, boys, often as young as twelve or fourteen, were quickly trained as operators for both the Union and Confederate armies. Although quick, responsible and enthusiastic, the young operators were treated shabbily. They had no army rank; they were underpaid for the dangerous work they had to perform; and they received scant consideration should they be compelled to vacate their camps during a retreat, not being considered worth a horse, and had to make their own way to the new lines.[4] Worst of all, in the event of death – and more than three hundred telegraph operators died from disease or were killed in battle during the war – their relatives received no pension, which often left families dependent on charity [16].

Almost all of the fighting took place within the Confederate states over a fluid and constantly changing front, or rather several fronts, geographically dispersed. During 1862–63 a number of bitter battles took place with large casualties for both sides at Shiloh, Manassas, Antietam, Fredericksburg, Chancellorsville, Gettysburg and Vicksburg, with no definite outcome [17]. The telegraph continued to play a vital part in most of these encounters. More lines were laid, instruments were purchased from abroad and communication posts were established, but no new techniques were developed. For a short time McClellan experimented with telegraphing information from captive balloons – the first time this had been attempted in battle – but this service was not to last, particularly after the general was removed from office by Lincoln in late 1862. The turning point came in 1863 with the battle of Chickamauga, south of Chattanooga, where the Union General Rosecraus was opposed by General Bragg's strong Confederate forces, and faced defeat. The problem was one of communication, and the need to transport troops from General Meade's Union army in Virginia – which within the short time necessary did not seem possible. Back in Washington, the telegraphers in their office at the War Department consulted maps and railway timetables, and by extensive telegraphing to railway centres and officers concerned with troop movements, commissariat and ammunition, brought about the rapid transportation of 20 000 fully equipped troops within a few days to Chattanooga. Catastrophe was averted and Chattanooga claimed by the Union forces. When it is realised that, not only clear railway lines would be required during the troop movement, but the correct transport wagons would be needed and the railway gauges changed we can understand why this has been described as 'the greatest transportation feat in the history of warfare up to that time, due entirely to the rapid communication available through the telegraph'.

After this Union victory there followed an unbroken succession of

---

[4] This was a consequence of Secretary of State for War Edwin Stanton's view that operators should not be members of the military services, so that they would be free to refuse orders from any officer. Unfortunately it also removed pension and other rights, a situation which was to some extent rectified by Andrew Carnegie's 'Honor Pension for needy survivors'.

Confederate defeats, until the final capture of Richmond in 1865 brought hostilities to a close. By the end of the conflict some 25 000 km of military lines had been constructed by the Union armies, and 8000 km of new commercial lines added within the Confederate States. This was a tremendous achievement, and many of the new lines were incorporated into the civilian network when it was restored to its owners. As described in the previous chapter, the commercial telegraph network took on a new form after 1865. Many of the telegraph companies had seen their lines and equipment destroyed, and only Western Union and the American Telegraph Company emerged as major companies able to continue providing a nationwide service in the post bellum United States.

4.4 European Conflicts

An early entry into the field of military telegraphy was during the Spanish civil war of 1838–40, when a Corps of Military Telegraphy was formed for visual signalling. Their use of lightweight material and transport by pack animals provided for tactical efficiency, and was carried forward to the introduction of electric telegraphy in 1848 and the African War of 1859–60, for which the same techniques were used. A Field Telegraph Corps was formed in 1873, whose personnel were popularly known as '*Torrecos*' from the word *torre*, meaning 'tower' [18]. In Italy also, armies were quick to apply the electric telegraph for their use. In 1860–61, during the struggles for unification, two army corps were despatched simultaneously from Tuscany and Romagna, to link up for the siege of Ancona. The two corps, although separated by the Apennines, were able to maintain continuous contact by means of telegraphic lines established in their rear. During the siege of the city, 20 km of telegraph cable was laid down around the city, and five telegraph stations set up to communicate with the fleet and the besieging army, using optical semaphores.

The French and Austrian armies both applied the electric telegraph to military operations during the Austro-French conflict of the early 1860s. Direct military operations carried out by the Austrians against the French telegraph lines damaged many of their suspended air lines. As a rapid solution to the replacement problems, the French were particularly successful in laying cable from wagons 'on the run' in the field, to link the general-in-chief with the head of every army corps. This was probably the first time this technique was used in warfare, and it became common in later campaigns. The success of their military telegraph operations in this war decided the French to form a Telegraph Brigade in 1868, and a School for Military Telegraphy in France in the same year [19].

In Germany towards the end of the nineteenth century the telegraph was also being incorporated into military communications. The engineer principally noted for these developments was Herr von Fischer Treuenfeld, then a major in the Engineer Corps, who later became an enthusiastic contributor to the *Proceedings of the Society of Telegraph Engineers* in Britain and to the *Military and Telegraph Bulletin* prepared by the Royal Engineers. He had spent some time in South America, and was appointed Director of Telegraphs for General Lopez during the war between Brazil and Paraguay which took place between 1864 and 1869 (in which Germany had an interest), a conflict which saw the telegraph used by both sides for tactical purposes. With the aid of the telegraph General Lopez was enabled to direct the movements of battle remotely from his tent, and he somewhat extravagantly claimed sole responsibility for his success in evicting the Paraguayans from their hold on Angostura and Lima [15]. Nearer home, von Treuenfeld used the telegraph to minimise the disadvantages of using combined armies operating from divergent bases in the 1864 war between Prussia and Bohemia. General von Moltke maintained telegraphic control of the three armies that formed the Prussian advance until they converged safely on their common goal. He asserted at the time that it 'was a triumph for the telegraph' [20], and this view was echoed by a speaker before the Society of Telegraph Engineers in London in 1872, who in referring to the Prussian conflict noted 'that modern warfare was almost an impossibility without the aid of the electric telegraph in the field' [21].

By the time of the Franco-Prussian War of 1870–71, Prussia had set up three organisations to aid the campaign: the Etappen (base) telegraphs, the state telegraphs and the field telegraphs. The civilian lines were incorporated into the first two of these organisations, as were the lines captured from the French. At the close of the war a vast network of state telegraphs was in place, covering the entire country and much of its neighbour, France. The field telegraphs were connected to the state or the various headquarters stations, often as part of a military action. According to General von Chauvin, appointed Director-General of Telegraphs, who wrote about these organisations at some length [22], 'The line was laid quickest when the telegraph detachments accompanied the advanced guards; very often this event preceded the most advanced troops . . . In such cases the telegraph stations were actually exposed to fire, and sometimes had to retire before an advancing foe.' Despite their perceived value, the establishment of special telegraph companies in the army lagged behind those of other European countries, possibly because of the difficulty of persuading civilian telegraph operators to volunteer for this somewhat aggressive use of their skills!

4.5 African Colonial Wars

The first active engagements for the British army in Africa in which telegraphy played a significant role were a number of punitive wars fought between 1868 and 1898. In the first of these, the Abyssinian War of 1868, electric telegraphy was not at all successful, mainly because no consideration was given to the tropical conditions under which the lines were to be installed or the terrain needed to be traversed, compounded by difficulties in the provision of technical equipment.

At that time every regiment had its own corps of signallers who, although experienced in flag signalling, were deficient in electric telegraphy. Where such expertise was needed, as we saw earlier with the Crimean campaign, it was the practice to employ civilians. The difficulty was lack of additional military funding for training, which the Royal Engineers had been seeking in vain for a number of years [23]. Funding became available in 1870, when a sudden outbreak of war between Prussia and France caused the British Government to order the military authorities to place everything in their respective departments on a footing for immediate action and, significantly, asked parliament to defray the cost. A scheme for the addition of a Telegraph Troop to the Royal Engineer Field Train was put forward and sanctioned in 1870; it was to be known as the 'C' Telegraph Troop, responsible for providing signals for an army corps. As discussed later, two further companies of the Royal Engineers, the 34th and 22nd, were also raised for service, mainly with postal telegraphs and paid for by the General Post Office.

The first opportunity to apply this new group of telegraph companies arose in 1873, during the First Ashanti War, when Lieutenant Jekyll RE commanded a detachment in a region which proved unsuitable for the horses of 'C' Troop. The difficulties of providing signal communications for this campaign, under the conditions of jungle warfare that prevailed, were severe. The wires were suspended from bamboo poles erected along the road being built from the Cape Coast to Prahsue (Pramiso) and Kumassi in advance of the main force (Figure 4.6). The work was hampered by damage caused by many violent thunderstorms and frequent electric shocks sustained by those handling the wire. These setbacks had one useful consequence, however – it engendered considerable respect among the native labourers employed, and persuaded them to refrain from their usual practice of pilfering material from the lines! The line eventually reached its goal at Accra, a distance of 137 km, with five intermediate telegraph offices. Morse sounders were used, and an operational hazard was experienced by the unexpected deafness of some of the operators caused by the heavy doses of quinine they needed to take [11].

The Zulu War of 1879 was the first campaign in which mounted telegraph troops of 'C' Troop took part. The ability to ride a horse became

*Figure 4.6   'Fixing the cable' by the Royal Corps of Signals in the first Ashanti
        War*
Source: *The Military Telegraph Bulletin, by courtesy of the Royal Corps of Signals*

essential for telegraph personnel during the campaigns that followed, and did much to improve the speed with which forward telegraph stations could be set up, often in advance of the attacking troops. In the Transvaal War of 1881 the signal companies were again successful in establishing good communications between the forward troops and their base headquarters in Natal, through a line hastily set up through to Pretoria. A full cable detachment included two mounted linesmen and a four-horse cable wagon with two drivers. In action, the cable could be laid at a fast pace by using a 'crook-stick' to guide the rapidly unreeling cable to the side of the road or track. Every two minutes a telegraphist on the wagon would press his key and report to the detachment commander, also mounted, if an answering signal was not received [24].

For the Egyptian War of 1882, the troops were landed at Ismailiyah and the telegraph company was able to make use of the railway wire up to a point within a few kilometres of Tel-el-Keber station, where the Arabs held an entrenched position. The company completed its assignment simultaneously with the attacking force by laying cable at night along the railway line, a task it completed at a rate of 10 km an hour, some of it under shellfire. After establishing a field telegraph station in a saloon carriage at the now-captured railway station, the company was able to make immediate connection directly to Ismailiyah, and hence to London via submarine cable, to announce the success of the campaign to the Queen and the War Office. The dispatches 'were sent off at 0830 hrs and at 0915 we received Her Majesty's reply' [25]. This direct and rapid communication from a campaign headquarters abroad to the War Office in Whitehall was unprecedented, and served to establish telegraph communication as a vital part of future military actions.

4.5.1 The Nile and Egyptian campaigns

Egyptian misrule in the Sudan resulted in a number of military expeditions by British troops from Cairo to Khartoum during 1884 and 1885, primarily to quell revolts initiated by a self-styled Mahdi, a spiritual and temporal leader, named Mohammed Ahmed. It was during one of these revolts that General Gordon was killed, a serious setback for the military in Victorian Britain. These expeditions provided the first major involvement for the newly formed Telegraph Battalion, formed from the amalgamation in 1884 of the 22nd and 34th companies, Royal Engineers, with the 'C' telegraph troop. The battalion was initially commanded by Colonel Webber, the Director of Telegraphs, and the operation was significant for the variety of problems that arose in establishing and maintaining telegraph communications in a desert location. For the Nile expedition the basis of the signal plan was to use the civil telegraph route between Cairo and Khartoum, which had been built in 1870. It started out as a four-wire system as far as Asyut, a distance of 350 km, but reduced to two and then to one uncertain line to Debba (Dabbah); beyond this point the line was non-existent (Figure 4.7). The first task of the telegraph companies was to restore the line to a reasonable working condition, with the assistance of the Egyptian Telegraph Department. Beyond Merawi a new line had to be installed over 600 km of rugged and desert territory, eventually reaching Khartoum after overcoming many difficulties en route. At one point an experiment was made to lay bare wires along the ground, over a distance of 37 km, and to operate a 'vibrating sounder', there being no poles available to erect an air line. The extreme dryness of the terrain made the line resistance to ground high, the circuit worked well, and thus was initiated a technique which was used thereafter in desert conditions, including later operations in the Western Desert during the Second World War.

This was the first campaign in which the new 'vibrating sounder' was used. The device was introduced by a Captain Cardew, who was responsible for a number of inventions and improvements to telegraph apparatus within the Royal Engineers [26]. This robust and sensitive device was used as a transmitter, having the mechanism of an electric bell but without the bell dome – what we would now recognise as an electric 'buzzer'. The system derived its sensitivity from the use of a telephone as the receiving instrument, which had not previously been done before by the military for telegraphic work.

The wire used for field telegraphs was stranded galvanised iron wire, less than 5 mm in thickness. For long lines No. 14 hardened copper wire was used, but it proved difficult to manipulate, and frequently broke. Insulated cables, carried on drums, consisted of three strands of 20 BWG[5] copper wire, insulated with vulcanised India rubber, round which a layer of canvas was wrapped. Later cables had an insulating layer of gutta-percha. During the long marches across the desert, the cable was reeled off an accompanying wagon, enabling contact to be made with headquarters during halts. For installations along railway lines and in certain fixed locations, bamboo and light iron tubular poles were erected for air lines. The standard equipment carried by the Telegraph Battalion of the Royal Engineers in peacetime consisted of four 'air-line' and two 'cable' sections, consisting of enough material to lay 32 km of wire and equipment for three telegraph offices. This equipment was however varied in wartime according to the demands of different campaigns [27]. The telegraph instrument used for long distances was the ubiquitous Morse sounder, supplemented by the Vibrating sounder, for bare lines laid across the ground.

In addition to military messages, press messages were handed in by special correspondents attached to the attacking force, enabling *The*

[5] BWG stood for Birmingham Wire Gauge; later it was replaced by the Standard (or Imperial) Wire Gauge (SWG).

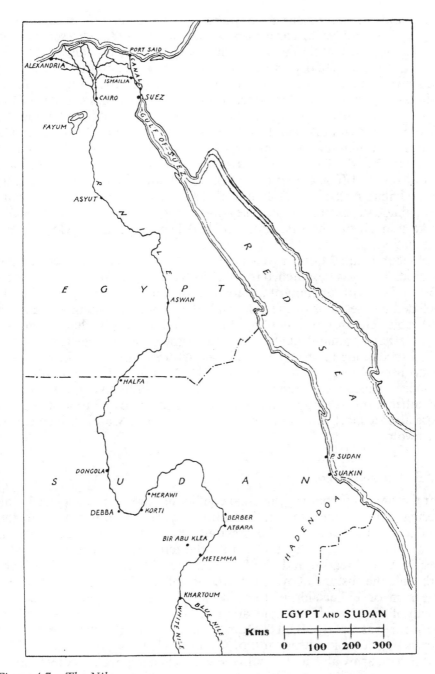

Figure 4.7   The Nile
Source: Nalder, 'History of the Royal Corps of Signals', 1958, reproduced by permission of the Royal Signals Institute

Times and the *Daily News* to keep the public informed in an immediate way, not possible with earlier operations abroad [28]. The scrutiny of private messages, although an urgent responsibility, was also a difficult one. More than 6000 messages in Arabic were examined during the campaign. Among other problems was the case of the Mudir of Dongola (Dunqulah), who always began his messages with a chapter of the Koran! Altogether, 29 new telegraph stations were established along the 1000 km route from Aswan to Khartoum. The amount of traffic carried by this joint military and commercial line was significant and amounted to several hundred telegrams per day. The Cairo office is said to have recorded over 17 000 cipher groups of information in a single night, plus some civilian traffic [12]. To handle this augmented traffic, arrangements were made to install duplex key-speed repeater equipment between Cairo and Kurti, the first time this had been used in a field military installation.

A second expedition proved necessary before Khartoum was finally reached. This was despatched in 1885 via Suakin (Sawakin) just south of Port Sudan, with the aim of approaching Khartoum through Berber. It included two sections of the Telegraph Battalion, advancing with airline and cable. The troops, including the telegraph operators, had to engage in close hand-to-hand fighting to defend their fortified telegraph posts established along the route. For this late African campaign the battalion made use of recorders, sounders and telephones, but not vibrating sounders, which were found to be the cause of inductive interference in the other forms of communication. The campaign came to a premature close after a short time, following the death of the Mahdi, which ended the revolt.

4.5.2 The Boer War

For the signal companies, this final conflict in Africa to take place in the closing years of the nineteenth century was totally different from the experience they gained in the previous limited punitive expeditions in northern and central Africa. Generally the action took place over a wider area, both in breadth and depth, unlike the Nile expeditions in which, although the distance covered by the installation teams was great, its extension on either side of the Nile was limited. In South Africa the length of the lines of communication was over 1500 km and the width at the front was up to about 800 km along a frontier subject to raiding parties which took a heavy toll of installed telegraph lines (Figure 4.8). The British army also had to contend with an enemy whose command of technology was at least equal to its own, and which possessed a cadre of skilled telegraphists capable of working and maintaining telegraph lines within their own domain.

A major feature of the campaign, certainly in its early phases, was the

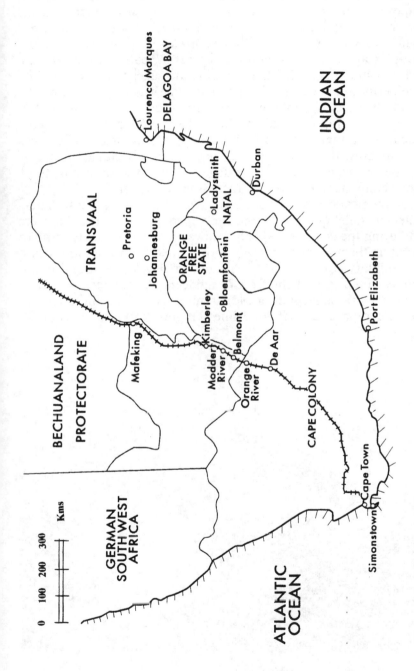

Figure 4.8   Southern Africa
Source: IEE Conference Paper No. 411, p. 50, 1995

use of cable carts, tried out during the Nile expeditions, which accompanied all the marches of Field Marshall Lord Robert's troops, so that at each temporary halt the leader was able to establish contact with his headquarters. The equipment used reached a level of sophistication not attained in earlier campaigns. At the Jacobsdal headquarters, double-current duplex apparatus was employed, and Wheatstone's automatic fast speed transmitters enabled the outgoing messages to be prepared by an off-line perforator process. This latter facility was much used by the Director of Telegraphs, Colonel R.L. Hippisley, who considered it inadvisable to send important dispatches at key speed in case the enemy tapped the line. Tapping of the British lines by the Boer telegraphers was frequent, and Lord Roberts is said to have taken advantage of this by sending false messages in plain language about future movements of troops, correcting them by a following message in cipher.

Most communications took place using Morse and vibrating sounders over the field installations and on the civilian commercial lines that were also used during the conflict [29]. The railway network, already in place and connecting the major centres in South Africa, formed an important link in the military communications routes, its development and maintenance an essential task of the Royal Engineers. To protect the railway lines from enemy raids a chain of blockhouses were built and interconnected by telegraph and telephone lines (Figure 4.9). On the railway

Figure 4.9   A Boer War railway blockhouse
Source: Author's photograph, taken in 1999

alone the telegraph companies erected 15 000 km of wire to serve both purposes. Armoured trains were an important defence feature of the railway network. Each train carried two linesmen and an operator, who as soon as the train stopped, established communication by connecting onto the adjacent telegraph line.[6] The telephone was not considered suitable for long-distance communication in the field, and its use was limited to blockhouses and some internal lines, such as those employed during the siege of Ladysmith. A few years earlier, General Alison had written to General Fielding expressing the view that

I should not like to supersede our present system of telegraph by telephone until I was more convinced of its being an advantage than I am now. It might be tried between brigade offices and regiments perhaps where now no communication telegraphic or telephone exists.

Although the war was nominally over with the flight of President Kruger and the fall of Pretoria in 1901, a prolonged period of guerrilla warfare continued for nearly two years after this date. The whole of the telegraph service in South Africa was put under military control, its operation shared by the civilian staff and the Royal Engineers Telegraph Service. Extensive telegraph and telephone networks were installed by the military in the Pretoria and Johannesburg area. Extensive use was made of the Van Rysselberghe telephone system, developed a few years earlier, in which the addition of an earth return enabled the twin metallic telephone lines to act as one line of a telegraph system, thus allowing both Morse transmission and speech to be carried over the same set of wires; a filter introduced into the telephone lines minimised interference from induced Morse generated currents [30, 31].

Throughout the campaign, visual communications by heliostat and heliograph were used very effectively, made possible by the prevalent sunshine and exceptionally clear atmosphere of South Africa. They were generally used as the principal form of tactical communication, forward of divisional headquarters, but the standard rate of transmission was slow compared with the electric telegraph, between 4 and 8 wpm and, apart from the use of ciphers which slowed the transfer of information even further, the visual transmissions could easily be read by the enemy. The heliograph proved valuable during the siege of Ladysmith, however, and was the only possible form of communication with the relieving forces during a major period of the siege.

Wireless played virtually no part in the campaign. Some attempts were made during the closing stages of the campaign to make use of wireless communication by the army, using equipment provided by the newly formed Marconi Company. However, this was not successful, and did not

---

[6] A restored armoured train, together with its telegraph equipment used in the Boer War, may be seen at the Railway Museum at York, England.

play any part in operations. A discussion of its use by the military and of the equipment concerned is given in a later chapter.

4.6 Early British Army telegraph training

The initial training received by officers and men of the Royal Engineers at the Royal Engineers School of Military Engineering at Chatham included some telegraph training. Following the formation of the Telegraph Battalion, described below, a more complete training schedule was put in place. The training in electricity and telegraphy was severely practical, and commenced with a 54-day course which covered the making-up and testing of batteries, galvanometers and resistance coils, resistance measurement, firing of mines, electric light, signalling and the use of various kinds of telegraphic instruments. More elaborate courses were developed later under the enthusiastic guidance of the Chief Instructor in Electricity, Major Phillip Cardew, mentioned earlier and dubbed 'the Edison of the Royal Engineers', who introduced several new devices, including the vibrating sounder and the 'separator', which were still in use in the First World War. The separator was a simple instrument, consisting of a combination of an electromagnet and two condensers, which enabled one line to be used for two separate circuits – one with vibrating sounders and the other an ordinary circuit with Morse sounders, without mutual interference. He was awarded a gold medal at an inventions exhibition in London in the 1880s for his working hot-wire a.c. voltmeter, the first practical voltmeter to apply a thermal effect, and which led to his appointment as Electrical Advisor to the Board of Trade in 1889.

4.7 The Telegraph Battalion

In 1870, following the nationalisation of the telegraph companies in Britain discussed in the previous chapter, expansion of the commercial network in Britain was held up by a serious shortage of technicians, particularly for maintenance and construction. At about the same time the army in India had arranged for detachments of sappers and miners to work with the civilian Indian Telegraph Department. In an interesting example of lateral thinking of the day, the War Office suggested that the General Post Office in Britain could benefit by a similar loan of sappers to mitigate their technician shortage, and at the same time enable a pool of trained telegraphists to be set up for the military.

It soon became an established army routine to extend practical training in telegraphy by sending groups of officers and men to St John's Wood barracks in London for duty with the Post Office. This included linesman training in the renewal of the telegraph roadline from Uxbridge to Oxford, and Major Beresford in his *Records of the Postal Telegraph Companies, Royal Engineers* commented on 'the novel sight of redcoats working in the streets of the City . . . carefully with a soldering iron and spirit lamp with an eager crowd of spectators watching'. Later work by the Royal Engineer telegraph companies included renewing a number of badly deteriorated underground lines, in particular those laid earlier by the Electric Telegraph Company. These had to be repaired while still in use, with minimum interruption to the service, which was therefore a difficult operation. The work eventually became so extensive that the Royal Engineers produced their own *Handbook for Underground Work*, which continued to be used as guide to the maintenance of such systems for a considerable number of years by the Post Office as well as the Telegraph Battalion [32]. Its popularity encouraged the Royal Engineers to produce other handbooks derived from their experiences with the public telegraph service. One of these, *The Telegraph Office*, was written for 'country postmasters, unacquainted with the principles of electricity and who may not be provided with galvanometers or other appliances for the detection of faults', and was 'Compiled by Officers of the Royal Engineers employed in the Postal Telegraph Service' [33].

Concern expressed by Post Office employees at having 'soldiers in the workplace' prompted a writer to the *Military Bulletin* for March 1884 to comment:

To allay the little nervousness in the Department which actually existed at the military invasion, the Headquarters of the Company did not actually march into St Martin's le Grand [then the Post Office Headquarters in London], but did, one lovely afternoon in May, arrive at Victoria Station, march through Hyde Park at the height of the season, and take up its quarters at St John's Wood Barracks [34].

Two companies of signallers formed from the Royal Engineers for this purpose were known as the 'Post Office Companies' (they were actually the 22nd and 34th Companies of the Royal Engineers) to join the original 'C' Telegraph Troop. In 1884 all three telegraph companies were amalgamated to form the Telegraph Battalion of the Royal Engineers, with distinctive uniform and badges, who were called upon to provide telegraph signallers for a number of campaigns in Africa, as described earlier.

4.8 The Society of Telegraph Engineers

It is to the 22nd Company of the Royal Engineers that we owe the formation of the Society of Telegraph Engineers in London, later to become the Institution of Electrical Engineers. For some time professional telegraphists and electrical experimenters within both civilian and military establishments had felt the need for a formal institution similar to the Institution of Civil Engineers and Institution of Mechanical Engineers, which would act as a qualifying body and learned society, and where they could meet, read papers and carry out all the functions of a professional organisation for their particular branch of science – for, as Sir William Preece was later to say, 'The engineer must be a scientific man' [35]. One of the founding members of the Society of Telegraph Engineers was (then) Captain Charles Webber, in command of the 22nd Company, and he is reputed to have discussed the idea during repairs of the telegraph route along the Uxbridge–Oxford road, with the Post Office Divisional Head, just after the 'Post Office Company' had been formed; Webber himself claimed that a colleague, Major Frank Bolton, was the catalyst for this new society [36]. Certainly it was shortly after this discussion that steps were taken and the society was formed in 1871, containing at its inception a high proportion of military telegraph officers [37]. Indeed, one of the first subjects of debate at its inaugural general meeting, proposed by Captain Dawson RN, was the application of electricity for the purpose of war. The first President of the Society of Telegraph Engineers was Charles William Siemens, already an FRS and one of the leaders of the telegraph industry in England. The second was Frank Scudamore, Director of the Post Office and architect of the nationalisation of Britain's telegraph services. The list of subsequent presidents for the next two decades reads like a roll-call of the greatest names in telegraphy in the nineteenth century, several of them having close military connections. For the first ten years of its existence 80 per cent of its members were engaged in telegraph occupations of which about a quarter were from the military. Towards the 1890s, when the society broadened its scope to include other branches of electricity, the proportion engaged in telegraphy fell, so that by 1889, when the society changed its name to the Institution of Electrical Engineers, it was only a few per cent, with an emphasis on learned society activities and civilian developments.

4.9 British Army signalling

With the formation of the Telegraph Battalion in 1884, signalling procedure had reached an established form which was used for the rest of the century and into the next. The first British *Manual of Instruction in Army Signalling* was produced in 1880, followed by frequent reissues in subsequent years. Messages to be transmitted were not only classified in items of priority, but also carried considerable sets of coded information on the origin of the message, its time of transmission and any modifications, such as corrections and requests for repetition. Important messages were transmitted in a number cipher, generally consisting of sets of four numbers, and passed through a cipher section for decryption before being sent to the addressee. Priority was indicated by a two-letter prefix attached to the message indicating the order of precedence [38]:

	Messages to be delivered to the receiving station	Messages to be transmitted from the receiving station
OHMS with priority	SB	XB
Messages connected with the working of the line	SG	XG
Ordinary messages OHMS	SM	XM
Private messages	S	X

Time was coded in a simple but effective way, again in the form of a set of letter codes, illustrated in Figure 4.10. The twelve letters from A to M denoted the hours, and also the minutes at 5 min intervals (0, 5, 10, etc.). The intervening four minutes were denoted by the letters, R, S, W, X. Letters sent singly indicated hours; sent in combinations of two, they represented the hours and periods of five minutes; sent in combination with the letters RSWX they represented hours and minutes; a.m. and p.m. are also included where necessary. So, for example, M was 12, MF was 12:30, MFS 12:32 and MFSAM 12:32 a.m. Each station was given a distinguishing call sign consisting (at that time) of two letters, for example NC was Newcastle and PR Pretoria, with field stations designated as PA, PB, PC, etc. These various prefixes (and others) were not sent with the message but as part of a telegraphic dialogue between the

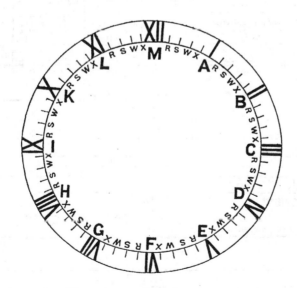

Figure 4.10   Army signalling: time coding
Source: War Office, 'Manual of Instruction' 1887

two station operators, in order to establish the characteristics of the message correctly before its content was transmitted. Corrections and requests to repeat doubtful words were also indicated by certain code letters, although later this procedure was rationalised to follow common telegraph codes established by international agreement (see later). Army telegraphs conformed to a fairly formal recording procedure, epitomised in the Army Telegraph Form shown in Figure 4.11. After a period of instruction, which occupied typically 40 to 50 working days, the operator would be able to complete the form directly on the message pad as the message was transmitted, tear off the form, and enter into a dialogue with the distant operator relating to the next message. A high level of efficiency was aimed for and attained by the new Postal Corps of the Royal Engineers. Its operators were usually volunteers from the civilian post office or railway who had generally received previous training and experience. A glimpse into the rapidly evolving technology of the period and the requirements it placed on the new recruits is caustically expressed in the cartoon shown in Figure 4.12, which appeared in the *Military Bulletin* at the end of the nineteenth century.

ARMY TELEGRAPHS AND SIGNALLING.

(A.S.) 119A. No. of Message

Prefix *X B* Code *M S* *p.m.*	Words. *15*	Sent	For Stamps or Signature of "Franking Officer."	Dated Stamp of Office or Received.
Office of Origin and Service Instructions.		At *10.25 p.m.*	X. This message is on account of Service.	At *P H D*
P H D	Charge.	To *P H F*		From
		By *S K*	(Sig.)	By

Please write distinctly, curtail addresses, and see instructions on back.

To *G. O. C., Portsmouth.*

If necessary, insert here the time at which the message was written, ___m.

(Day of month)	(Sender's number)	(In reply to number)		
Please	*wire*	*me*	*when*	*Jumna*
arrives				

From *A. A. G., Horse Guards.*

D 2

Y. The above message may be forwarded as now corrected. Censor of Army Telegrams.	Z. Signature of Addressor, or other person duly authorized to send this message in name of Addressor.

[See over.

Figure 4.11 Army signalling: code form
Source: War Office, 'Manual of Instruction' 1887

CREATES GREAT DISSATISFACTION TO THE POSTMASTER GENERAL, AND THE DEPARTMENT GENERALLY, THAT WHILE THEY CAN DUPLEX, QUADRUPLEX, MULTIPLEX, AND GENERALLY PERPLEX THE TELEGRAPHIC INSTRUMENTS. THE SAME SYSTEM DOES NOT APPEAR TO BE APPLICABLE TO THE CLERKS · WITH A VIEW TO REMEDYING THIS OBVIOUS DEFECT A NEW INSTRUMENT WILL SHORTLY BE PRODUCED — *The Pedal-dexter Needle!*
TELEGRAPHISTS WILL BE EXPECTED TO PASS AN EXAMINATION IN THE ABOVE DUPLEX WORKING (ONE BY THE HANDS AND THE OTHER BY THE FEET SIMULTANEOUSLY) IN THREE MONTHS · ANY TELEGRAPHISTS NOT FOUND PROFICIENT IN SIX MONTHS FROM THE COMMENCEMENT WILL BE IMMEDIATELY DEGRADED TO THE SORTING BRANCH ·
By Order

Figure 4.12   Cartoon from The Military Telegraph Bulletin
Source: By courtesy of the Royal Corps of Signals, Blandford, Dorset

References

1   SIEMENS, W. von: 'Lebersinnerungen' (Prestel-Verlag, Munich, 1890)
2   SIEMENS, E. von: 'Recollections of Werner von Siemens' (Lund, Humphries, London, 1983)
3   BERESFORD, C.F.C.'The first telegraphs in war', *The Military and Civil Service Telegraph Bulletin*, 15th April 1887, (38)
4   HAIGH, K.R.: 'Cableways and submarine cables' (Adland Coles, London, 1968)
5   EARDLEY, S.M.: 'The life of Vice Admiral Edmund, Lord Lucas' (London, 1898)
6   NALDER, R.F.H.: 'History of the Royal Corps of Signals' (Royal Signals Institution, London, 1958)
7   PEMBERTON, C.B.: 'Battles of the Crimean War' (Batsford, London, 1962), p. 184
8   LUKE, P.V.: 'Early history of the telegraph in India', *Journal of the Institution of Electrical Engineers*, 1891, **20**, p. 111
9   MOON, Sir PENDEREL: 'The British conquest and dominion of India' (Duckworth, London, 1989)
10   SHRIDHARANI, K.J.: 'The story of the Indian telegraph' (Posts and Telegraph Department, New Delhi, 1956)
11   BATES, D.H.: 'Lincoln in the Telegraph Office' (University of Nebraska Press, Lincoln, Neb., and London, 1996)
12   GREELY, A.N.: 'Reminiscences of adventure and service' (Scribners, New York, 1927)
13   HARLOW, A.F.: 'Old wires and new waves' (D. Appleton Century, New York, 1936)
14   PLUM, W.R.: 'The military telegraph during the Civil War in the United States' (Jansen McClurg, Chicago, 1882)
15   WOOD, D.L.: 'A history of tactical communication techniques' (Ayer Co., London, 1974)
16   GREELY, A.N.: 'The photographic history of the Civil War' (Review of Reviews, New York, 1911)
17   FARMER, A.: 'The American Civil War 1861–65' (Hodder & Stoughton, London, 1996)
18   FISCHER-TREUANFELD, R von: 'Military telegraphy in Spain', *The Military Telegraph Bulletin*, 15th May 1886, (27)
19   'Military telegraphy', *The Electric Telegraph Review*, 1870, **1**
20   BERESFORD, C.F.C. 'The field telegraph and its use in war ... ', *The Military Telegraph Bulletin*, 15th June 1886, (28), p. 215
21   MALCOLM, Captain: 'Army telegraphs', *Journal of the Society of Telegraph Engineers*, 1872, **1**, p. 188
22   CHAUVIN, Major General von: 'The organisation of the electric telegraph in Germany for war purposes' (E.S. Miller, Berlin), translated by Captain Hare RE, *Proceedings of the United Services Institute*, 1884, (126/7)
23   PORTER, WHITWORTH: 'History of the Corps of Royal Engineers' (Longman Green, London, 1889), p. 120
24   'Instructions for laying and maintaining field cables', War Office Manual (War Office, London, 1907)

25 MACKWORTH, Sir ARTHUR 'The Field Telegraph Corps in Egypt', *The Military Telegraph Bulletin*, 16th February 1885, (12), p. 90

26 'Captain Cardew's vibrating sounder', *Journal of the Society of Telegraph Engineers*, May 1886

27 'P.V.L.':'Military field telegraphs in England', *The Military Telegraph Bulletin*, 15th May 1886, (27), p. 231

28 BERESFORD, C.F.C.: 'The field telegraph with the expedition to the Eastern Sudan, 1885', *The Military Telegraph Bulletin*, 15th January 1886, (23), pp. 175–8

29 HIPPISLEY, Lt Col R.L.: 'History of the telegraphic operations during the war in South Africa 1899–1902' (London, 1903)

30 MOURLON, M.C.: 'The Van Rysselberghe system of simultaneous telegraphy and telephony', *The Electrician*, 1885, **14**, pp. 459–60, 476–7

31 BEAUCHAMP, K.G.: 'Exhibiting electricity' (Institution of Electrical Engineers and Peter Peregrinus, London, 1996), p. 143

32 BERESFORD, C.F.C.: 'Records of the postal telegraph companies', *The Military Telegraph Bulletin*, 1884, 15th April, p. 11; 15th May, p. 19; 15th August, p. 6

33 'The Telegraph Office – compiled by the Royal Engineers' (Royal Engineers, London, 1877)

34 BERESFORD, C.F.C.: 'Records of the postal telegraph companies, R.E.', *The Military Telegraph Bulletin*, 15th March 1884, (1), p. 2

35 FITZGERALD, G.F.: 'The application of science – lessons from the nineteenth century', *Journal of the Institution of Electrical Engineers*, 1900, **29**, pp. 394–400

36 WEBBER, C.: Note in *Journal of the Society of Telegraph Engineers*, 1893, **21**, p. 543

37 READER, W.J.: 'A history of the Institution of Electrical Engineers' (Peter Peregrinus, London, 1987), p. 10

38 'Manual of instruction in army signalling' (War Office, London, 1887)

Chapter 5

Submarine cables

With the interior of Britain, America and most of Western Europe provided with a network of telegraph lines linking principal centres of commerce, the contrast between rapid transfer of information by overland telegraph lines with the leisurely onward progression of messages by packet-ship or steamer to overseas locations proved stifling to business enterprise. The demand for submarine telegraph cables providing a similar service to their terrestrial counterparts, but across the seas and oceans, became pressing. Many companies were formed (and often rendered bankrupt!) to initiate experimentation in cable manufacture and laying on the ocean bed, a process which was to lead to a number of successful connections between land stations on different shores during the second half of the nineteenth century.

5.1 Leaving the land

Technically the problems of guiding telegraph signals on land and under the sea were quite different, and this is reflected in the lengthy solution to the problem of submarine transmission: it was 14 years after Cooke and Wheatstone installed their first land telegraph route along the railways before a satisfactory submarine cable was evolved. Several initial attempts were made by enterprising individuals before the problems involved began to be even partly understood. Charles Wheatstone was one of the earliest of these, using cable insulation techniques he had applied successfully to buried underground wires, and made a number of trials in Swansea Bay in 1844 [1]. He used a multiple cable of seven copper wires covered with tarred hemp, and even described a machine to manufacture the cable. Signal leakage, however, proved unacceptable. Learning of the insulation properties of gutta-percha (of which more later), he proposed to apply this around the wires and to sheathe the insulated wires in a lead tube. He failed to devise a method of using gutta-percha effectively, however, and little progress was made. His contemporary on the other side of the Atlantic, Samuel Morse, had also studied the problem and laid an insulated cable across New York harbour in 1842. Like his contemporary – another American named Ezra Cornell, who successfully laid a cable 20 km in length, linking Fort Lee with New York city along the Hudson river – he used rubber insulation which was found not to be very satisfactory when immersed in water over a long period of time.

5.2 Gutta-percha

The lack of a suitable insulation material for submarine cables was a major stumbling block for the early telegraph experimenters. India rubber appeared to be ideal, but in practice it was subject to corrosion in sea water and its insulating properties broke down after a short period of immersion. There was a further problem: a suitable insulating material must be impervious, not only to the penetration of sea water at the great pressure prevailing in the ocean depths, but also to biological attack which was known to affect many forms of insulation, chiefly destructive boring worms found in all the world's oceans. A solution which proved very satisfactory was found in coagulated latex obtained from a certain group of trees, principally the percha tree, and subsequently known as gutta-percha (Malay, *getah percha*). Dr William Montgomerie, a resident surgeon of Singapore, was instrumental in bringing this to the attention of the Royal Society of Arts in 1843 [2], and brought a number of samples with him to London the following year. It was subsequently tested by Michael Faraday, who pronounced it ideal for use as an electrical insulator, although its initial use in Britain was to make patent bungs for soda water bottles! [3]. The latex is obtained from a range of palaquium trees, all of which are found in South-East Asia. It is not as elastic as rubber but can be worked easily by heating without impairing its insulating properties. What does affect these properties is the proportion of resin and other impurities present in the derived latex [4]. Purchasers of the raw material from eastern traders quickly learnt that its purity varies widely with the kind of tree tapped, and that the best trees were confined to the southern end of Malaya and on Singapore island.

The method of obtaining the latex by cutting a V-shaped notch in the bark, tapping the tree and attaching a cup to catch the liquid as it flows from the cut is not so effective as with rubber latex (Figure 5.1). The Malays had used small quantities of the material to make useful devices, such as riding whips, and for this purpose used to cut down the entire tree and bleed its trunk and branches in several places at once to obtain the latex. This was the method used by native gatherers of the material for

Figure 5.1   'Tapping the tree'
Source: By courtesy of the Singapore Botanical Gardens

European manufacturers, and led to wholesale tree-felling in Malaya in the mid-1840s which was to become disastrous in ecological terms [5]. The trees of Singapore were felled first, starting in 1847; by 1857 there were none left. The search moved to Johore and other areas of Malaya, to Sumatra and Borneo, and even the Philippine Islands. Sir Hugh Low, then Governor General of the Straits Settlements, realised that trees were being cut down in large numbers and recommended that the export of gutta-percha should be banned completely [6]. Although trade was reduced, this prohibition lasted only until 1889, when the rate of exploitation again increased rapidly. Only about 1 kg of latex is obtained from tapping a single tree, but this rises to about 5 kg when the entire tree is cut down. It has been estimated that in 1845–46 alone the quest for gutta-percha led to the felling of about 70 000 trees, and some estimates put this much higher.[1] Cultivation was tried in jungle clearings, and efforts by the French to extract the latex from crushed leaves met with some success. It was not until the first decade of the new century that these efforts were rewarded, and a stable and renewable production of this valuable material was secured for the telegraph industry.

The principal company importing gutta-percha into England was the Gutta Percha Company, founded by Thomas Hancock in 1845, shortly after Montgomerie brought his samples to London. While initially the company processed the material for various purposes, it quickly became a major manufacturer of gutta-percha insulated cables, following the invention of a machine for the extrusion of gutta-percha tubing by Henry Bewley, a Dublin chemist. The company had a monopoly on the supply of gutta-percha insulated core for many years, and nearly all the cables made before 1865 were manufactured at their works on the banks of the Thames. (The armouring of the cables was carried out by other companies, mainly those in the wire rope industry.) The favourable position of Britain for imports from Malaya, the high cost of gutta-percha for non-British manufacturers, and its occasional scarcity, made the search for an alternative cable insulation material an important one. In the 1920s a new coagulated latex called balata was procured by tapping the 'bullet tree', a native of the Guianas of South America. Its electrical properties were not as good as those of gutta-percha, but with suitable processing to remove its high resin content it could be applied to cable production and was used extensively in the period leading up to the second World War, until the formulation of polyurethane and similar artificial dielectrics replaced organic compounds completely in cable manufacture [7].

Werner von Siemens was one of the first to see the potential of gutta-percha in telegraph communications. In 1843, two years before Henry

---

[1] A single long cable, such as one across the Atlantic, could require hundreds of tonnes of high-grade gutta-percha for its completion.

Bewley in Britain had introduced his extrusion machine for the Gutta Percha Company, Siemens constructed his own machine in Germany for 'seamless' wire covering, similar in principle to the machine for making macaroni, and not very different from the way insulated cables are produced today. This was a significant improvement on previous methods of manufacture since the presence of the smallest hole in the cable covering, which would allow penetration by water, would be fatal to the performance of the telegraph cable, especially when immersed in sea water at great depth. Siemens laid some gutta-percha covered cables in Kiel harbour, but used them not for telegraph communication, but to electrically fire submersed mines for the military. This had previously been carried out by Schilling in 1812, when he laid cables in St Petersburg, and again in the River Seine in Paris in 1814, to fire mines for his regimental commanders. Siemens first used gutta-percha insulation for line telegraphy when, at the request of the army, he laid a long subterranean wire from Berlin to Grossbeeren in 1847, a task which led directly to the formation of the firm of Siemens & Halske and to contracts for several further government telegraph lines. The first of these new contracts was a telegraph line from Berlin to Frankfurt-am-Main, where the German National Assembly was sitting. This enabled the election of the Emperor that year, taking place in Frankfurt, to be known within the hour in Berlin. Further lines in the following year linked Berlin with Köln, Hamburg, Breslau and the Prussian frontier at Verviètes. Underwater cables formed part of these undertakings, with crossings of the rivers Elbe and Rhine. To protect the gutta-percha insulated cables from fouling by ships' anchors and fishing tackle in these busy waters, Siemens provided a chain of wrought iron tubes, the first in a series of armoured undersea cables, later to be manufactured in his cable factory established at Woolwich, England.

The first really successful underwater cable in the open sea, which established the commercial value of the submarine telegraph cable, was laid in 1849 by Charles Walker of the South-Eastern Railway Company. His cable, which used gutta-percha insulation and was about 3.5 km long, connected a ship, the *Princess Clementine* anchored offshore, to a temporary station linked to the telegraph system of the South-Eastern Railway by overhead wires. With its aid he exchanged telegrams with the chairman of the railway in London, some 134 km away [8].

5.3 Crossing the Channel

A much needed cable across the English Channel was laid by the brothers Jacob and John Watkins Brett, who were to fund the General Oceanic and Submarine Telegraph Company in 1846 for this purpose, and hopefully to support a further grandiose scheme 'to establish a telegraph connection from Britain across the Atlantic to Nova Scotia and Canada'. But their first aim was to establish a submarine telegraph between France and England, for which they ordered a cable 40 km long, having its central copper conductor covered with 12 mm of gutta-percha, to be manufactured at the works of the Gutta Percha Company in London. By 1850 the company was fully engaged in supplying covering material and making submarine cables, having built a thriving factory on the north side of the Thames. The finished cable was assembled on a horizontal drum mounted on the deck of a small steamer, the *Goliath*, chartered by the enterprise (Figure 5.2). The cable was successfully laid in 1850 and connected to the telegraph network of the South-Eastern Railway at Dover, the other end located at Cape Gris Nez near Calais. The shore ends of the cable were protected by enclosing them in thick lead tube [9].

Figure 5.2   Laying the cross-Channel cable
Source: Bright, 'Submarine Telegraphs', 1898

Unfortunately there was no protection from ships dragging their anchors on the bottom of this busy and shallow section of the Channel, and the link was broken after only a few hours. However, the principle had been established, and in 1851 a second cable was laid, this time in quad (four-core) cable, manufactured by R.S. Newall and Company to a much heavier armoured design (Figure 5.3). R.S. Newall were originally specialised in the production of wire ropes and equipment for collieries, and flourished as a manufacturer of deep-sea cables. In 1840 the firm took out a patent for the production of a wire rope with a soft hempen centre, and this design was to prove ideal for the manufacture of armoured submarine cable having a soft inner core of gutta-percha extruded onto one or more copper wires. The cross-channel cable was laid in October 1851, only just after Queen Victoria had declared closed the 1851 Great Exhibition in London, at which much interest had been shown in the plans for the enterprise [10]. Its success was immediately followed by a number of lines laid between Britain, the Continent and Ireland, and brought with it large orders and a virtual monopoly to R.S. Newall on the manufacture of armoured submarine cables.

In 1852 no fewer than three unsuccessful attempts were made to establish telegraph communications between Britain and Ireland, the first

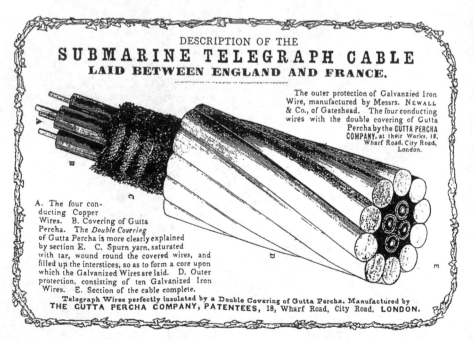

Figure 5.3   Submarine telegraph cable laid between Britain and France, 1851
Source: White, 'Handbook of the Royal Panoptican', 1854 (IEE copyright in Exhibiting Electricity, Beauchamp, 1997)

between Holyhead and Howth and then twice between Portpatrick and Donaghadee. Finally, on the fourth attempt, a line was laid between Portpatrick and Donaghadee. Other lines were laid between Britain and Belgium, Holland and Germany, while in the Mediterranean cables were laid between Corsica and Sardinia, and Italy and Corsica. The last of these cables, 177 km in length, completed the connection of France with its colony in Algeria via Italy in 1854. These cables were laid by the Mediterranean Electric Telegraph Company under the direction of J.W. Brett, who went on to lay many other cables linking Greece with Turkey and Egypt, and with extensions of the cables from Italy to reach Malta and Corfu [11]. Figure 5.4 shows the cables that had been laid between Britain, Ireland and the Continent by 1854. A year earlier, the St Petersburg branch of Siemens & Halske had succeeded in laying a Newall submarine cable from Oranienbaum to Kronstadt in the Baltic, a distance of 7.5 km, to join with the underground telegraph line from Oranienbaum to St Petersburg and the rest of the Russian telegraph network. It was working by the autumn of 1853 and continued to operate with minor interruptions until 1928 [12]. (Table 5.1 gives a fairly full list of submarine cables laid in the last two decades of the nineteenth century.)

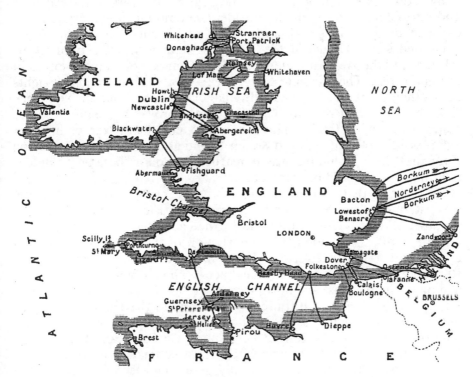

Figure 5.4   Cables laid between Britain, Ireland and the Continent in 1854

A British Engineer, Francis. N. Gisborne, was successful in laying the first undersea cable in North America, from New Brunswick to Prince Edward Island in 1852. He planned a cable from Newfoundland to New York, and actually obtained from the Newfoundland Legislation a 50-year charter to carry this out. However, his project ran out of capital and he transferred the charter to a wealthy paper merchant in New York, Cyrus Field, an action which was to have a major impact on the future of transatlantic cables.

5.4 The Siberian Telegraph

In the early 1850s many telegraph pioneers thought that worldwide telegraph communications would be accomplished only by taking the cables across predominately land routes, with only short sea crossings – a task which appeared to be within practical realisation. A major plan to link Europe to America across the Bering Strait was evolved by Perry McDonald Collins in 1854, and became known as the Siberian Telegraph [13]. Although this did not subsequently materialise in its conceived form, its consideration led to the formation of a formidable cable company and an extension of the Russian telegraph network to the Far East and into China and Japan. A considerable amount of exploration was carried out by Collins [14], and several of his surveying reports were submitted to the United States Congress. Samuel Morse supported his scheme, and the Western Union Telegraph Company became involved in the enterprise. The Russian government gave Collins exclusive rights across their territories together with financial support, and the New York Chamber of Commerce gave its unreserved approval. Some progress was made in 1865 when the Russians began to extend their telegraph line east from Irkutsk to Ulan Ude and Sretensk, and about half the required line through British Columbia was constructed. Collins became confident that the project would be completed.

However, in that year a working submarine cable across the Atlantic was established between North America and Europe, and its success prompted Western Union, who had by then acquired all of Collins' concession rights, to abandon the construction of the Bering sea line. The lines already constructed within Russia were eventually extended by the Russian Government, and continued across the Gobi desert into China and on to the Far East and into Japan. The Siberian line in this modified form provided a northern route from Europe to the Far East and was finally open to public use in January 1872 – a line almost 13 000 km long running eastwards from St Petersburg, via Omsk and Oudinsk, to Vladivostock; the line was connected via a submarine cable to Japan (Figure 5.5). The company eventually assuming responsibility for the completion and operation of this massive enterprise was the Danish

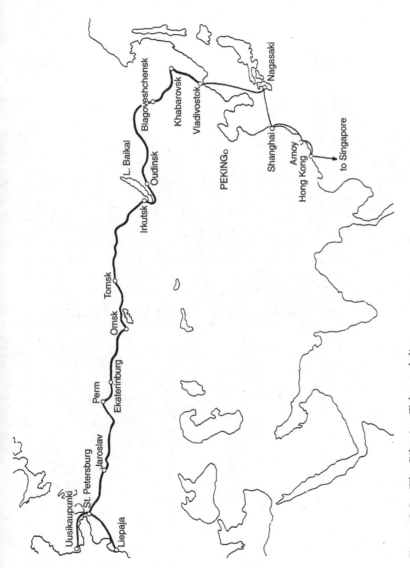

Figure 5.5   The Siberian Telegraph line
Source: Ahvenainen, 'The Far Eastern Telegraphs', 1981, reproduced by permission of the Academia Scientiarum Fennica

Table 5.1 Principal submarine cables to 1903

Date	Route	Length (km)	Company	Promoter
1842	New York Harbor	—	—	S. Morse
1849	*Princess Clementine* off Folkstone	3.2	South-East Railway	C. V. Walker
1850	Dover–Cap Griz Nez (Calais)	f	English Channel Submarine Telegraph Co.	J. W. Brett
1851	Dover–Cap Griz Nez (first successful channel cable)	44	Submarine Telegraph Co.	J. W. Brett
1852	Portpatrick–Donaghadee	f	English & Irish Magnetic Telegraph Co.	C. Bright
1853	Portpatrick–Donaghadee	64	English & Irish Magnetic Telegraph Co.	C. Bright
1853	Holyhead–Howth (Ireland)	f	English & Irish Magnetic Telegraph Co.	R. S. Newall
1853	Dover–Ostend	130		R. S. Newall
1854	Sweden–Denmark	58	Glass Elliot & Co.	W. Thomson
1854	Genoa–Corsica	145	Mediterranean Telegraph Co.	J. W. Brett
1854	Corsica–Sardinia	26	Glass Elliot & Co.	W. Thompson
1855	Orfordness–Hague	185	R. S. Newall & Co.	L. Clark
1855	Cape Breton–Newfoundland	137	Electric Telegraph Co.	J. W. Brett
1855	Varna–Balaclava	550	Electric Telegraph Co.	R. S. Newall
1855	Sardinia–Algeria	f	Mediterranean Telegraph Co.	J. W. Brett
1855	Italy–Sicily	9	Glass Elliot & Co.	R. S. Newall
1856	Prince Edward Is.–New Brunswick	20	Atlantic Telegraph Co.	J. Pender
1857	England–Netherlands	184	Electric Telegraph Co.	R. S. Newall
1857	Cagliari–Bon North Africa (first deep-sea cable)	290	Siemens Co.	W. S. Siemens
1858	Ceylon–India	77	Glass Elliot & Co.	R. S. Newall
1858	Valentia–Newfoundland	f	Atlantic Cable Co.	C. Bright
1858	England–Hanover	45	Glass Elliot & Co.	R. S. Newall
1858	Australia–King Island	225	W. T. Henley	
1859	Folkestone–Boulogne	38	Glass Elliot & Co.	R. S. Newall
1859	Toulon–Corsica	97	Glass Elliot & Co.	R. S. Newall
1859–60	Suez–Aden–Muscat–Karachi (Red Sea route 1st India cable)	4991	Red Sea & India Telegraph Co.	L. Gisborne

Year	Route	Distance	Company	Engineer
1861	Malta–Tripoli–Bengazi–Alexandria	2471	Glass Elliot & Co.	L. Gisborne
1861	Newhaven–Dieppe	129	W. T. Henley	
1864	Gwadur–Fao (Turkish route, 2nd India cable)	2334	Indo-European Telegraph Co.	R. S. Newall
1865	Biserte–Marsela	266	Siemens Co.	
1865	Fao–Baghdad–Bombay (Turkish route, 2nd India Cable)	2898	Indo-European Telegraph Co.	Col. Stewart W. Siemens Col. Stewart
1866	Valentia–Newfoundland (1st successful Atlantic cable)	4495	Anglo-American Telegraph Co.	J. Pender
1866	Buenos Aires–Montevideo (across River Plate)	38	River Plate Telegraph Co.	J. Proudfoot
1866	Florida–Havana (Cuba)	280	International Ocean Telegraph Co.	J. Scrymser
1868	Alexandria–Malta	1500	Anglo-Mediterranean Telegraph Co.	J. Pender
1869	Brest–St Pierre	5300	French Atlantic Cable Co.	E. d'Erlanger
1869	Batabanó–Santiago	837	Cuba Submarine Telegraph Co.	C. Bright
1870	Marseilles–Algiers	1300	Marseilles, Algiers & Malta Telegraph	J. W. Brett
1870	Falmouth–Gibraltar–Malta	5632	Falmouth, Gibraltar & Malta Telegraph	J. Pender
1870	Jarraica–Cuba	210	West India & Panama Telegraph Co.	C. Bright
1870	Teheran–Black Sea–Karachi (Black Sea route, 3rd India cable)	11 000	Indo-European Telegraph Co. (Siemens–British project)	Siemens
1870	Suez–Aden–Bombay (all-sea route, 4th India cable)	5787	British India Submarine Co.	J. Pender
1871	Madras–Penang–Singapore	4190	Eastern Extension Co.	J. Pender
1871	Singapore–Hong Kong	2737	Eastern Extension Telegraph Co.	J. Pender
1872	Singapore–Jakarta–Darwin–Adelaide	6250	British–Australian Telegraph Co.	C. S. Osborne
1872	Buenos Aires–Valparaiso	1300	River Plate & Brazil Telegraph Co.	J. Reid
1873	Puerto Rico–Jamaica	1120	West India & Panama Telegraph Co.	C. W. Bright
1873	Rio de Janeiro–Maldonado	1652	Platino–Brasiliera Telegraph Co.	W. Siemens
1873	Lisboa–Madeira	1000	Brazilian Submarine Co.	V. Monck
1873	Madera–Cape Verde Islands	188	Brazilian Submarine Co.	V. Monck
1874	St Vincent–Pernambuco (Brazil)	5386	Brazilian Submarine Co.	V. Monck
1874	Ballinskelligs–Nova Scotia	4130	Direct United States Cable Co.	Siemens Bros
1874	Barcelona–Marseilles	336	Direct Spanish Telegraph Co.	J. Pender
1875	Valenta–Halifax	6000	Direct United States Cable Co.	H. Labouchere
1876	Para–Demerara	1600	Central American Telegraph Co.	C. M. Hooper

Table 5.1 Contd.

Date	Route	Length (km)	Company	Promoter
1874	St Vincent–Pernambuco (Brazil)	5386	Brazilian Submarine Co.	V. Monck
1874	Ballinskelligs–Nova Scotia	4130	Direct United States Cable Co.	Siemens Bros
1874	Barcelona–Marseilles	336	Direct Spanish Telegraph Co.	J. Pender
1875	Valentia–Halifax	6000	Direct United States Cable Co.	H. Labouchere
1876	Para–Demerara	1600	Central American Telegraph Co.	C. M. Hooper
1876	Sydney–Nelson (New Zealand)	2272	Eastern Extension Telegraph Co.	J. Pender
1877	Rangoon–Penang	1511	Eastern Extension Telegraph Co.	J. Pender
1877	Aden–Bombay	3345	Eastern Extension Telegraph Co.	J. Pender
1879	Brest–St Pierre (Canada)	4130	Cie Française de Télégraphe de Paris à New York	Siemens Bros.
1879	Durban–Mozambique–Aden	6822	Eastern & South African Telegraph Co.	C. W. Stronge
1880	Brest–Cape Cod–Porthcurno	5885	Cie Française du Télégraphe de Paris Ponyer-Quertier	
1880	Aden–Zanzibar–Durban	6822	Eastern Telegraph Co.	J. Pender
1881	Cornwall–Causo	4155	Western Union Telegraph Co.	Siemens Bros
1884	Halifax–Hamilton (Bermuda)	1700	International Cable Co.	W. T. Henley
1886	Bathurst–Capetown	3680	African Direct Telegraph Co.	
1889	Dakar–Bathurst	9060	West Africa Telegraph Co.	J. Pender
1889	Capetown–Mossamedes	2805	Eastern & South African Telegraph	J. Pender
1889	Porthcurno–Cape Town (Boer War cable)	4745	Eastern Telegraph Co.	J. Pender
1890	Bermuda–Halifax	1700	Halifax and Bermuda Cable Co.	Henleys Ltd
1890	Zanzibar–Mombasa	250	Eastern Telegraph Co.	J. Pender
1891	Bacton–Borkum	338	British and German Governments	
1895	Paris–Manars	2639	Amazon Telegraph Co.	Siemens Bros
1900	St Vincent–Ascension–St Helena–Cape Town	7700	Eastern Telegraph Co.	J. Pender
1901	Durban–Mauritius	2787	Eastern Telegraph Co.	J. Pender
1901	Rodriguez–Cocas Island–Freemantle	6235	Eastern Telegraph Co.	J. Pender
1902	Bamfield (Vancouver)–Fanning Island–Fiji–Norfolk Island (trans-Pacific cable)	11 480	Pacific Cable Board Telegraph Construction & Maintenance Co.	S. Fleming
1903	West America–Philippines	9864	Commercial Pacific Cable Co.	J. Pender

Great Northern Telegraph Company. This company had earlier obtained important concessions from Russia to operate in its territory, and indeed the Russian Government depended on the Great Northern for technical know-how, and never failed to provide financial support for the company and its enterprises. Landing permission was easily obtained by the Great Northern from the Japanese, but it had problems in obtaining full cooperation from the Chinese until they had established their own internal network in the 1880s, under the direction of their telegraph pioneer, Li Hung-chang. Thereafter the Great Northern was an energetic promoter of telegraph lines in the Far East and a serious rival to the only other company operating in the region, John Pender's Eastern Telegraph Company.

5.5 Oceanic cables

Up to about 1855, submarine cables were laid only in relatively shallow waters. The New Brunswick cable was just 22 fathoms (14 m) below the surface, and no experience had been gained of laying cables in the deeper waters that comprised the major oceans of the world, a factor which had a major influence on the consideration of the Bering Strait route discussed above. This experience was about to be acquired, but often at a great financial cost. It was becoming apparent to the ambitious telegraph entrepreneurs of the time that three major challenges awaited their energy and ingenuity:

1   the Atlantic crossing,
2   the route to India and the Far East, and
3   the long route across the Pacific ocean from the East to the Americas,

all of which would have to be laid in deep waters.

The first of these was the long-established dream to lay a cable over 3000 km long across the Atlantic, where not only the distance, but a depth of many hundreds of fathoms presented a daunting task. The commercial desirability of linking the Old World with the New was apparent to men of business in both continents, and entrepreneurs and financiers appeared ready with energy and money to initiate the project. One of these, the American businessman Cyrus Field, was inspired by the achievements of Francis Gisborne in New Brunswick and began to examine the possibility of an Atlantic crossing with Samuel Morse and a Lieutenant Maury of the US Navy. In Britain, J.W. Brett, fresh from his experience in laying cable across the English Channel, was persuaded to join Field's syndicate in a new company, the Atlantic Telegraph Company, founded in 1896 with C. Field, J.W. Brett, E.O. Whitehouse and C.T. Bright as directors. Soundings of the ocean bed between Ireland and Newfoundland had been undertaken earlier by the US Navy, and

Lieutenant Maury was able to report that 'the bottom of the sea between the two places is a plateau, which seems to have been placed there especially for the purpose of holding the wires of a Submarine Cable and of keeping them out of harms' way'. The 'plateau' lies between the 48th and 55th parallels of latitude, and had been known to mariners for some time. In his reply to Field's enquiry, Morse indicated that a magnetic current could be conveyed through a length of cable stretching across the Atlantic (some doubt had been expressed by many 'experts' that the current would be dissipated long before it reached the other end) [15]. The subsequent efforts of Field, Brett, Bright and others of the cable-laying syndicate have been adequately reported in the literature and only a brief outline is included here for completeness [8, 14]. However, in order to appreciate the operational electrical problems encountered by the telegraphists at the time it will be useful to digress a little and consider these in some detail.

5.6 Theory and techniques

It was not until transoceanic cables were considered or attempted that the twin problems of line resistance and cable capacitance began to assume significant dimensions. The variability in cable resistance with the purity of the copper metal used was not at first considered important, and it was some time before the practice was begun of measuring each section of the line with a sensitive galvanometer, as it was produced, enabling higher-resistance sections to be rejected before they could be incorporated into the finished cable. With the cable resistance in each uninterrupted span known and reduced to a minimum, the joint effect of the low-pass filter formed by the series resistance and parallel cable capacitance could be clearly seen and measured. Morse signalling consists of a series of rapid changes in voltage level at the sending end, but the effect of transmitting these changes through a low-pass filter results in a current at the distant receiving end of little more than a gradual change in level, confused by a steady reduction in peak value as the signal progresses down the line, caused by the inevitable leakage of current through the imperfect insulation of the cable.

To understand this effect, consider what happens to the current at the receiving end when (a) a step function of voltage is applied at the sending end and (b) when a finite voltage pulse is applied, as happens when a Morse dot or dash is transmitted. The plot of current against time at the receiving end of a long cable is known as an 'arrival curve', and its shape was first calculated by Professor William Thomson (later Lord Kelvin) in his seminal paper 'On the theory of the electric telegraph', read before the Royal Society in London in 1855 [16]. A set of such curves is shown in Figure 5.6. Here the horizontal axis represents time, and the vertical

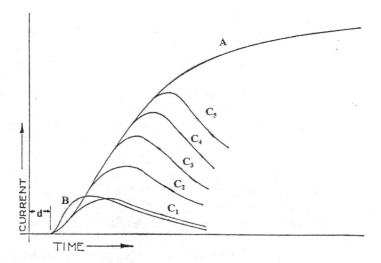

Figure 5.6   Telegraph arrival curves
Source: Bright, 'Submarine Telegraphs', 1898

axis represents the current at the receiving end when a potential is applied at the sending end for different lengths of time. Curve A is the arrival curve when a step-function voltage V, is applied. After a short interval of time, shown as the transmission delay of the cable, d, the current begins to increase and approaches asymptotically the value V/R, where R is the cable series resistance. Curve B shows the effect of an infinitely short pulse of applied voltage at the sending end: a small rise in current after the transmission delay, and then a gradual reduction to a quiescent value. Curves C_1, C_2, C_3, etc. represent a more practical situation when a Morse dot of finite length is sent. Curve C_1 corresponds to a dot or pulse length equal to d, C_2 to $2d$, C_3 to $3d$ and so on. These curves illustrate clearly that if an attempt is made to transmit a series of Morse symbols too rapidly, then the resulting changes in current at the receiving end will be superimposed on one another, and any distinguishing features will be lost. Thomson analysed this effect for the known characteristics of a transatlantic telegraph and concluded that signalling was feasible: that the individual Morse characters would be distinguishable from one another, but that the speed of transmission would be low. He calculated that the time required for the potential of the distant end of the cable to reach a given fraction of its maximum value would be proportional to rcL^2, where L is the cable length, c the capacitance per unit length and r the resistance per unit length. Hence, for a given design of cable, signalling speed is inversely proportional to the square of the cable length. This was subsequently confirmed by C.F. Varley, who performed experiments on great lengths of cable in the East India Dock,

London, for the Joint Committee of Inquiry (see Section 5.10). Thus, over a very long length of cable the differences between consecutive signals (dots and dashes) would become difficult to identify with the crude detection methods available at the time, unless the transmission speed were kept very low.

Several changes in operational methods were put into effect in the 1860s to ameliorate these difficulties. One consisted of a change in the implementation of the Morse code. Instead of using pulses of equal polarity but different length, the dots and dashes of the Morse code became pulses of equal length but of opposite polarity. Figure 5.7 shows the variation in receiving current with this new arrangement. Although a distorted version of the original square-topped Morse characters is received, they can still be distinguished from one another. This new code, termed the 'cable code', soon became universal for all transmissions over submarine cables.

Two further techniques came into use with cable systems. In a simple telegraph system a single Morse code symbol, a dot or a dash, will result in a long 'tail' to the symbol as described above, and the symbol will be easier to recognise if this tail is reduced to zero as quickly as possible. Various proposals were made to achieve this rapid reduction, principally by applying an opposite potential after the required signal had been received, a technique then dubbed 'curbing the signal' [17]. This worked extremely well for long terrestrial lines where the marking signal, either a dot or a dash, could be one polarity, and the space signal between the dots and dashes could be of the opposite polarity. For the submarine telegraph cable code, where the dots and dashes were of equal length but opposite polarity, a simple technique of curbing the signal by reversed polarity was no longer effective, except for the single-polarity Morse letters such as e, i, s, h, t, m and o. As a consequence a number of elaborate curbing routines were developed for use with the cable code, each having its own enthusiastic following. A popular example was William Thomson's 'automatic curb sender', designed to impose extremely accurate proportions in the lengths of the signal currents and spaces, so that by sending

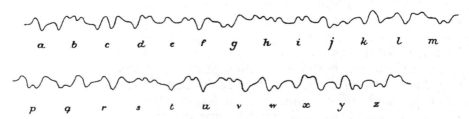

Figure 5.7   The telegraph 'cable code', first used for the transatlantic cable in the 1860s

automatically an accurate reversed current after every marking current a significant improvement in clarity could be achieved [18].

Many believed that these curbing systems would be completely ineffective on submarine telegraph cables, and advocated an operational technique known as 'signal shaping'. Here the rectangular shape of the transmitted signal was modified in such a way as to assist in the reconstruction of a near-rectangular current waveform at the receiving end, which enabled the rate of transmission as well as the clarity of the received signals to be considerably increased. Signal shaping reached complex forms in the early 20th century, (particularly on the Pacific cable, which we consider later); in its initial form it consisted simply of series capacitors inserted at each end of the cable. These also acted to block low-frequency earth currents, which were regarded as an impediment to achieving a high operating speed.

5.6.1 Loading

While William Thomson was the first to explain the signal distortion that occurred on very long telegraph lines, it was Oliver Heaviside who developed a more complete theory of signal propagation and suggested a solution. In a paper published in *The Electrician* in 1887, he showed that the inductance of the line, previously neglected in performance calculations, could play a significant role in determining the shape of the transmitted signal [19]. If the series inductance of the line could be increased, there would be an improvement in the shape of the received signal. Heaviside suggested that this could be done either by loading the cable dielectric with finely divided iron filings, or adding inductance at a large number of distinct points along the line. Experiments with 'loading coils' were carried out by G.K. Winters, Silvanus Phillips Thomson, G.A. Campbell and others. Gradually a number of techniques were evolved which not only increased the efficiency of submarine telegraph cables, but were also responsible for the later rapid development of the telephone system, using existing air lines over long distances, to provide the enhanced frequency performance needed to meet the wider bandwidth requirements of speech circuits. Winters, in a paper published before Heaviside's, had described the use of lumped inductive shunts across the ends of cable systems, in a way suggested earlier by Varley with his capacitance-coupled cables, but failed to relate this to a process of continual inductive loading [20]. G.P. Thomson, however, took the process further and described in several patents methods of including inductors at intervals along a cable, by dividing the cable into sections with separate inductors shunted across the line conductors or by the use of transformers [21]. In the United States, M.I. Pupin also performed experiments on loading coils, and his criteria were put into effect in some long-distance terrestrial lines [22]. Their value in telephone lines was demonstrated by

G.A. Campbell at about the same time. He made use of series-loaded lines, and developed an exact expression for the propagation coefficient of a series-loaded cable which is still used today in telephone system design.

These various ideas on loading were not taken up immediately by the submarine telegraph companies, probably because of the difficulty of measuring the insulation resistance of a cable from the cable end when it was equipped with loading coils, and similar problems which could confuse the determination of fault location by the resistance method. The original distributed loading, suggested by Heaviside, was more practical for this application and in 1902 was first put into effect by a Danish engineer, C.E. Krarup, who produced achieved distributed loading by placing a closed spiral of iron wire around the central conductor. The first loaded cables of this type were laid in waters around the Danish coast from Elsinore to Helsingborg. Later, distributed loaded cables were laid across the Straits of Dover by the Telcon Company in 1911–12 and proved successful in extending the limits of telephone transmission from Britain to towns in France. Iron alone was not, however, very satisfactory since it gave only a limited increase in inductance. By the 1920s loaded submarine telegraph cable was being produced by wrapping a narrow thin tape of permalloy spirally around the central copper conductor. Permalloy has a very high permeability, high resistance and low hysteresis loss at the low values of magnetic intensity associated with telegraphic currents and, when used for the first time on the New York to Azores cable in 1924, it enabled speeds of 400 wpm to be realised [22, 23]. The technique is routinely used today on lengthy cables, such as those across the Pacific, and made possible the development of telephone as well as telegraph communication by submarine cable.

5.6.2 Sensitive detectors

While the various operational changes in 'cable code' and 'signal curbing' helped to differentiate between the two constituent symbols of the Morse code, the problem of leakage remained, the current signal becoming progressively weaker as it travelled down the line. The ideal solution is to divide a lengthy cable into short sections and reconstitute the signal by retransmission at each section. A number of attempts were made to design instruments that would relay the message automatically and reduce the error often introduced by repeatedly reading and retransmitting the message. To assist this process, several ingenious 'amplifying repeaters' were constructed – this was before thermionic valve amplifiers became available. The most successful of these repeaters was the Western Union 'magnifier', developed by E.S. Heurtley and patented by him in 1909 (British Patent number 17555). A light moving coil carried the current and, since the coil was situated within a permanent magnetic field, it was free to rotate according to the strength and direction of the current flowing in it through two attaching wires. These wires were heated by the electric current flowing through them and were cooled by two small blowers, one for each wire. When the coil turned, one wire moved closer to its blower, and the other moved farther away. This affected the resistance of the wires which formed part of a bridge circuit, the balance of which was disturbed to produce an enhanced electrical signal, large enough to operate a relay. The Heurtley magnifier was a complex affair, difficult to construct and adjust, and was fairly quickly supplanted as soon as valve amplifiers came into use [23].

Where the terrain was suitable, a long submarine cable could be divided into smaller sections and a repeater or 'magnifier' inserted to enhance the signal before transmission over the next section. This could be carried out for example by 'island hopping' where suitable landings could be found. In a long run, such as crossing the Atlantic ocean, such a solution was not possible. What was needed in the 1850s was a more sensitive form of detector than the magnetic relay, one which could be operated by the very small currents reaching the end of a long line. This was found, again by William Thomson, with his highly sensitive mirror galvanometer recorder. Here minute movements of a magnetised needle were magnified by the action of a beam of light falling onto a mirror suspended in its supporting wire, and displayed on a curved scale (Figure 5.8) [24]. The magnets used were small flat pieces of watch spring, cemented onto the back of the mirror, the combination weighing less than 0.1 g. An external magnet could be rotated about a vertical axis to provide a controlling force to bring the reflected spot of light to a convenient point on the scale. Later, in 1863, Thomson introduced a more sensitive astatic version of his mirror galvanometer (see Chapter 2), which

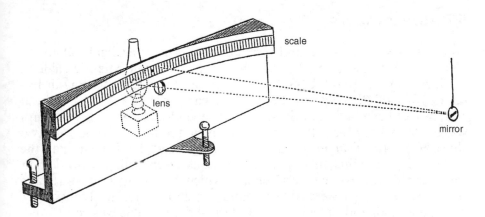

Figure 5.8   Thomson's mirror galvanometer
Source: Thomson, 'The Life of William Thomson', 1910

had four coils arranged in pairs on either side of the magnetic needles. A problem with this technique was the effort required to interpret the rapid movements of the light spot and, at the same time, to write down the character so represented. It was customary to employ two operators, a 'mirror operator' to interpret the movement of the mirror and call out the character it represented, and a second operator to write down the characters, a procedure which did not appeal to station managers!

William Thomson was again able to provide an ingenious solution: in 1867 he patented a sensitive indicator capable of providing a permanent record. This was the 'siphon recorder', which in the following decades became the standard cable receiving device [25]. It resembled a modern moving-coil microammeter, having a coil suspended between the poles of a powerful magnet which moved when current flowed through it. This movement was conveyed to a capillary tube, one end of which moved across a paper ribbon, the other end dipping into a well of ink. The writing end of the tube did not touch the paper, so the device was virtually 'friction free'; the ink was caused to emerge by maintaining an electric charge on the tube with the paper ribbon held at earth potential as it passed over a metal roller. The ink was ejected onto the paper as a series of closely spaced dots, giving a straight line in the absence of a signal, but moving to the left or right as the signal went positive or negative, to indicate a dot or a dash. The first public demonstration of the siphon recorder took place in June 1870 at the London home of Sir John Pender on the completion of the British Indian submarine cable, although it had been used operationally for several years before then. In the United States, and as marketed by the Automatic Telephone Manufacturing Company, it became known as the 'undulator' and was in wide use for terrestrial as well as submarine cable operation until well into the 1930s.

5.7 Atlantic crossing

None of these techniques and instrumentation were available when the Cyrus Field syndicate made its first attempt to lay an Atlantic cable in 1857. A single ship, the USS *Niagara*, started from Valentia Bay in Ireland with half the cable on board. The cable parted after only about 550 km had been laid, and the effort was abandoned. A second attempt was made early in 1858, this time commencing in mid-Atlantic with two ships, HMS *Agamemnon* and the USS *Niagara*, each loaded with half the cable and proceeding in opposite directions after having spliced their cables together. Despite several breaks, which necessitated a restart each time, the two ships were able to land the shore ends in Trinity Bay, Newfoundland, and Valentia Bay, and the Atlantic cable was completed. The date was 5th August 1858. This success proved to be short-lived, for after only a few weeks of working an electrical insulation failure put the cable out of action. One report (never completely substantiated) attributed the failure to Wildman Whitehouse, the engineer in charge, who by the use of induction coils increased the operating potential at the sending end to 2000 V in an attempt to generate sufficient current at the other end to operate the insensitive relays used.[2] Before this failure, the line had demonstrated its value by passing over 700 messages, including an important one from the British Government countermanding an order for the embarkment of two regiments to leave Canada for England, thus saving the Treasury about £50000 in transportation costs.

For a number of reasons, considered later, a third proposal for an Atlantic cable was deferred until 1864, when the Atlantic Telegraph Company contracted out to the Gutta Percha Company the task of manufacturing the cable required. Earlier attempts had revealed the problems that arose when two ships attempted to meet in mid-ocean and to lay a spliced cable simultaneously. This time it was decided to load the entire cable on board a single ship for the attempt. Only one ship afloat could accommodate this amount of cable, together with all the coal required for the crossing – Brunel's *Great Eastern*, a vessel of 22000 tonnes displacement and over 200 m in length, at that time lying idle and waiting for a role. It found it in cable-laying, for which it was ideally suited, and subsequently carried out a number of tasks of this kind, mainly for John Pender, who was to become a major figure on the telegraph scene. In July 1865 the *Great Eastern* spliced her cable with a heavily armoured shore cable, previously laid by SS *Caroline*, and began her voyage to the United States. Again, difficulties beset the operation of laying the cable, mainly faults in the cable picking-up gear installed on the *Great Eastern*, and on 11th August the work had to be abandoned. Financially this failure proved disastrous, and in order to attract sufficient funding for a further attempt, the Atlantic Telegraph Company amalgamated with the newly formed Anglo-American Telegraph Company, further support coming from Daniel Gooch, at that time Chairman of the *Great Eastern* Company and a Member of Parliament. More cable was taken on board, and on 30th June 1866 the *Great Eastern* arrived at Valentia Bay for another attempt. The ship arrived at Trinity Bay, Newfoundland, on 27th July, its mission completed, and a working Atlantic cable finally in place. As a bonus the *Great Eastern* successfully recovered the 1865 cable, spliced it with new cable and proceeded to Trinity Bay for the second time, thus providing not one but two working cable links across the Atlantic.

By the end of 1873 five cables had been laid across the Atlantic, linking Britain with Canada. Onward connection to the United States was by landline, a situation which was not completely satisfactory to the

[2] On present-day long cables of transatlantic length the usual applied working voltage is 50 V, and no cables are operated at greater than 70 V. This limitation is solely to protect the cable insulation from any possible damage.

American authorities. In 1873 a new company was formed, the Direct United States Cable Company, in which the Siemens & Halske Company played a major role. Its objective was to link the United Kingdom directly by submarine cable to the United States. This would require an extremely long cable, so the plan was modified so as to lay a cable to Tor Bay in Nova Scotia, where the signals would be received and immediately relayed over a second submarine cable to Rye Beach in New Hampshire, a total length of 4927 km. The cable was to be manufactured by Siemens in Woolwich, England and laid by Siemens' own cable ship, *The Faraday*, which had been built and launched in Newcastle in 1874. This was especially designed for cable-laying, having a gross tonnage of 4800 tonnes, 108 metres in length and incorporating twin screws for close manoeuvrability. The cable was laid in September 1875 and, before it was open to the public, was tested by Sir William Thomson,[3] who later wrote that the cable

proved to be in perfect condition as to insulation and showed its electrostatic capacity and copper resistance to be so small as to give it a power of transmitting messages, which, for a transatlantic cable of so great a length, is a very remarkable as well as a valuable achievement [26].

The company went on to use *The Faraday* for a large number of cable-laying and repair operations, including eight transatlantic cables, one of which was commissioned by the Compagnie Française de Télégraphie de Paris in 1879 to link Paris with New York.

5.8 Links to South America

At about the same time as the Atlantic was being spanned by submarine cables, telegraph technology was being introduced to the South American republics, which later benefited enormously from cables connecting the New World with Europe. One of the earliest of these new ventures in South America was the River Plate Telegraph Company, established by a Glasgow engineer, John Proudfoot, who obtained concessions from Argentina and Uruguay, initially to lay a 39 km cable across the mouth of the river Plate and to operate a telegraph service through a connecting landline between Buenos Aires and Montevideo. This was in 1865, and a key item in his equipment for this service was the Wheatstone ABC telegraph, designed to be used by untrained operators. This was the beginning of intensive telegraph operations in South America, which were later to be developed by, among others, the West India Cable Company, the Cuba Submarine Cable Company, the Brazilian Telegraph Submarine Company, the Central & South American Telegraph Com-

---

[3] It was during these tests that Thomson discovered that a 1 per cent change in the impurities in the copper could cause as much as a 50 per cent change in conductivity, a factor which caused the cable users to become much more alert to the techniques of cable manufacture than before.

pany, John Pender's Eastern Telegraph Companies and Siemens & Halske [28, 29].

An important goal for the early telegraph pioneers in this area was to establish connections between the several islands of the Caribbean and, through Panama, to link the continents of North and South America. Many colonial powers were involved in this enterprise, mainly Britain, France and Spain, in addition to the neighbouring countries of South America and financiers in the United States and Canada. The questions of concessions, landing rights and tariffs proved to be complex, and led to the endless formation and disbanding of cable companies as the responsibility (and anticipated profit) passed from one company to another. A full history is described comprehensively by Jorma Ahvenainen in his *History of the Caribbean Telegraph Before the First World War* [27]; only the outcome of the operations is noted here. A key figure in the exploration and laying of submarine cables in this region was Sir Charles Tilston Bright, who had been active in laying the first Atlantic cable and a number of Mediterranean cables [28]. Much of the Caribbean was in British hands, and the West India Cable Company and the Cuba Submarine Cable Company, both British firms, were responsible for laying cables linking the islands of Cuba, Jamaica, Trinidad and several smaller intermediate islands with the Brazilian mainland at Georgetown in 1870–72. These routes were not easily achieved or maintained afterwards. The Caribbean sea-floor was marked with great and sudden variations in depth, and had a rocky, coralline surface which damaged the cables and, in certain areas, a considerable layer of soft alluvial mud, which made the recovery of cables in the event of damage an almost impossible task. Linking Cuba to Punta Rasa in Florida had already been achieved by the International Ocean Telegraph Company in 1867 and 1869, using cables manufactured by the Gutta-Percha Company in Britain (there were no submarine cable manufacturers in the United States at that time), the work being supervised by F.C. Webb and Charles Bright. By 1875 the Cuba Submarine Telegraph Company had succeeded in laying working cables between Jamaica and Colón (Panama). Other companies took over the laying of cables along the west coast of South America, principally the Central & South American Telegraph Company, an American company, which laid most of the cables from Panama north up to San Josi, and by land to Mexico City and south to Paita and Collao in Peru during 1882.

While telegraph connections could, in the 1880s, be made from the Caribbean and South America to Europe, this was by no means a direct process. For example, in communications between the West Indies and Europe the message had to be transmitted via the lines of at least five companies, and this caused high tariffs to be imposed. Furthermore the British, French and North American governments were anxious to have their own direct routes which would be completely under their own control. The British were the first to achieve this, through the Halifax to Bermuda Telegraph Company, registered in England. A cable for this route was manufactured and laid by W.T. Henley's Telegraph Works in 1889 and extended, after several lengthy negotiations, to link Bermuda, Jamaica and several of the smaller islands in the Caribbean in 1898. An all-British line was now in operation between Great Britain and the West Indies, and proved vital to maintain communications later when war erupted between the United States and Spain, blocking all the links to Spanish possessions in the Caribbean. The French, through the activities of the Société Française des Télégraphes Sous-Marins, established a direct line from New York to Haiti, the West Indies and Venezuela in 1896 and, with the Brest–Newfoundland cable laid earlier in 1869, were in a position to maintain a French-operated line to their Caribbean possessions, albeit via a short land line from Newfoundland to New York. The United States already had a cable connection linking Cuba to Florida, laid by their International Ocean Company in 1867. This was, however, regulated by concessions granted by the Spanish authorities, but at the conclusion of the American–Spanish war, Puerto Rico was ceded to the United States, while Spain relinquished its sovereignty over Cuba, thus enabling the telegraph traffic over the Cuba–Florida route to be placed solely in American hands. Only two further lines were added later: one by the Commercial Cable Company from New York to Havana, and one by a subsidiary of Western Union from New York to Colón. Both were laid in 1907 thus ensuring that the two all-American companies would have control of United States operations in the Caribbean.

5.9 Cable-laying technology

Laying oceanic cables also invariably meant establishing a facility to recover from the ocean floor a cable which had parted company with the cable carried aboard ship. This could (and did) occur several times during a single cable-laying exercise, which often took place in very deep water. The British excelled in this technique. At the 1878 Paris Exposition was shown, Andrew Jamieson's British patent, a self-relieving grapnel for raising submarine cables. Detection of cable lying on the sea bed reached almost an art form: 'an officer sitting on the taut grapnel cable where it passes over the bow sheave and, by the feel of vibrations through his posterior, can tell the nature of the ground over which the grapnel hook is passing and can feel change in tension when the cable is hooked' [29].[4] In busy sea communication areas where the ocean floor may be criss-crossed with many working cables, the recovery of a given

[4] A more scientific technique is to detect the small e.m.f. that is generated as the cable is lifted through the Earth's magnetic field.

cable is not quite so simple, and a graphic description of the problem was given by Captain Holmes of the Royal Signal Corps just before the Second World War. He cites dragging ships' anchors and typhoons as major operational problems in keeping cable communications working in Hong Kong waters:

In a wind of typhoon force, ships at anchor have to go full steam ahead in order to keep their position and not drag their anchor; if the anchors once start to move, nothing can stop them and the ship is out of control. . . . the Colony was struck by a Typhoon in August 1936, and by a very severe Typhoon in September 1937, the submarine cable repair work encountered was considerable; all the damage was caused by shipping. [30]

Recovery in the event of an electrical fault developing was also required, and for this the exact location of the fault (generally a short circuit or direct contact with sea water) was needed. Numerous methods were used, many of which depended on the measurement of cable resistance, and since this depends, among other factors, on the sea temperature in which the cable is immersed a comparative test was used involving a Wheatstone bridge. An alternative test was to use a differential galvanometer in which the currents passing through the cable and a variable resistance were balanced to give a zero indication when the two were equal. To minimise the effects of induced currents such as earth currents, the polarity of the testing voltage source was made reversible, and the two readings were averaged. Other measurements were of cable capacitance and insulation resistance, both of which were necessary if a complete fracture was found [31]. The capacitance of a uniform cable is proportional to its length, so the ratio between the total capacitance and that up to the fracture gave a measure of the length of the measured portion, again using a bridge method. This was complicated by the leakage of the charge applied to the cable, and by the time taken for this leakage to be observed as a change in the resistance of the cable. Similar tests were carried out by measuring insulation resistance, which was inversely proportional to the length of cable measured. Again, the measurements were complex and time-consuming, requiring two separate measurements, one with the polarity of the testing source potential reversed, the two measurements being averaged to minimise the effect of induced currents. Each measurement was recorded only after a steady reading was obtained (this could take several minutes for a long cable). When the fault was a partial earthing of the cable without fracture – a breakdown of insulation at one point – and both ends of the cable were available – as in factory testing, or when a second well-insulated cable was available for the return path – then a loop test could be employed. In a loop test the two sections of the cable formed two sides of a Wheatstone bridge, so that the ratio of the bridge resistances on the variable arms at balance gave the ratio of the resistances of the two sections of the cable, and hence their length

[32]. An absolute measure of the insulation resistance of an undamaged cable could be obtained by charging the capacitance in the cable to a high potential, disconnecting it, and then determining, from its measured potential voltage decay to half its initial value, the time constant and hence its insulation resistance component. All these tests were often carried out on board ship, and to compensate for the rolling of the vessel special forms of 'marine galvanometer' were used, fitted with controlling magnets designed to cancel out any displacement of the galvanometer coils in the Earth's magnetic field as the ship rolled.

To minimise the possibility of laying a faulty cable, techniques of continuous testing of cable insulation as the cable was paid out over the ship's stern were evolved, following a lead set by Willoughby Smith in 1879 [33]. He monitored a signal sent continuously through the cable from a shore station to which it was linked by connecting a sensitive galvanometer through a plate surrounding the cable at its exit point. This acted as one plate of a capacitor, the other being the moving central wire, thus effecting a measurement with no direct connection to the cable.

5.10 A Committee of Inquiry

The mid-1850s saw a series of spectacular failures in laying submarine cables, not only the early Atlantic cable attempts, but also the first line to India and various lines in the Mediterranean.[5] These failures led to heavy financial losses for both industrial backers and the British Treasury, prompting the latter to recommend the formation of a Committee of Inquiry in the construction of submarine telegraph cables. This was convened in 1859, chaired by Captain Douglas Galton RE and with four members each from the Board of Trade and the Atlantic Telegraph Company, one of whom was Charles Wheatstone. It was an important investigating committee, and its conclusions had far-reaching implications for subsequent developments in all aspects of cable production and laying [34]. The report, published in April 1860, was entitled 'Report of the Joint Committee appointed to the Lords of Commons of the Privy Council for Trade and the Atlantic Telegraph Company to inquire into the Construction of Submarine Telegraph Cables' [35]. It was a comprehensive document, 80 000 words in length and containing the testimonies of more than 43 experts, including all the leading practising telegraphists of the time – Bright, Jenkins, Preece, Wheatstone, Latimer Clark, William Thomson, Siemens, Varley – and various representatives from the Navy and industry.

William Thomson's evidence was probably the most valuable. In 1858

---

[5] The gravity of this situation was revealed in 1861, when it was recorded that of 18 296 km of cable laid, only 4830 km was actually working.

he had described his mirror galvanometer, whose action depended on changes in current strength, so that it was no longer necessary to wait for the entire line to be cleared of its electrical charge before the next message could be sent. Indeed, the insensitive relay system used previously had been shown to be virtually ineffective if the cable length was more than 1000 km in length. Many of the conclusions of the Committee of Inquiry were taken into account in subsequent cable production and laying operations, considerably reducing the number of failures. Among these failures, the report noted:

1   Variations in conductivity caused by impurities in the copper used.
2   Porosity of the gutta-percha insulation resulting from faults in its manufacture or the presence of air bubbles.
3   Bad joints where cables were spliced.
4   Small punctures in the casing leading to the ingress of sea water into the gutta-percha coating, destroying its insulation properties.
5   Eccentricity of the conductor, which could have been caused by a high storage temperature for the finished cable, either in factory storage tanks or in the hold of a cable ship.

The techniques then being used to manufacture cable, and the conditions in which the finished cable was stored, were considered as contributing most to the failures of the Atlantic and the Red Sea Indian cables. Specifically, Galton wrote 'The account which we have given shows that [the Atlantic cable's] failure was in great measure owing to the absence of a proper preliminary experimental inquiry into the conditions required in the construction of such a cable.' He added that 'the storage of the finished cable left a lot to be desired'. The poor storage of a finished cable at the Glass & Elliot works is often quoted as a cause of subsequent failure, for it was left uncovered in the open during an unusually hot summer. This softened the gutta-percha and displaced the copper core from its central position in the cable, the effect of which was felt when the cable was subjected to sharp bends during storage or laying. Measurements on the cable as it was being manufactured became an essential requirement for submarine applications. To check for continuity of the copper wire a simple galvanometer test sufficed. A method of charge storage was used to check continuously for insulation, in which the rate of leakage of charge applied to the finished cable proved an effective indicator. A satisfactory method of testing for the many splices that formed part of the manufactured cable was also evolved and applied to all cable manufacturing after 1891; it was based on the techniques described by Willoughby Smith, who also gave evidence at the Inquiry [36].

To facilitate the sharing of knowledge concerning the technology of telegraphy brought to light in this and other reports, John Pender and others founded *The Telegraph Journal*, which appeared weekly, priced 6d (2½p) from 1861, becoming *The Electrical Review* in 1899.

5.11 A Cable to India . . .

The second great challenge to the telegraph entrepreneurs was to find a
way to connect Great Britain with its largest possession, British India.
For some time the government's inability to communicate quickly with
the Far East had been a major handicap in its handling of such events as
the swiftly developing Indian Mutiny in 1858, and in its many China
trading interests. One consequence of Galton's Commission of Inquiry,
coupled with a growing appreciation by the government of the value
of rapid communications with its overseas empire, was that the Treasury was now fully prepared to lend financial support to a suitable
enterprise leading to improving such communications. An initial step
was to ask Charles Bright to prepare a report for the Treasury on the
possibility of a line to India. The outcome was a parliamentary blue
book, *The Establishment of Telegraphic Communication in the Mediterranean and with India*, published in 1860 [37], which took the form of
a list of all those British companies with recognised capability to
establish submarine cables and their detailed financial probity. The
report thus made clear which companies could be safely approached
to undertake such a major project if government funding was to be
involved.

Before Galton's and other government inquiries had published their
findings, private enterprise (with some government help) had already
begun the first attempt to construct an Indian telegraph, along what was
known as the Red Sea route, initiated by the Red Sea & India Telegraph
Company. Figure 5.9 shows this and three other routes attempted or
completed. The intention with the Red Sea route was to use the projected
link from Malta to Alexandria, existing land lines from Alexandria to
Suez, in which the company already had a monopoly, and to build new
lines along the bottom of the Red Sea to Aden and then across the
Arabian Sea to Karachi. The Red Sea submarine cable from Suez was
laid by the Newall Company, rather hurriedly and with poor supervision,
and reached Aden in 1859. It failed repeatedly and was abandoned in
1861, just before the completion of the Aden–Karachi line, which itself
failed after a few weeks. Through-connections from Egypt to Karachi
had been possible for a few days only, and messages were sent, but the
entire project was poorly thought out and executed, and occasioned
financial loss to the Treasury – a major factor in triggering the government inquiry described earlier.

The second route, known as the Turkish route, was laid along a mainly
overland course, taking advantage of existing cable routes across Europe
to Constantinople, before crossing into Turkey to reach the Persian Gulf
and then, again by land, across Baluchistan to its destination at Karachi.
New lines beyond Constantinople would have to be built, and the
European & Indian Junction Telegraph Company was formed to carry

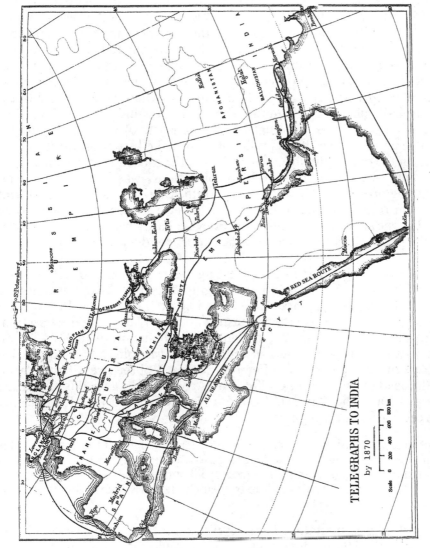

Figure 5.9 Cable routes to India by 1870
Source: After Goldschmid, 'Telegraph and Travel', 1874

this out. Cable manufacture was contracted out to W.T. Henley's works in London, and was far better executed than the ill-fated Red Sea cable. William Henley, who founded the firm in 1837, was one of the earliest manufacturers of insulated wires for electrical purposes. After 1859 he moved his works to large premises at North Woolwich, which was eventually to extend over six hectares to include a wharf for loading submarine telegraph cable into his own cable ships. His works became noted for their ability to produce cables for the exacting submarine cable contracts, and in 1865–66 manufactured the shore ends of several Atlantic cables as his was the only works possessing the sufficiently powerful armouring machines required [38]. The Indian cable contained a single core of copper wire covered by four layers of gutta-percha, several layers of asphalt silicon compound, a hessian dressing, twelve layers of armoured steel wires and finally yet more layers of asphalt mixture. In producing this cable the manufacturers benefited from previous experience on the desirability of high-purity copper for long submarine cables, and only wire proven to have high conductivity was used in the manufacturing operation. The total length of the cable was 2335 km, and with a weight of 5000 tonnes it was by far the heaviest length of cable to be used in one submarine cable-laying exercise, partly because of the heavier armour plating, but also because the copper conductor was much thicker [39].

A British colonel in the Royal Engineers, Patrick Stewart, at that time Director-General of Indian Telegraphs, assumed responsibility for surveying the route, under the direction of Sir Charles Bright, and eventually for supervising the construction work also. The first section, a landline, was built from Karachi, across Baluchistan to Gwadar on the north coast of the Gulf of Oman, and was joined to a submarine cable terminating in Fao (Al Faw) at the mouth of the river Shatt al Arab at the head of the Persian Gulf. It was no easy landing for the submarine cable at Fao. The *Illustrated London News* described the efforts to get the cable ashore, through

thick mud the consistency of cream . . . Sir Charles Bright set the example to his staff and the men and was the first to get out of the boat and stand up to his waist in the mud – an example followed by all of the officers and men, who were soon wallowing in the soft mud up to their chests, but still dragging the end of the cable.

This effort was maintained for several kilometres until firm land was reached [40]. The landline section was completed in 1863 without too much difficulty. Laying the submarine cable section was achieved using sailing vessels, towed by steamers supplied by the Indian Navy. This combination appeared to work well, separating as it did the tasks of carrying the cable and of carrying the coal used for fuel for the steamers, while retaining the manoevrability of a powered ship. Two intermediate junction landing places were made at Mussendam and Bushire, where the signal could be reconstituted to avoid the attenuation which would otherwise be experienced. At the Bushire junction an alternative route was planned by landline across Persia, northwards to Teheran to connect with the Turkish telegraph network at Khanikas. The line from Fao, leading westwards to Baghdad, where it joined the Constantinople line, was not completed until 1864, and on 27th January 1865 the first unbroken telegraph connection between India and Europe was opened. Sadly, Colonel Stewart, who energetically oversaw the completion of the line, did not live to see this: he died in Constantinople at the early age of thirty-three, a week before the line opened.

While the Indian connection continued to operate for several years, it was not operationally a great success. Telegrams took days rather than hours to reach their destination, and frequently arrived mutilated – and there was no opportunity to gain access through the intervening operators to determine the original text. The problems were attributed almost entirely to the Turkish section of the route, where the variety of scripts used by the Turkish operators and their unfamiliarity with European languages led to retransmission errors. The situation was scathingly denunciated by Sir James Anderson in his book *Routes to India*, in which he states that

no steady improvement [on existing lines] has yet been affected, nor is there any reason to anticipate anything better than a continuous and unprofitable outlay of large sums of money even to keep open the present spasmodic unreliable communication . . . there are several lines through Europe and Asia, and one through Persia, meeting the lines of the government at the Persian Gulf; delay and inaccuracy characterise the whole of them [41].

Three years after its inauguration, a select committee of the House of Commons, worried by continual complaints, decided that the enterprise was 'thoroughly inefficient and that to a great extent [the cable operation] was at the mercy in Asia of a set of incapable and unimprovable Arabs, Turks and barbarians', and recommended that any further line to India should be 'administered by a single management'.

This challenge was taken up by the German firm of Siemens & Halske, which had strong English connections. It proposed a route to be developed for the Indian Government from the English Channel across Europe and Russia to Teheran, where it would link with the existing line to India via Bushire and the Persian Gulf, thus avoiding Turkey completely. A double line was proposed to run from Lowestoft under the North Sea to Norderney, via Emden and Berlin to Thorn on the East Prussian frontier, and then by way of Odessa and Kertsch to Teheran, where it would connect to the existing Indian line to Karachi, Bombay and Calcutta, previously constructed by the British Government. This route was accepted by the Indian Government after some heart-searching (it had up to then objected to private commercial companies operating its internal telegraphs), and the Germans formed a company in 1868, the Indo-European Telegraph Company, to begin this new route, known later as the Indo-European or Black Sea route. Much of the construction work was carried out by British sappers of the Royal Engineers under Major Bateman Champion. The overland portion of the route from the German frontier to the Black Sea passed through wild uncivilised country, and a strong form of construction was adopted using very thick iron wire and iron posts. A new submarine line was laid under the Black Sea using, surprisingly, rubber insulation in place of gutta-percha, and manufactured by W.T. Henley's company in Gravesend and shipped to Bombay.[6] Again, a combination of sailing vessels to hold the cable and steamships to tow them was used, and the cable laid after a number of difficulties had been overcome (described by J.M. Adams, a grandson of one of the British telegraphists who worked the line [42]). The entire line, including a duplicated section from Teheran to Karachi, was completed in January 1870 and was extremely successful, messages taking fewer hours than they had taken days over previous routes to India, with an average accuracy surpassing any previous records.

Messages to be transmitted were prepared as punched paper tape using the double-hole code, already in use with Wheatstone's recorder on other submarine cables, and only at Teheran was it necessary to decode and retransmit the information between sections. This process considerably shortened the transmission time of messages along the route and reduced the number of errors. The telegraph equipment for the line was designed and manufactured by Siemens of Berlin. To automate the process of relay transmission, Siemens designed a punched ribbon telegraph or 'inker'. It was fitted with a cranked-handle electromagnetic generator such that the armature rotated once as the punched paper tape advanced by one hole position. Thus a current was transmitted, depending on the hole positions, as a series of alternating positive and negative impulses. A polarized relay was made responsive to these currents. For a dot (that is, two holes in line across the tape), a positive pulse resulted. For a dash, (two holes in the strip spaced apart) the next *negative* half-cycle of current generation was also received, resetting the relay and a longer transmission ensued [43]. The symbols were recorded on a moving paper tape at the receiver, using an inked disc pressed onto the paper by the action of a relay. The polarized relay could be made to repeat the impulses for onward transmission to the next telegraph station as required.

Although Britain now had several alternative cable routes to India, there was still disquiet about their continued availability in times of

---

[6] This may have been because gutta-percha insulation was known to lose its shape in the high temperatures found in tropical climates.

conflict or war. *The Electrician* in April 1885 noted that all the present long-distance telegraph cables passed at some point through foreign countries, notably Russia, China and France, any of which would be in a position to block transmission of government messages between Britain and the Far East [44]. A route was needed which would pass mainly under the sea, be controlled entirely from Britain and, incidentally, be operated without the payment of way-leaves to foreign states, which could always be revoked in times of stress. This was to be considered as the fourth route to India, the all-sea route, to pass by sea from Porthcurno in Cornwall to Gibraltar and Malta, under the Mediterranean to Alexandria, by landline to Suez, under the Red Sea to Aden and finally across the Arabian Sea to Bombay.[7] A direct cable between Malta and Alexandria had already been laid by the Anglo-Mediterranean Telegraph Company in 1868. This was a deep-sea cable, 1500 km long, manufactured by the Telegraph Construction & Maintenance Company (Telcon), a new company formed by John Pender in 1864 through the amalgamation of the Gutta-Percha Company and the cable-makers, Glass Elliot of Greenwich. Several other sections of the line already existed; the missing piece in the jigsaw was a cable from Britain to Malta, the construction of which now had to be considered.

John Pender, then chairman of the British-Indian Submarine Telegraph Company, was invited to undertake the installation of this new cable. He formed a company, the Falmouth, Gibraltar & Malta Telegraph Company, which suggested that the landing point in Britain would be Falmouth, but in the event he was to choose Porthcurno in Cornwall, which had been a busy station on the mail packet service and was later to become an important cable and wireless station for Britain. In addition, a doubling of the line from Suez to Bombay was contemplated to maintain the reliability of communication. This new line from Suez to Bombay was also to be completed by Telcon. The company directors had no reservations about being able to complete the task efficiently and on time, a considerable store of experience having by now been amassed by the various cable companies. They proposed to use the *Great Eastern*, a ship with an impressive cable-laying history – it had laid the first successful transmission cable – and big enough to carry the 3800 km of cable in its holds as well as the necessary coal required for the operation. The cable was paid out and the line made operational in just over two months. A subsidiary company then began the laying of a cable from Porthcurno to Alexandria, to complete the all-sea route to India by June 1870. Although three companies, and not one overall controlling company, carried out the project, it had a common leader in John Pender, who succeeded in making this route

---

[7] The decision to build a new cable route was no doubt influenced by the prospects of an expansion in trade following the completion of the Suez Canal, then under construction by Ferdinand de Lesseps (it was completed in 1869).

independent of foreign influence and, since it was almost entirely carried by sea, the question of way-leaves did not arise. Celebrations were now in order and the Lord Mayor of London entertained the Prince of Wales (later King Edward VII) to a banquet at the Guildhall in honour of the occasion:

From the table where HRH sat there was flashed messages of congratulation to the Viceroy of India at Simla whose reply was received by the Prince in less than one hour ... This was speedily eclipsed a few months later; the result of the English Derby reaching Reuter's Office at Bombay with the short space of four minutes [45].

Porthcurno, which had been established as the terminal station in Britain for this new Far East route, became an important centre for training telegraph operators for long-distance telegraph routes. It was a well-equipped if lonely establishment. A description of life there is given by John Pender's second wife, Emma, on a visit in 1877:

There is a telegraph station near to the Land's End, put down in a secluded bay, and no doubt tryingly lonely ... The Company have furnished it with a billiard table, library etc. and now it is a kind of school for preparing young men for the telegraphic positions abroad [46].

The success of the training routines established in this telegraph school was a result of the efforts of the first superintendent of the Porthcurno Station, Edward Bull. He had the difficult task of training young men – many of whom arrived at the school barely able to write – not only in the technology, but also to accept the long working hours required. In order to be tolerable operators, working an expensive long-distance cable route, they had to be proficient at sending and receiving Morse code at a rate of 10 wpm, and be able to use a mirror galvanometer or siphon recorder accurately without continually being asked for repetitions – which could be expensive for the company at the going rate of £4 per message. Not all the young men at Porthcurno were 'operators in training'; some were 'probationers' who were willing to work alongside trained operators for no pay while they acquired the techniques of the trade. They were content to do this in the expectation of employment elsewhere, since it became something of a cachet to be able to describe oneself as a 'Porthcurno-trained operator', as graduates of the school were in demand for the many new submarine telegraph stations established in the closing years of the nineteenth century.

5.12 ... and further East

In extending the reach of the telegraph network from Europe towards the Far East in the 1870s there were only two contenders, the Swedish Northern Telegraph Company, which had built and managed the

Siberian landlines through Russia to China and Japan, and the British Eastern and Eastern Extension Telegraph Companies working submarine lines under the management of John Pender.

In John Pender telegraphy found a new kind of pioneer. A trader in cotton in Glasgow and Manchester, he was essentially a financier and politician (he became Member of Parliament for Wick in 1872 and 1892), providing much needed financial support for the second attempt to lay an Atlantic cable by becoming a director of the English & Irish Magnetic Telegraph Company in 1852. The eventual success of this venture encouraged him to invest further in telegraphy and to initiate the construction of submarine cables in many parts of the world. To this end he founded a number of telegraph companies, most of which were eventually incorporated into The Telegraph Construction & Maintenance Company (Telcon), which dominated overseas cable construction in the early part of the twentieth century. His achievements in over thirty years of telegraphy were recognised by the state in the honours list of 1887 (the golden jubilee of Queen Victoria's accession to the throne), in which he was made a Knight Grand Cross of the Order of St Michael and St George [47]. After the Atlantic cable and his successful India cable project, his companies participated in almost a frenzy of cable-laying throughout the world along almost any route where they could profitably be laid. Trade with South-East Asia and China was booming, and cables were laid to link Singapore, an important trading centre, with Penang in Malaya and to Madras in India. Pender's first links with China came with the establishment of the China Telegraph Company in 1869, formed to link Singapore and Hong Kong via Shanghai with a 2700 km cable completed in 1871. Later, also from Singapore, cables were laid to Batavia and Java in Indonesia.

The success of these ventures prompted the amalgamation of several of the John Pender companies to form two new companies in 1872, the Eastern Telegraph Company and the Eastern Extension Telegraph Company, to link all of Britain's Asiatic and Australasian possessions [48]. This was just after the inland telegraph companies in Britain were nationalised, and although fears were expressed that the same fate would overtake the overseas cable companies, the Eastern Telegraph Company pressed ahead with its complex programme, financed entirely by private capital. The success of the Swedish Great Northern Telegraph Company in providing an overland telegraph service from Europe somewhat mitigated the monopoly that Pender sought with his amalgamation of all the relevant telegraph companies to form his Eastern and Eastern Extension Telegraph Companies. A compromise was reached in 1883 with an agreement between the Swedish and British companies to share both the traffic and the building of new lines in the Far East, and this continued after the British cable policy was handed over to the Postmaster General towards the end of the century. In 1900 the companies laid a joint submarine cable from Shanghai across Chefoo (now Yantai) to Taku in northern China. From there a landline extended to Peking via Tientsin, and a complex network of telegraph lines began to spread over northern China. At this time the Great Northern Telegraph Company made an important and unique contribution to this Far East development, quite apart from laying the cables and operating the telegraph stations. It created and printed a Chinese Telegraph Dictionary in which Chinese written symbols were represented by numbers transmitted in Morse code; the recipient transcribed the numbers back into Chinese by reference to the dictionary, making the system available to the public at all telegraph offices serving their routes [49].

Other links established by the Eastern Telegraph Company were those connecting Britain with South America and with Africa. Two important new links with Porthcurno were established in 1873 to connect to Portugal's capital via Vigo and Carcavelos, a total distance of 1430 km. These two new cables soon began to carry the majority of the traffic for the Eastern, Eastern Extension and Brazilian Submarine Telegraph Companies. A branch of the Eastern Telegraph Company, subsequently established in Vigo, served as a connection point for the cable links to Brazil, where a number of telegraph lines from several South American republics met. Through this Porthcurno–Vigo link, Pender's companies were able to participate in the huge expansion of telegraph traffic that occurred between the Old World and South America described in Section 5.8.

The first cable to connect Africa with the growing international telegraph network was completed in 1880. It was laid from Aden along the east coast of Africa to Zanzibar, Mozambique and Durban. In East Africa, the Eastern Telegraph Company and a number of associate companies connected many of the east coast ports to the submarine telegraph systems of the Mediterranean. Although the submarine lines connecting these ports were laid in shallow waters and hence were vulnerable to damage, this was considered preferable to constructing landlines through difficult jungle terrain. The whole of the African coast was now the monopoly of the Telcon Company, which was responsible for cable manufacture, line construction, maintenance and operation of all telegraph communication from Cairo to the Cape. The dependence of the country on 'John Pender's cables' was demonstrated by the part they played in the Boer War, during which a Pender cable was the only means of communication between the Colonial Office in London and President Kruger's government at the outset of negotiations in 1899. The difficulty of 'explaining everything by cable' has been cited as one of the reasons for the eventual breakdown of these negotiations, leading to the commencement of hostilities, and was regarded in some circles at the time as a negative recommendation for such 'high technology'.

The growth in the activities of the Eastern Telegraph Company was remarkable during the 15 years that John Pender remained chairman. In 1872 it owned 14000 km of submarine cable, owned or rented a further 2000 km of landlines, had 24 stations, two cable repair ships and hired others (including the *Great Eastern*) for cable-laying operations. By 1887 it had increased its holding to 36000 km of submarine cable and 64 stations, all but eight of them abroad. By 1900 the Eastern and its associated companies held a virtual monopoly on communications throughout the Empire by the medium of telegrams which was to remain unchallenged until the emergence of long-distance wireless communication in the 1920s.

5.13 The Australian connection

An important task for the Eastern Telegraph Company soon after its formation was the extension of the London–Bombay line to Australia. A landline was laid by the company from Bombay to Madras, where the submarine cable recommenced, eventually reaching the northernmost tip of Australia, with a cable laid from Java to Darwin by the British Australian Telegraph Company in 1871.

Telegraphy arrived in Australia a few years after it had been firmly established in America and was beginning to provide an accepted public service in Europe. The first landline was laid between Melbourne and Williamstown in 1853, a distance of 11 km, by S.W. McGowan, who claimed to have been a pupil of Samuel Morse in America (but whether he studied art or electricity is not recorded). He went on to install other short lines linking towns in Victoria, and sufficiently aroused the interest of the neighbouring colony of South Australia for its governor to formally request the appointment of a superintendent of telegraphs in the colony by the London Colonial Office. The choice made by London was a significant one. In 1858 it appointed Charles Todd, a Londoner then in charge of the Galvanic Department of the Royal Observatory, Greenwich, and formerly on the staff of Cambridge Observatory. He wasted no time on reaching South Australia, quickly setting up telegraph lines from Adelaide to Melbourne in 1858, followed shortly thereafter by lines to Sydney, Brisbane and Hobart in Tasmania. With the major cities of the continent in telegraphic communication with one another, using the Morse sounder and printer, business interests turned to the possibilities of similar communication with Britain.

The initial response came from Brett and Carmichael, fresh from their exploits in the Mediterranean attempting to link Italy with Corsica, Sardinia and North Africa. Their Mediterranean Submarine Cable Company wrote to the Australian Colonial Government in 1860 with plans to lay a cable from Ceylon (Sri Lanka) to the west coast of Australia, via the Cocos Islands. This suggestion was premature, particularly in view of the later mishaps that befell the company in the Mediterranean and ended in severe financial failure, and no more was heard of this scheme. An entrepreneur with more experience of eastern submarine cable links, Francis Gisborne, made determined attempts to interest several governments in the possibilities of a cable from India to Australia. He planned to link Bombay via a landline to Calcutta, from where a submarine cable would lead to Rangoon, Singapore and Java, finally passing through the Torres Strait to Brisbane on the eastern coast of Australia, and so connecting with existing telegraph links to Sydney, Melbourne and Adelaide. While successful in some sections of this route, Gisborne met with severe opposition from the Australian Government, which was not prepared to meet the cost of an extra 3000 km of submarine cable, implied in Gisborne's route from Java to Brisbane, and visualised a shorter link to the northern coast of Australia and its continuation by landline to the main cities in the south. Charles Todd expressed this view in his 1859 recommendation to the Governor of South Australia that a cable should be brought 'from Timor to the northern coast of Australia, near to Cambridge Gulf, and thence overland to Adelaide'. This would save the laying of over 3000 km of submarine cable, but would require the construction of a similar length of landline through country only partly explored. It was ten years before this view finally prevailed and the overland line was laid.

The exploration and search for a telegraph route from Adelaide to the north coast occupied many pioneers over several years. One of these was a scientist, Benjamin Herschel Babbage, son of Sir Charles Babbage, well known in England as the inventor of the first mechanical computer. Benjamin Babbage was employed by Charles Todd, Superintendent of Telegraphs, to survey part of the proposed route and advise on the geology of the central terrain – and also to keep an eye open for signs of gold, as the discovery of a goldfield would be a very desirable outcome of a telegraph route survey! The major burden of exploration fell on John McDonall Stuart, who in the face of much hardship eventually found a way through the mainly arid country of central Australia [50].

The landline was to be laid between Port Augusta, north of Adelaide, and Darwin on the northern coast, a distance of about 3000 km, Darwin being the terminal of the Java submarine cable. A new company was formed by Captain Sherard Osborne RN to lay the submarine cable from Singapore, where it would join an existing cable from Rangoon and the India network, and continue via Java to Darwin. This was the British Australian Telegraph Company. It undertook to lay this cable for the South Australian Government on condition that the government 'pledge themselves to have a landline open for traffic by the first of January 1871, connecting Port Darwin with the present system of colonial telegraphs'. It also imposed a penalty clause for late completion of the land connection. A similar provision was available to the South Australian

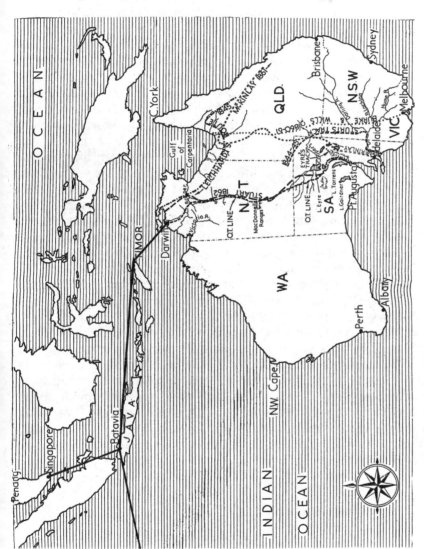

Figure 5.10 Austrailian telegraph connections
Source: Clune, 'Overland Telegraph', 1955

Government in the event of non delivery or operation of the submarine connection – which would prove fortunate for the final outcome.

The challenge was taken up by Charles Todd on behalf of the South Australian Government, and work commenced in December 1870 with the arrival in Adelaide of 3200 km of galvanised wire from Johnson and Nephew Ltd of Manchester, together with iron poles, insulators and other equipment for the venture. This was as big an undertaking as the earlier American project to link their Atlantic and Pacific coasts with an overland telegraph line, but without the detailed surveying that preceded the American construction. To a large extent the Australian transcontinental line was erected on a 'survey as you go' principle, so that major routing decisions were made by the teams responsible for the construction of the several sections of line (Figure 5.10). Work was commenced from both ends of the route, and the first pole of the line was planted at Darwin on 15th September 1870. The linesmen had to contend with extremely variable conditions along the route, on which there was no supporting settlement of any kind. For the builders of the overland telegraph line the problem was not only to find a convenient route for the line, and secure supplies of water and food for the construction parties and their livestock, timber for poles and access for wagon transport, but also to maintain the project during the alternating dry dust and heat of the summer and the monsoon conditions of the 'wet' season. The line was completed in August 1872, later than the contract allowed, but ahead of a satisfactory operation of the submarine cable, which although laid successfully from Banjoewangi (Java) to Darwin in November 1871, had developed a fault in June 1872 which was not cleared until October.

In the first five days of the complete cable service from London to Darwin, 153 messages were sent from Australia to London and 148 messages received. Celebratory banquets were held simultaneously in Sydney, Adelaide and London. Charles Todd was informed of his knighthood by a telegraph message from Queen Victoria at Windsor, whereupon the superintendent of telegraphs joined the many cable pioneers who now held this honour for their work in spanning the Empire with a continuous telegraph network.

5.14 The world encompassed

To complete the chain of submarine telegraph cables linking the Empire, the remaining task was to connect Australia and New Zealand with the American continent across the Pacific Ocean. This presented the third and largest of the challenges facing the Victorian telegraph engineers. The project had been discussed for a number of years, various commercial solutions having been proposed and later discarded [51]. It was not until 1887, at a Colonial Conference called by the government in

London, that a resolution was finally passed recommending a thorough survey of the Pacific by the Admiralty to establish the practicability of such a cable. This was by far the longest submarine cable yet proposed, over 11 000 km in length. Many considered that the retarding effect of the electrical capacitance for the cable would be too great: 'No signals would come out at the other end!' they said. The survey was held up for many years while successive governments wrangled over who should pay for the link to be built. When the survey was finally undertaken in 1901, the bed of the Pacific Ocean, where the cable would lie, was found to be soft, and a safe resting place for the cable, not the hard coral bed that many had expected. Soon afterwards a Pacific Cable Board was formed and funding agreed. This was the first occasion on which countries of the Empire worked together on a telegraph project, and the first instance of direct government activity in a submarine cable-laying exercise, all previous cables having been financed by private enterprise. The cable was owned jointly by the governments of Australia, Canada, New Zealand, the United States and the United Kingdom [52]. The chief architect of the project was Sir Sandford Fleming, Chief Engineer of the Canadian Pacific Railroad Company. So that the landing points would lie within the British Empire (for strategic and defence purposes), he decided to launch the cable from Southport, on the Gold Coast of Australia and terminate it at Bamfield on Victoria Island, off Vancouver in Canada. Intermediate stations were to be established at Norfolk Island, Suva in Fiji and Fanning Island in mid-Pacific, all of which at that time formed part of the British Empire. An alternative route from Sydney to Auckland in New Zealand and from there to Fiji was also planned.

Laying the Pacific cable commenced in 1902 from the Australian end to its termination at Bamfield in Victoria. Telcon, the new amalgamated company of John Pender's empire, provided a cable ship, the *Anglia*, and the cable was manufactured in Telcon's works in North London. The first section, linking Australia with New Zealand, was completed on 25th March, and was followed by the 1500 km cable from Norfolk Island to Suva at the beginning of April. The longest section, between Fanning Island and Bamfield, required special consideration. The 6500 km of armoured cable needed weighed 8000 tonnes, and to accommodate this (a bigger load than even the *Great Eastern* could carry) a special cable ship, the *Colonia*, was built in Newcastle, England, capable of carrying the complete cable. This was laid in a single voyage, and remains the largest length of cable ever put down in one piece. This task was completed without incident in 18 days on 18th September. The final section of 3200 km between Fiji and Fanning Island was laid by the *Anglia*, and the entire line was working by 31st October 1902. The following year saw the completion of another Pacific cable by the American Pacific Cable Company, but this linked the west coast of the United States with China and did not have any serious impact on the Empire's communications at the time [53].

FIRST WORLD-CIRCLING TELEGRAPH MESSAGE

The message delivered to His Excellency Lord Minto at Government House, Ottawa, Nov. 1st, 1902,
after circling the globe.

Canadian Pacific Railway Company's Telegraph

All messages taken by this Company are subject to the conditions printed on our Form 2.

```
10. RA.          MD. RO.          (39 Words)          8.35 A.M.
          (via Commercial Eastern Australia Pacific)
To Governor General,
            Ottawa.
          Receive globe encircling message via England,
South Africa, Australia and Pacific Cable congratulating
Canada and the Empire on completion of the first segment
state controlled electric girdle the harbinger of  incal-
culable advantages, national and general.
From Ottawa, Oct. 31st, 1902          Sandford Fleming.
```

Figure 5.11   The first world-encirling telegram, sent from Sir Sandford Fleming to Lord Minton, the Foreign Secretary, in 1902

The encircling of the globe by telegraph cable was now complete, and the governments of Britain and the Empire, who were mainly concerned with this enterprise, had achieved the long sought-for goal of accomplishing the task with every cable landing on British-controlled land. The completion of the cable was duly celebrated in various capitals of the Empire, initiated by the receipt of the 'first world-circling telegraph message', from Sir Sandford Fleming in Ottawa to Lord Minton, the Foreign Secretary in London, via Australia and South Africa (Figure 5.11). Although the operating speed of the completed cable was better than expected, the high capacitance of the Pacific section placed a considerable limit on the rate of communications, the initial rate being about 8 to 20 wpm, and even then some skill was required to interpret the results [54]. After some experience an operator working these lengthy submarine cables would be able to identify a given symbol from the characteristic length of the signal, in addition to the number of dots and dashes it contained. At a later stage, when the Pacific cable had been in use for several decades, the Eastern Telegraph Company developed a method of regeneration of the signal at a relay station, such as Fanning Island, which considerably shortened the time for message transmission. The technique involved the equalisation of the attenuation and phase of the transmitted signals up to at least a frequency equal to the fundamental frequency of reversals at the highest signalling speed used. Today we would call this 'signal processing', employed universally for signal shaping on digital networks by the use of solid-state devices, but in the 1920s this was achieved by accurate timing, 'sampling' of the distorted signal received and retransmission to the next relay station, with which it was kept in synchronisation by an electromechanically controlled tuning fork [55]. In the four years from 1923 to 1927 this regenerator system was installed at 120 stations working over 230 000 km of cable, allowing as many as 14 cable sections to be joined in series, with a significant increase in overall transmission rate [56]. Later, loaded cables laid across the Pacific enabled the transmission speeds to be increased to 300 wpm [57].

Operator training for the Pacific cable route took place mainly in Australia, at a training school which formed part of the cable station at Southport. As in earlier telegraph organisations, it was found advantageous to employ young people who were adept at hand-keying. Advertisements such as the following appeared regularly in the Sydney newspapers:

WANTED: *boys 15 years of age to learn submarine telegraphy and serve overseas. Apply Pacific Cable Board 77 Pitt Street, Sydney.*

A detailed and descriptive account is given by Bruce Scott of what happened to the selected candidates:

The trainees wore blue uniform, trimmed with red embroidery and with a red crown emblazoned on the breast pocket, the insignia of the Imperial Service. A full training at the School was a prize to the young people and not all the boys survived the course. It included sending and receiving Morse code with a key and brass sounder. The speed required was not high (the cable, at least initially, could not respond at speeds greater than 12 wpm) but the sending had to be nearly faultless and maintained for several hours at a time.

The cable code over the Pacific route required manipulation of a pair of cable keys controlling positive and negative currents for the dots and dashes, a skill quite different from normal Morse code keying using a single key. In addition the operator needed to be familiar with the operation of the mirror galvanometer, and later the siphon recorder, which eventually superseded all other forms of receiving and recording the coded signals. Still later in the development of the system the Wheatstone automatic transmitter was used, but the input signal still needed to be prepared separately on punched paper tape, as described earlier, and a new skill had to be learnt in the preparation of off-line input tapes.

The cables used for the route consisted of a single armoured core with a sea return. This return path had a far lower resistance than would be obtained with the thickest return cable extending over thousands of kilometres. However, this technique did lead to interference by stray earth currents, which was particularly severe during the time of high solar activity. Duplex transmission on the Pacific cable was generally applied using a Wheatstone bridge technique. The cable needed electrical balancing separately for each route, and this led to a 'bridge balancing room' being established at each relay station, filled with accurate resistance and capacitance boxes, where correct balancing for each route could be achieved.

The telegraph transmission facilities, now available on a worldwide basis, were by no means completed with the opening of the Pacific route. From Porthcurno in Cornwall in 1901, at the beginning of the now-completed 'Imperial route', a new feature began to be seen on the landscape which was to affect telegraph communications to an even greater extent than the use of submarine cables. This was the complex of aerials being erected at Bass Point for Marconi's attempt to signal across the Atlantic, by repeating the Morse letter 's' continuously. The eventual transformation of telegraph signalling through the advent of wireless telegraphy forms the subject of the second part of this book.

References

1 DILLWYN, L.W.: Diary (1844), National Library of Wales, Aberystwyth
2 HUDSON, D., and LUCKHURST, K.W.: 'The Royal Society of Arts 1754–1954' (John Murray, London, 1954), p. 142
3 HAIGH, K.R.: 'Cableways and submarine cables' (Adland Coles, London, 1968), p. 26
4 BURKILL, I.H.: 'A dictionary of the economic products of the Malay Peninsula' (Ministry of Agriculture, Kuala Lumpur, 1966)
5 CLAPHAM, J.H.: 'An economic history of modern Britain', vol. 2 (Cambridge University Press, Cambridge, 1963), p. 43
6 *Journal of the Royal Society (Straits Branch)*, 1883, **12**, p. 208
7 COLGRAIN, W.: 'The Telcon story' (Telegraph Construction and Maintenance Company, London, 1950), p. 99
8 BRIGHT, C.: 'Submarine telegraphy' (Crosby Lockwood & Son, London, 1898), p. 5
9 BEAUCHAMP, K.G. 'Exhibiting electricity' (Peter Peregrinus and the Institution of Electrical Engineers, London, 1996)
10 'Mediterranean electric telegraph', *Illustrated London News*, 1855, **27**, p. 423
11 KUNERT, A.: 'Telegraphen Seekabel' (Karl Glitscher, Cologne and Mülheim, 1962)
12 AHVENAINEN, J.: 'The Far Eastern telegraphs' (Suomalainen Tiedeakatemia, Helsinki, 1981)
13 COLLINS, PERRY McDONOUGH: 'Overland exploration in Siberia' (New York, 1864)
14 RUSSELL, W.H.: 'The Atlantic telegraph' (David & Charles, Newton Abbott, 1972; 1st edn 1865)
15 THOMSON, W.: 'On the theory of the electric telegraph', *Philosophical Magazine*, 1856, 4th series, **2**, p. 152
16 SMITH, WILLOUGHBY: 'Working of long submarine cables', *The Electrician*, 1879, **3**, p. 258

17 EWING, J.A.: 'Thomson & Jenkin's automatic curb sender', *Journal of the Society of Telegraph Engineers*, 1876, **5**, pp. 213–48

18 HEAVISIDE, O.: 'Electromagnetic induction and its propagation', *The Electrician*, 1887, **19**, p. 71

19 WINTER, G.K.: 'On the use of electromagnetic induction in cable signalling', *Journal of the Society of Telegraph Engineers*, 1874, **3**, pp. 103–6

20 THOMSON, G.P.: 'Improvements in means for use in or in connection with the conveyance of varying electrical impulses, applicable to electric signalling for telegraphic, telephonic or other purposes', British Patent No. 22^{F}304 (1891)

21 PUPIN, M.I.: 'Propagation of long electrical waves', *Transactions of the American Institute of Electrical Engineers*, 1899, **16**, pp. 93–142

22 'The loaded submarine cable', *Electrical World* (NY), 1925, **86**, p. 353

23 BUCKLEY, O.E.: 'The loaded submarine cable', *BSTJ*, July 1925, **4**, pp. 355–6

24 MALCOLM, H.W.: 'Future progress of cable telegraphy – use of magnifiers', *The Electrician*, 1915, **74**, p. 526

25 THOMSON, W.: 'Improvements in testing and working electric telegraphs', British Patent No. 329 (1858)

26 THOMSON, W.: 'Improvements in receiving or recording instruments for electric telegraphs', British Patent No. 2147 (1867)

27 THOMSON, W.: 'Sir William Thomson's report to Messrs Siemens Bros and Company Ltd on tests of the direct U.S. cable' (Nichols & Sons, London, 1875)

28 BARTY-KING, H.: 'Girdle round the Earth – the story of Cable & Wireless, 1929–1979' (Heinemann, London, 1979)

29 SCOTT, J.D.: 'Siemens Brothers 1858–1958' (Weidenfeld & Nicolson, London, 1958), pp. 87–8

30 AHVENAINEN, J. 'History of the Caribbean telegraphs before the First World War' (Suomalainen Tiedeakatemia, Helsinki, 1996)

31 'Sir Charles Tilston Bright', *in* 'Dictionary of national biography', Vol 2, pp. 271–3

32 SCOTT, R.B.: 'Gentlemen on imperial service: a story of the trans-Pacific telecommunications cable' (Sono Nis Press, Vancouver, BC, 1994)

33 HOLMS, K.E. 'Submarine cables at Hong Kong', *Royal Signals Quarterly*, April 1939, **7**, (24), pp. 99–110

34 SMITH, W. 'The rise and extension of submarine telegraphy' (London, 1891), pp. 109–31

35 MAVER, W.: 'American telegraph systems: apparatus and operating', (Wm. Maver, New Jersey, 1898)

36 SMITH, W.: 'The working of long submarine cables', Journal of the Society of Telegraph. Engineers, 1879, **8**, pp. 72–4

37 GALTON, D.: 'Ocean telegraphy', *Edinburgh Review*, 1861, **113**, pp. 113–43

38 Parliamentary Paper 27, vol. 62, 591 (1860), Microfiche Record 66.495–502, Public Record Office, Kew, Surrey

39 Parliamentary Paper 2605/425, vol. 62, 269.461 (1860) Microfiche Record 66.491–5, Public Record Office, Kew, Surrey

40 'Henley's centenary – 100 years of W.T. Henley's, *The Electrician*', *The Electrician*, 1937, **68**, p. 3

41   BRIGHT, C.: 'Presidential address to the Society of Telegraph Engineers', *Journal of the Society of Telegraph Engineers*, 1887, **16**, p. 27
42   'The telegraph cable to India', *Illustrated London News*, 8th July 1865, pp. 532–3
43   ANDERSON, Sir JAMES: 'Telegraph routes to India considered' (Metchin & Son, London, 1868)
44   ADAMS, J.M.: 'Development of the Anglo-Indian telegraph', *Engineering Science & Education Journal*, 1997, **6**, (4), pp. 140–48
45   SIEMENS, W.: 'Wissenschaftliche und technische Arbeiten', Vol 2 (Berlin, 1891), pp. 260–75
46   'War and submarine cables', *The Electrician*, 1885, **14**, (23)
47   COLLINS, H.M.: 'From pigeon post to wireless' (Hodder & Stoughton, London, 1925)
48   PENDER, Mrs J.: from 'History of Cable & Wireless', CD-ROM (Cable & Wireless, London, 1997)
49   'Pender, Sir John', *in* 'Dictionary of national biography', Suppl. III, pp. 258–9
50   'Danish watchmaker created the Chinese Morse system', *Morsum Magnificat Journal*, 1997, (51), p. 14
51   CLUNE, F.: 'Overland telegraph' (Angus & Robertson, Sydney, 1955)
52   GARRETT, G.R.M:. '100 years of submarine cables' (1930), p. 34
53   PACKER, J.E.: 'The history of Cable & Wireless', IEE PGS7 Summer Meeting, Swansea, July 1989, p.15.5
54   WOOD, K.L.: 'Empire telegraph communications', *Journal of the Institution of Electrical Engineers*, 1939, **84**, pp. 638–61
55   KINGSBURY, H., and GOODMAN, R.A.: 'Methods and equipment in cable telegraphy', *Journal of the Institution of Electrical Engineers*, 1932, **70**, p. 477

PART 2

AERIAL TELEGRAPHY

Chapter 6

Marconi and the experimenters

The history of practical wireless communications, as it applies to wireless telegraphy, could be said to commence with Marconi's work. However, this would be to ignore a vast hinterland of events leading up to Marconi's achievement in 1896, which, although not in themselves productive of applicable results, allowed a practical commercial system to emerge through Marconi's enterprise. First to appear was a sound theoretical backing provided by James Clerk Maxwell, a physicist then holding a professorship in natural philosophy at King's College, London, with his seminal paper of 1864, 'A dynamical theory of the electromagnetic field', in which the electromagnetic wave theory was mathematically expounded. Maxwell was able to prove by experiment that the rate of propagation for these waves agrees closely with that of light, and with considerable prescience was able to state, 'we have strong reason to believe that light itself – including radiant heat and other radiations if any – is an electromagnetic disturbance in the form of waves propagating through their electromagnetic field according to electromagnetic laws'. This view did not receive widespread acceptance, not least because it was couched in difficult mathematical terms. One of the few who took note of Maxwell's theoretical ideas at the time was Hermann von Helmholtz, Professor of Physiology at Heidelberg, who in turn excited the interest of his pupil Heinrich Hertz in the possibility of the generation of electromagnetic waves. Hertz's interest led to the first practical demonstration, designed initially to test Maxwell's theory, but which quickly became an exercise in electrical communication at a distance without the aid of wires [1].

6.1 Beginnings

Two forms of distant wireless transmission[1] were being considered in the late 1880s: inductive field transmission using large loops of wire or direct conduction through the ground – a method favoured by Sir William Preece in his experiments of 1884 [2]; and true radiation field transmission, initiated by Hertz in his classical experiments with dipoles in 1888. Preece and the Post Office did succeed in communicating by inductive field transmission between Anglesey and the Skerries lighthouse off Holyhead, and later transmissions across the Bristol Channel in 1900. However, the rapid reduction in field strength with distance compared with that possible using a radiation field led to all wireless systems not primarily employing a radiation field being discarded, except for certain specialised short-distance communication techniques used in the First World War and described in Chapter 8. In Hertz's early radiation experiments, electromagnetic waves were produced by the extraordinarily simple mechanism of an electric spark across a gap between two metal electrodes, and they were detected at a distance with an even simpler device of a loop of wire, which Hertz named a 'resonator'. These experiments caused considerable interest among physicists of the day, including Sarasin and de La Rive in Geneva, Slaby, Garbasso and Aschkinass in Berlin, Branly in Paris, Lodge and Jackson in Britain, Bose in Calcutta, Popov in Russia, Tesla in America and Righi in Italy. Professor Righi, whose lectures were attended by the young Marconi in about 1890, probably had the most direct and long-term influence on the new technology, although a more immediate effect resulted from the demonstration of Hertz's invention by Sir Oliver Joseph Lodge, at that time Professor of Physics and Mathematics at the University of Liverpool.

Hertz died at the early age of thirty-six in 1894, and Lodge, in a commemoration lecture entitled 'The work of Hertz' given at the Royal Institution, London, in that year, demonstrated Hertz's invention and at the same time introduced a new device for detecting electromagnetic waves, discovered by Professor Édouard Branly of Paris in 1890, to replace Hertz's spark resonator [3]. He named it a 'coherer', which in Lodge's version consisted of a glass tube loosely filled with iron filings and closed at each end with a metal plug (Figure 6.1). The device normally exhibited a high resistance, but when affected by an electric spark, its resistance was reduced considerably, so allowing a control current to

[1] The term 'wireless' was used generally to indicate transmission from point to point without the use of intervening wires until about 1906, when the term 'radio' began to take its place, introduced by the 1906 Radiotelegraphic Conference. In this book the term 'wireless' is reserved for aerial communication using Morse telegraphy, i.e. wireless telegraphy (W/T), and 'radio' for the aerial transmission of telephony, i.e. radio telephony (R/T), although the latter term may also embrace the transmission of any form of coded communication.

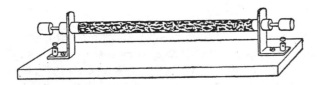

Figure 6.1   Lodge's coherer
Source: Lodge, Royal Institution Lecture, 1894

pass and activate a relay or other device to indicate the reception of an electric wave. A slight mechanical shock was then necessary to restore the normal high resistance, and Lodge achieved this by using a converted bell mechanism arranged to tap the glass tube periodically. He later repeated his talk, together with Hertz's experiments, at a British Association meeting in Oxford, and published it later in book form [4]. Like Hertz, Lodge was a scientist, not an entrepreneur, and his desire was to understand the processes of electromagnetic transmission and detection, not to pursue its wider development: he 'neglected the prelude to what has now developed into wireless telegraphy' [5].

Aleksandr Stepanovich Popov, a professor at the Imperial Torpedo School in Kronstadt, Russia, was one of the many European scientists who became interested in Lodge's interpretation of Hertz's work and made use of a similar apparatus for his own work on the study of atmospheric electricity. For reception Popov used a long vertical wire insulated at the upper end and connected to earth via a coherer at its lower end. This proved to be a considerable improvement on Lodge's elevated plates, and was later adopted by Marconi in his commercial telegraph work. Popov's primary objective however, was not telegraphy, as he stated clearly at the end of a paper published in 1896 [6]: 'In conclusion, I may express the hope that my apparatus, with further improvement, may be adapted to the transmission of signals to a distance by the aid of quick electric vibrations as soon as a means for producing such vibrations possessing sufficient energy is found.' A later publication in 1906 by John Ambrose Fleming, the first Professor in Electrical Engineering at University College, London, noted the lack of progress in these developments:

We are left, then, with the unquestionable fact that at the beginning of 1896, although the most eminent physicists had been occupied for nine years in labouring in the field of discovery laid open by Hertz, and although the notion of using these Hertzian waves for telegraphy had been clearly suggested, no one had overcome the practical difficulties . . . or given an exhibition in public of the transmission of intelligence by telegraphic signals by this means.

6.2 Marconi

It is clear that these experimenters who were contributing to the early days of wireless telegraphy had other interests or responsibilities which limited the time they could apply to the development of Hertz's original ideas. Several were professors with departmental and lecturing responsibilities; Jackson had his duties with the Navy;[2] both Popov and Lodge had other pressing research interests, Popov with the use of electromagnetic detectors for storm detection and Lodge with optical investigations.

The young Guglielmo Marconi alone was unencumbered with other responsibilities and had the time (and finance) to be able to devote all his energies to this fascinating work. He was himself amazed that no one had grasped the commercial possibilities of the generation of Hertzian waves in view of the easy availability of its principle components the required equipment – the spark gap, the induction coil, the resonator and the coherer. He remarked, 'I could scarcely conceive it possible that their application to useful purposes could have escaped the notice of such eminent scientists' [7]. He carried out experiments in the garden of his father's estate at the Villa Grifone near Pontecchio in the spring of 1895, and soon added a vital fifth ingredient, used earlier by Popov for his storm detection work – the elevated aerial. This was regarded by some as Marconi's principle contribution, since he was, in essence, a practical implementer of existing technology, rather than an innovator, much as Morse had been with terrestrial telegraphy.

He repeated Hertz's experiments and soon achieved a range of several metres using Branly's coherer in place of Hertz's resonator, as Lodge had done. Attaching a 'grounded aerial' to his equipment for both transmitter and receiver produced much better results, and he quickly saw that there was a direct relationship between the height of his new aerial and the distance attainable [8]. With even the modest aerial height in his garden at the villa, Marconi was able to obtain reliable transmission over about 2.5 km. As several other experimenters had already found it was hardest to obtain good results with the coherer, because of the difficulty of automatically reducing it to a non-conducting state after the reception of a signal. So he introduced a relay, of the kind already used for receiving the signal, to actuate a tapper (as Lodge had done) so as to repeatedly disturb the coherer, and this appeared to work well. He also added a siphon printer, readily available as a standard item in use for submarine telegraphy, to provide a permanent record of his results.

Marconi now had a system capable of demonstration. He chose in the first instance to exploit his invention in Britain, in the knowledge that Britain was a maritime nation whose dependence on overseas trade

---

[2] The significant contribution made by Sir Henry Jackson of the Royal Navy is considered in detail in Chapter 9.

seemed to offer a good market for such communication devices. He was fortunate to secure a contact with Sir William Preece, then chief engineer to the Post Office, and arranged a number of meetings with him in 1896, and while in England he filed a specification for his wireless system with the British Patent Office (British Patent No. 12039) which was recorded on 2nd March 1897. Preece arranged for a demonstration of Marconi's equipment to take place between the Post Office building at St Martins-le-Grand and a station erected on the roof of the Savings Bank Department in Queen Victoria Street in London. Further demonstrations were carried out, primarily for Army and Navy observers on Salisbury Plain, and a range of about 3 km achieved. The naval observer was Sir Henry Jackson, who had carried out his own wireless experiments during the previous year. Other tests followed, including an important one between Lavernock Point and Breamdown, across the Bristol Channel, using tall masts 33 m high [9], the results of which led directly to the formation of the Wireless Telegraph & Signal Company in 1897 to manufacture transmitting and receiving sets.[3] Initially the sales of his equipment were poor. Some equipment was sold to the British Army in 1898 for use in the Boer War (see Chapter 4) but, as we shall see, it was not until 1900, when the Admiralty signed a contract with Marconi to supply equipment and to train operators, that the company found its feet. Success was assured after a contract with Lloyd's of London was finalised in 1901, which established that any shipping line using Lloyd's vast network of shipping agents would also have to use Marconi equipment if they wanted to maintain contact with the Lloyd's bases and other Lloyd's ships.

Earlier, Marconi had carried out a number of significant experiments for Trinity House in London to establish wireless contact between South Foreland Lighthouse, near Dover, and the East Goodwin lightship, some 20 km distant. This was in 1898, and for over two years it performed an important role in ensuring the safety of vessels, and saving a number of lives, through this rapid means of communication. The East Goodwin lightship was the scene of the first public demonstration of the value of wireless in protecting the lives of seafarers. In dense fog on 28th April 1899, a steamer, SS *R.F. Mathews*, ran into the lightship and inflicted serious damage. The lightship was able at once to communicate with the South Foreland Lighthouse and obtain prompt assistance. Fleming was the first British scientist to report on the operation of this telegraphic link. He commented on its rapidity and efficiency in a letter to *The Times* dated 3rd April 1899. After remarking on the speed of communication at 12 to 18 wpm, and the automatic prints of the Morse code signals on paper tape, he wrote:

[3] In 1900 the company's name was changed to Marconi's Wireless Telegraph Company.

Marconi has placed a lightship on the Goodwins in instant communication, day and night, with the South Foreland lighthouse. A touch of a key on board the lightship suffices to ring an electric bell in the room at the South Foreland, twelve miles away, with the same ease and certainty with which one can summon the servant to one's bedroom at a hotel . . . If awakened by a bell rung from the lightship, he is able to ring up in turn the Ramsgate lifeboat, and if need be, direct it to the spot where its services are required within a few seconds of the call for help . . . Up to the present time none of the other systems of wireless telegraphy employing electric or magnetic agencies has been able to accomplish the same results over equal distances.

Marconi later established two stations, one on the Isle of Wight and the other in Bournemouth, some 20 km away, between which he carried out regular transmissions to support his experiments and provide publicity. Operations between these two stations were witnessed by Lord Kelvin and others, and were an essential preliminary to his first transmission across the English Channel in 1899. Marconi had obtained from the French Government permission to erect a mast at Wimereux, near Boulogne, and the receiving station was at the South Foreland Lighthouse 50 km away. The cross-Channel transmissions were successful and led to further demonstrations, including one during Fleming's British Association lecture in September 1899, when telegraphic communication was carried out between the BA in Dover and its counterpart, the French Association for the Advancement of Science, meeting at the same time in Boulogne. Extensive publicity attended these demonstrations, The *Times* on 8th April 1899 commenting that,

The apparatus, moreover, is ridiculously simple and not costly. With the exception of the flagstaff and 50 metres of vertical wire at each end, he can place on a small kitchen table the appliances, costing not more than £100 in all, for communicating across 30 or even 150 km of channel.

Marconi was not alone in carrying out wireless experiments across the water. While he was busy with the South Foreland transmissions, Professor Ferdinand Braun of Berlin University was transmitting across the bay at Cuxhaven with a spark transmitter incorporating a Tesla induction coil, and a coherer of his own design. Captain Jackson (as he was then) reported on Braun's results to the Admiralty, expressing the view that

Considering that an oscillating discharge between sparking balls in conjunction with a vertical wire, is used for transmitting signals, and that the principle therein involved is patented by the Wireless Telegraph Company, it seems improbable that Professor Braun could use his apparatus in England without payment of Royalties. [10]

Although Braun never attempted to deploy the system in England, he went on to cooperate with Siemens in producing a range of sets which, together with those manufactured by the Telefunken Company, offered a viable alternative to the Marconi Company's products in Continental Europe and America as the 19th century drew to a close.

A defect of the early Marconi transmitting apparatus was that the frequency of oscillation of the spark system in the transmitter could not be 'tuned' to match that of the resonant circuits in the receiving apparatus. Lack of tuning not only reduced the sensitivity of the system, but also rendered the receiver susceptible to a wide variety of interfering signals, and the transmitter to radiation over a wide interfering bandwidth. This need was recognised by Oliver Lodge in his patent application of 1897 for the introduction of tuning into such apparatus, which he termed 'syntony', and was to prove of fundamental importance in the following decade. Marconi was not slow to appreciate its advantages, and applied Lodge's ideas to his 'syntonic wireless telegraphy' in 1899–1900 [11]. In his arrangement, shown in Figure 6.2, tuning was provided for the primary of a coupling transformer containing the spark gap to match the secondary aerial circuit, with corresponding tuning arrangements at the receiving circuit. Including the spark gap directly in the aerial circuit has an inhibiting effect on the transmission and leads to considerable damping. The resulting wave train at discharge then contains only a few reversals and imposes a relatively long interval before another spark discharge could take place. If the spark gap is included in the primary

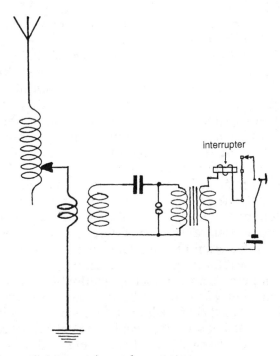

Figure 6.2   Marconi's 'syntonic' tuned transmitter

circuit of an inductively coupled circuit, the wave train is longer, less power is needed and the transmission is over a narrower bandwidth. Marconi referred to this coupling circuit as the 'jigger', which became a standard arrangement with all future spark transmitter systems. This system is described at length in his important British Patent No. 7777 of 1900, which covered the tuning of four circuits: the spark gap, the transmitting and receiving aerials, and the secondary of the receiving circuit. This patent became a cornerstone of the Marconi Company's activities for a number of years, and in effect controlled the use of wireless and prevented its development by any other company, until the patent expired in 1914. (Marconi had purchased the Lodge syntony patent in 1911.) In an earlier British Patent, No. 1862, appearing in the year 1899, Braun had also placed the spark gap in a primary circuit coupled to the aerial, but apparently had not seen the great advantage of loosely coupling the two circuits and tuning them to the same frequency, as Marconi had done, and so did not reap any significant commercial advantage.

6.3 Transatlantic attempt

Following the success of his cross-Channel transmission and the improvement obtained through the use of tuning, Marconi was now encouraged to attempt transmission across the Atlantic. In a preliminary experiment in 1901 he had already successfully transmitted a signal between the Lizard in Cornwall and St Catherine's point on the Isle of Wight, a distance of over 300 km, and thus established that the Earth's curvature would not limit signalling to visible lines of sight (a belief strongly held at the time), helping to ensure adequate collaboration for the Atlantic attempt. After several abortive attempts in which high winds brought down his 60 m masts – first at Poldhu, his transmitting station in Cornwall, and later at the receiving station at Cape Cod in Massachusetts – he was successful in using kites to support a 150 m aerial at a new receiving station located at St John's, Newfoundland. From there, on 12th December 1901, he and his assistant were able to hear distinctly the three Morse dots of the letter 's', being transmitted at intervals from Poldhu [12]. Commercial transmission followed in 1903 from Poldhu to a redesigned receiving station at Cape Cod, with the sending of the first commercial messages for a newspaper. Shortly afterwards Marconi started to erect a permanent commercial station for transatlantic traffic in Newfoundland, but was checked by the Anglo-American Telegraph Company, which claimed that such an action violated its landing rights and franchise in that country as a cable and telegraph company [13]. Marconi transferred his enterprise to Canada and was successful in obtaining permission from the Canadian Government to erect a station at Glace Bay to communicate with his Clifton station, already under construction in Ireland. Operation commenced in 1907 with a limited public transatlantic service using a transmitting power of about 80 kW.

By this time the siphon recording device at the receiver had been abandoned in favour of headphones in conjunction with a carborundum (silicon carbide) crystal detector. Crystal detectors had been discovered by Braun in about 1874, and very many pairs of metals and crystalline substances were subsequently found to exhibit the property of one-way conductivity, or rectification, although satisfactory adjustment was often frustratingly difficult. The carborundum detector was not always very sensitive, and a new detector appeared at the turn of the century which, although complex, received widespread support as a reliable receiving device. This ingenious invention, patented in 1902, was 'Marconi's reliable magnetic detector', known as the 'Maggie' to countless marine telegraph operators. It functioned by detecting the demagnetisation of a bundle of fine iron wires after exposure to weak electromagnetic radiation. An endless loop of such wires was made to revolve slowly by clockwork as it passed over the poles of a pair of horseshoe-shaped magnets, with a portion of the wire carried through two concentric coils of wire, as shown in Figure 6.3. One of these coils formed part of the

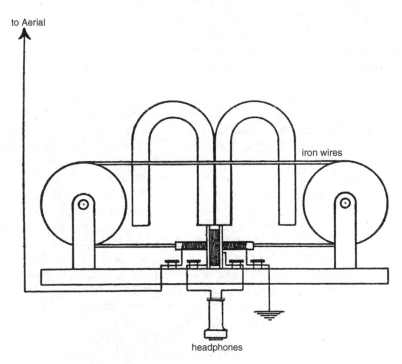

Figure 6.3   Marconi's reliable magnetic detector
Source: Baker, 'History of the Marconi Company', 1970

circuit of the receiving aerial, and the other was connected to a telephone headset. The demagnetisation, caused by the incoming signal from the aerial, induced a momentary increase in the induced electric current in the telephone circuit. This was repeated as each wave train reached the aerial, and a musical note was heard if the spark frequency was greater than about 100 Hz. The 'Maggie' became standard equipment in Marconi's marine installations, until replaced by the thermionic detector in 1910.

6.4 Spark and arc

Much effort was devoted to obtaining a reliable spark in the transmitters of this period. The aim was to produce a regular series of wave trains, as undamped as possible so as to approach a continuous waveform – sadly lacking in many of the sets of the time, partly because of poor design and often because of misadjustment of the equipment. Advice given to operators included 'an open spark must be kept white and crackling and have considerable volume' and 'if too long, it will be stringy; if too short, an arc will be formed' [14]. A difficulty with high transmitter powers was that some of the energy was transferred back from the aerial to the spark circuit after the first burst of oscillation, transforming the spark into a short-period arc, thus lowering the transmission efficiency and, under some conditions, causing transmission on two separate frequencies at the same time. Reducing the coupling coefficient between the two circuits could reduce this latter effect, with, however, an inevitable reduction in efficiency [15], but a better technique was to apply one of the various methods of rapid deionisation being developed in the 1890s to prevent an arc from being generated in the first place. The first of these to be applied was Elihu Thomson's 'magnetic blowout'. This applied a suitably timed magnetic field at right angles to the direction of the spark to dissipate the metallic vapour which formed immediately after the spark, and which in the normal way produced the continued conductivity of the spark [16]. Other methods of quenching the spark were to direct an air blast at the gap, achieving similar results to Thomson's magnetic blowout, or to modify the design of the gap itself. A successful technique was to use a rotary spark gap, consisting of one or more stationary members and a rotating member constructed like a wheel with projecting spokes. This ensured that the electrodes were close enough to create a spark (or arc) only intermittently and briefly. Figure 6.4 shows several techniques for producing a quenched spark train.

Where the energising power source was derived from an alternator, a higher efficiency was obtained by synchronising the appearance of the sparks across the gap with those moments when the alternating voltage was at a maximum potential. This was initially proposed by Nikola

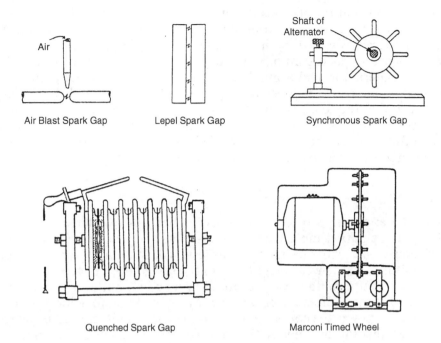

Figure 6.4   Various forms of quenched spark gap
Source: Robinson, 'Manual of Wireless Telegraphy', 1909

Tesla in 1896 in his US patent No. 20981, and subsequently improved by Reginald Fessenden in 1907 in his US patent No. 706787 [17]. Fessenden had made considerable progress in the development of wireless equipment while employed by the US Weather Bureau, and was later associated for a number of years with the National Electric Signalling Company (NESCO), a firm founded by two Pittsburgh entrepreneurs, Hay Walker and Thomas Given, mainly to support Fessenden's numerous inventions. His mechanically quenched 'synchronous rotary spark gap transmitter' used a three-phase alternator rotating at a rate of 125 Hz. This was arranged such that the gap was present at both positive and negative peaks of the alternating supply, thus giving a spark rate of 750 Hz. The sequence of short spark trains obtained was almost equivalent to continuous wave generation, and produced a musical note which was quite distinctive. Fessenden applied this signal as a carrier for the transmission of speech signals with some notable success [18].

At about the same time that Fessenden was designing his synchronous spark gap transmitter, Marconi was also developing Tesla's ideas, and produced his own version in 1907. His designs were used for marine communication, certain fixed shore stations and, most spectacularly, at the large transmitting stations at Caernarvon and Brunswick, discussed later. He produced several variations of design for a synchronous rotary disc discharger, typified by his 'timing disc', a simplified diagram of which is shown in Figure 6.5. A studded disc, running synchronously with the alternator, had a number of metal studs around its periphery corresponding to the number of poles of the alternator, and the design was such that the spark discharge took place at or near the peak of the generated voltage waveform. The predominant tone in the note produced had a frequency in the range of 600–750 Hz, which enabled an operator to distinguish the signal during long periods of atmospherics.

6.4.1 'Short spark' operation

With the application of these various methods of quenched and synchronous operation more transmissions could now be accommodated, and interference with one another was lessened. But not for long. Higher transmitting powers were sought, not only to avoid the interference problem as more transmitters were brought into action, but also to increase the range of the equipment. It was quickly realised that to achieve a high transmitted power the coupling between the two circuits should be close. However, this brought about the 'double frequency' effect, referred to

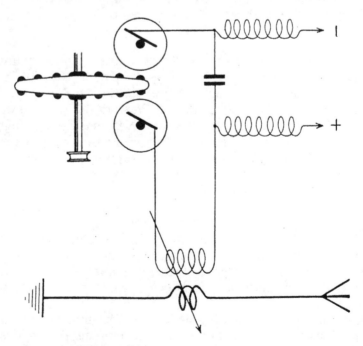

Figure 6.5   The Marconi timing disc
Source: With acknowledgement to Marconi Ltd

earlier, in which there were two peaks in the tuning curve, each responsible for radiating a part of the power. Further, the improvement in
bandwidth was not enough – each transmission was still taking place
over too wide a bandwidth because of the irregularity of the spark excitation of the aerial tuned circuit. These, then, were the two problems that
faced the wireless experimenters in the first decade of the twentieth
century:

1   How to generate a purely sinusoidal waveform at radio frequency,
    and
2   How to do this efficiently and radiate a large power.

The breakthrough came in 1906 with the publication of a paper by
Professor Max Wien of Danzig, Germany, in which he made a thorough
analysis of the mechanism of radio waveform generation by the use of an
electric spark [19]. He showed that it was not necessary to use a long
spark to generate powerful radio waves, and that a continuous signal
consisting of only one frequency could be achieved with a high degree of
efficiency by ensuring the close coupling of the spark energy source with
the aerial circuit. Part of this problem of efficiency arose from the action
of a powerful spark across the gap between two metallic contacts, which
was often made as long as 300 mm under the mistaken belief that in this
way a strong transmission could be obtained on the long wavelengths
then in use. In operation the first spark caused a certain amount of
ionisation to appear, which increased the conductivity of the gap, often
creating an electric arc which resulted in less power transferred to the
radiating aerial. The quenching methods available to avoid this were
complex, and their incorporation inevitably reduced the overall efficiency
of the spark system.

In his spark system Wien used instead a very small spark gap, less than
1 mm, and showed that as the gap is made very small then only a single
frequency is radiated (Figure 6.6). Since these small gaps passed little
energy, quenching occurred automatically. Further, he showed that his
short spark gap worked in quite a different way to the longer gaps that
had been used previously, and acted to 'shock' a tuned circuit into continual resonance, unaffected by the characteristics of the spark circuit.
This may be understood from the circuit arrangement he used, shown in
Figure 6.7. On applying a direct current potential across the capacitance-
inductive circuit, LC, shunted across the gap, the capacitance is charged
until its potential reaches a level at which a discharge takes place through
the gap G. A certain amount of ionisation could occur within this space
which, by its low conductivity, would maintain an arc while the capacitance discharges and transfers its stored energy to the inductance L. This
would then proceed to transfer its energy back through the gap to C,
repeating the process between C and L to give rise to an oscillatory
waveform across the LC circuit. In Wien's method the ionisation was

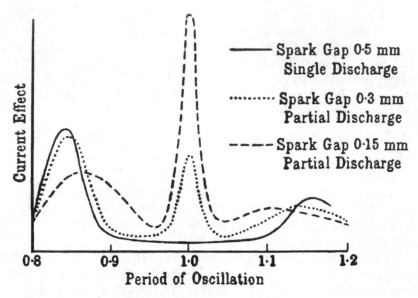

Figure 6.6   Performance of Wien's short spark gap
Source: Physikalische Zeitschrift, 1906

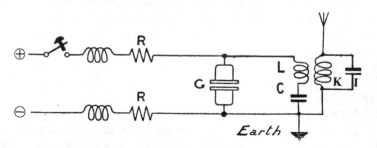

Figure 6.7   Arrangement of Wien's short spark circuit
Source: Physikalische Zeitschrift, 1906

easily removed by the cooling effect of the spark gap electrodes since the spark was so small, and the spark/arc was extinguished when the potential across C reached its minimum, and before it again became charged by the direct current supply. Resonance in the LC circuit did not occur, or at least did so for very few cycles. Instead, a quick surge of energy was transferred to the coupled secondary circuit IK, causing this to resonate – but at its own frequency, unaffected by the presence of the GLC circuit which now had no circulating energy. The IK circuit was thus 'shock excited' by regular unidirectional surges of current in LC as the capacitor C was repeatedly charged and discharged through the gap G, and a reasonably continuous waveform was obtained. To obtain a large radiated power it was necessary only to arrange for each of a series of such short gaps to supply its energy to the coupled aerial resonant circuit, rather than one long spark gap.

Following Wien's invention, a number of commercial versions of the 'short-gap' quenched spark transmitter began to appear. The first of these was constructed by Baron Egbert Von Lepel, with a spark transmitter which he designed and patented in 1908 (German Patent No. 17349), and for some time this enjoyed popular appeal. In his transmitter the spark was generated between two metal plates separated by a few millimetres of a paper disc, except at the centre where a small hole was cut. The spark, which appeared during the peak of each high-frequency oscillation at the edge of this gap, gradually burnt away the paper as the discharge proceeded. After some period of use the paper needed to be replaced, having been completely consumed. In operation the spark gap generated a fairly continuous waveform which could be modulated at an audio frequency by adjusting the repetition rate of the spark discharge [20]. A transmitter using a Lepel spark system of moderate power, with water-cooling of the spark plates, was tried out by the British Army between Slough and Hunstanton in Norfolk in 1909, a distance of 150 km, and found to be effective in producing an aerial power of 250 to 300 W from a simple inductively coupled transmitter. This was heard at the receiver as a clear high-pitched note, and, the wireless operators 'were at times giving the National Anthem and other airs played on a keyboard at Slough by means of which the circuits were adjusted to give a regular scale of spark frequencies' [21]. Lepel's system was not developed further in Britain, although it was used by the Compagnie Général Radiotélégraphique in Paris and at several other telegraph stations in France, and in Belgium. The Telefunken Company, after investigating the Lepel system, patented its own system of quenched spark transmission in 1909 which became extensively used. This was close enough to Lepel's design to cause a considerable patent battle between the two companies, which is recorded in the correspondence columns of *The Electrician* in the period just before the First World War [22, 23]. Other quenched spark gaps were developed by Marconi and several companies in the United States. In all cases a number of copper discs were separated by annular rings of mica forming a series of small spark gaps, following the design of the Telefunken company. The spark was confined to the airtight space inside the mica rings, and in operation shock-excited a coupled tuned circuit connected to the transmitting aerial.

A significant development of the short spark gap was made in 1911 by Roberto Galletti di Cadilhac who worked in Lyon, France, from 1910, where he developed his highly efficient quenched spark transmitter and was able to obtain an efficiency of up to 80 per cent, and a very pure transmitted waveform [22].[4] He used a very high applied potential, which he obtained initially through the Grenoble Power & Light Company, which had recently erected a transmission line from Moutiers in Savoie to provide the energy necessary for Lyon's tramway system. This operated at 60 kV, later increased to 150 kV – a level which he calculated would enable a transoceanic telegraphy system to be achieved. Galletti used this power in a new design of a multiple spark gap system in which each spark was initiated in sequence. His system is shown in Figure 6.8, taken from his patent application (British Patent No. 15 497, 1910), and consisted of a number of spark gaps operating in parallel, coupled by a common capacitor. By shock-exciting an aerial tuned circuit, through multiple discharge from these gaps he could obtain a very efficient and high-powered radiation [23].

Galletti's design of spark gap consisted of a series of metal discs separated by 1 mm rings of mica, compressed together to form an airtight inner region where the spark would form. He found that the resistance of the gas generated in the gap by the spark rose after the first discharge, and he considered this to be responsible for cutting off (quenching) the spark when the condenser discharge reached its peak, contrary to previous ideas which had attributed the quenching action to the cooling effect of the metal disc elements. In 1911 Galletti was able to develop his quenched spark transmitter for long-distance transmission. He entered into an agreement with the Indo-European Telegraph Company, which was anxious to add a transatlantic wireless communication service to its successful line telegraph operations across Europe and Asia to India, and looked to Galletti to design such a system. To establish this he formed The Galletti Wireless Telegraph & Telephone Company, with offices in London. The site he chose for his transatlantic transmitter was at Champagneux in the Savoie, where a huge cliff, 500 m high, served in place of an aerial mast. From its summit he was able to stretch a series of ten cables, each 950 m long, joined at the base and connected to his quenched spark transmitting system, transmitting at a wavelength of 4000 m. This operated to produce an aerial power of about 300 kW from a specially designed d.c. dynamo driving his multiple capacity system [24]. Following a series of tests within Europe, he was able to communicate by Morse and Baudot codes with telegraph stations in the United States and Canada. He established two-way communication with the powerful US Navy transmitter at Tuckerton, New Jersey, with a performance later confirmed by Alessandro Tosi (the co-inventor of the Bellini–Tosi wireless direction finder), as exceeding that of any other transatlantic transmission from Europe at that time [25].

---

[4] This achievement did not go unchallenged and a spirited correspondence developed in the pages of *The Electrician* during 1911 between Galletti, W.H. Eccles and A.J. Makower on the calculation of this efficiency and the results obtained by Galletti at Lyon. (See *The Electrician*, for February 3 page 673, March 24 pages 957–959, and April 14 pages 26–7.)

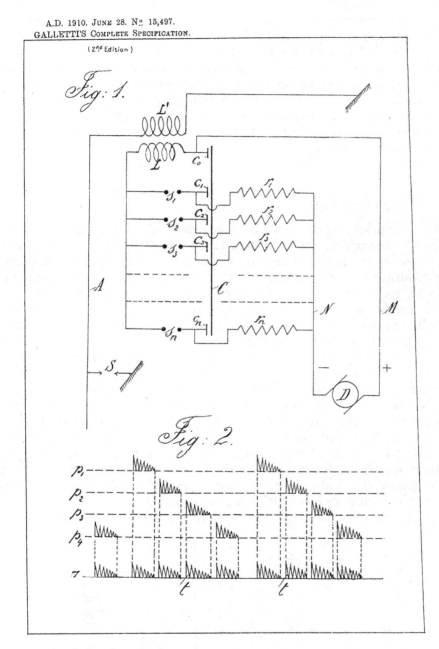

A.D. 1910. June 28. N.º 15,497.
GALLETTI'S Complete Specification.

Figure 6.8   Galletti's multiple spark gap transmitter
Source: British Patent No. 15 497, 1910

Its success, however, was short-lived. The transmission experiments from Champagneux took place in 1914, just at the commencement of the First World War. The station was closed down shortly afterwards by the French Government, the station dismantled, and its major components transferred to a government station in Bordeaux, where no practical use was made of them or of Galletti's research in wireless transmission. They were in fact much damaged in the move and, although they were returned to Galletti in 1919, it was no longer possible, or even practical, to continue his interrupted research. His company was eventually wound up in November 1925 [26].

6.4.2 The electric arc

All the spark systems described above were shock-excited systems in which a tuned circuit was put into a state of oscillation with a short burst of spark energy, and continued to resonate after the spark had subsided. By ensuring a continuous train of short sparks, as obtained by applying Wien's technique, the circuit could be maintained in a more or less continual resonant state. An arc transmitter achieves continuous-wave oscillation quite differently, by employing the 'negative resistance' characteristics of the electric arc, discovered earlier by those working on the use of the electric arc for illumination purposes. When the arc is shunted by an auxiliary circuit containing inductance and capacity, its negative resistance characteristic enables continuous oscillations to be sustained at the resonant frequency of the LC circuit (Figure 6.9). The most successful of these 'musical arcs', a term applied by William Duddell, a pioneer in this technique, was that designed by Valdemar Poulsen in 1902 [27]. For his arc he used a carbon cathode and a water-cooled copper anode, the electrodes being rotated for even wear, and subjected to a strong magnetic field to locate the arc correctly. He also enclosed the arc in an atmosphere of hydrogen, a major contribution to the technique of arc generation, which significantly increased the arc's intensity. These

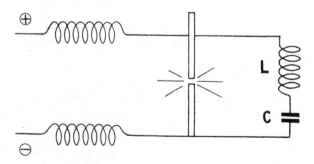

Figure 6.9   Duddell's arc generator
Source: Blake, 'History of Radio Telegraphy and Telephony', 1928

modifications considerably increased the slope of the negative resistance characteristic of the arc, and allowed the frequency of oscillations to be sustained within the band required for broadcast transmission. In order to render the Morse transmission of the Poulsen arc intelligible (a continuous wave when detected would result in a direct current, heard as an uninformative 'click' in the telephone), the aerial current was interrupted at an audio rate using a vibrating relay contact, referred to by Poulsen as a 'Pederson tikker' (ticking contact) [28].[5] The actual transmissions were thus interrupted continuous waves (ICW), rather than the continuous waves (CW) of an unmodified arc transmitter. In the more powerful Poulsen arc systems, continuous waves were radiated and the interrupting mechanism was included in the associated receiving equipment (see Section 8.2.2). A few Poulsen arc installations were used commercially in Britain, but they were used most seriously the United States, particularly by the US Navy. A modified version of the arc system was developed by C.F. Elwell for the Federal Telegraph Company and installed for the US Army at a station at Croix d'Hins near Bordeaux at the end of the First World War. This was a 1 MW station which continued to operate until replaced by a valve transmitter in 1936.

The arc possessed several advantages over various kinds of spark set. It was less efficient than the best of the spark systems, but its transmission range was at least three times that of a spark, it produced a true continuous wave and the operating cost was low. The arc systems were most effective when used to transmit long wavelengths in the range 1800–2400 m (125–167 kHz), but inefficient at medium waves, including the emergency and general calling wavelength of 600 m (500 kHz), and for this reason they were never used extensively by the Marconi Company. Although arc stations rated as high as 3.6 MW were built in the early 1900s, the Poulsen arc was not ideal as a generator of continuous waves, although it was considerably quieter in operation than the powerful spark transmitters, and the search for other methods of CW generation continued.

6.4.3 The high-frequency alternator

An interim development, before the use of thermionic valves supplanted all other methods, was the high-frequency alternator. It was not easy to construct although the technique appears theoretically simple. An existing design of alternator which had been in use for a decade to generate low-frequency alternating current power was modified to generate a higher frequency, ideally in the 10–100 kHz region, which could then be

---

[5] Pederson was by no means the first to make use of this idea. The technique had been embodied in several patents by Tesla and Poulsen, as well as Pederson, in the years between 1901 and 1907, and used by von Arco who referred to the device in 1907 as a *Tickerschaltung* – a 'ticking contact'.

used directly to provide radiating power in a transmitting aerial. There were a number of major practical difficulties. The frequency f in hertz of a conventional multi-pole a.c. synchronous alternator having both stator and rotor windings is given approximately by

$$f = NP/120$$

where P is the number of poles and N is the speed of rotation in revolutions per minute. Several designs were pursued by Thomson, Fessenden, Steinmetz, Lodge, Tesla and Duddell, all with large numbers of pairs of poles and armature rotation speeds of up to 30 000 rpm [29]. The most successful were those designed by Ernst Alexanderson, one of which is still capable of operation today.[6] Ernst Fredrik Werner Alexanderson was born in Uppsala, Sweden, and emigrated to the United States in 1901 having completed his postgraduate studies in electrical engineering in Germany. He was employed at the Commercial & Cable Electrical Company in New Jersey, and quickly became a leading authority on the design of high-power, high-frequency alternators for wireless transmission. During the First World War he produced his largest alternator, an inductor machine which employed a fixed armature consisting of multiple poles and a slotted iron rotor, which revolved at 20 000 rpm, achieving a frequency of about 100 kHz and a developed power of 200 kW. The constructional tolerances on such a machine were severe. An allowable width of only 3.2 mm was available for each pair of coils and their insulation. The rotor diameter was 300 mm, and the centrifugal force at its periphery was equivalent to several tonnes. Despite the design difficulties a total of 12 such machines were built by General Electric in America and operated by the US Navy in the summer of 1918. Other machines were constructed in various countries, most of them having a lower operating frequency than the Alexanderson machine, and making use of frequency multiplication to achieve a usable operating radio frequency.

6.4.4 Frequency multiplication

The difficulty of generating electrical power directly at very high frequencies focused attention on the possibility of using external circuits from which multiple-frequency components could be drawn. In the early 1900s very many of these circuit devices appeared, most of them relying on the harmonics produced when the generated high frequency is passed through a non-linear device, such as a flux-saturated iron-core transformer. One of the more widely applied systems was that developed jointly by Georg von Arco and Alexander Meissner of the Telefunken

---

[6] The last working 200 kW Alexanderson alternator is located at Varberg, Sweden, and was demonstrated in 1995, when its transmissions were received at the Institution of Electrical Engineers, Savoy Place, London, by an audience gathered together for the international conference on '100 Years of Radio' on 5th September 1995.

Company in 1911. Two iron-cored transformers were excited from a common d.c. source, but in opposite directions, up to the point where the iron is saturated. When an a.c. current was applied to both transformers, the core of one increased in saturation in one direction, while the other decreased, and vice versa. The saturated core thus varied only slightly in flux, while the other behaved as an unsaturated core during each half-cycle. Two secondary windings were connected so that the corresponding voltages induced in each winding were subtracted one from the other. The two distorted voltage curves produced a resultant which, if the primary d.c. and a.c. currents were correctly related, contained a sine wave of twice the original frequency. The effect was to produce full-wave rectification, the waveform of which, upon Fourier analysis, was found to lack the original frequency, but showed a predominant second harmonic, together with a set of even harmonics of decreasing intensity. A similar commercial system was developed independently by Kujerai in Japan. In order to obtain a reasonable efficiency from such an arrangement it was essential to follow the doubler by a tuned resonant circuit to limit the applied frequencies to only that part of the spectrum which is intended to be active. A system devised by A.M. Taylor made use of a complex system of delta-connected chokes, in which the separate cores become saturated early in each cycle, and a transformer carrying three primaries in series with each choke, the common core of which never became saturated. The resulting waveform distortion was such that a triple frequency conversion could be obtained from a three-phase supply. Such a system was implemented in Telefunken's alternator transmitter, located at Nauen, described later in this chapter.

An alternative and remarkable method of frequency multiplication was also being developed concurrently with that of Telefunken. This was the Goldschmidt system, in which, by incorporating several tuned circuits adjusted to higher harmonics of the fundamental generator frequency across the stator and rotor windings of the alternator, it became possible to combine the functions of generation and frequency transformation. With careful tuning the energy contained in the lower and intermediate frequencies of the process could be transferred to the highest harmonics generated, and a frequency multiplication of four or five times the basic excitation generator frequency could result [30]. Several stations were constructed incorporating the Goldschmidt generator, the most successful being that at Tuckerton, New Jersey, by the German firm of Hochfrequenz Machinen Aktiengesellschaft für Drahtlose Telegraphie for the German Government, and taken over by the US Navy at the outbreak of war in 1914. This produced aerial powers of 100 kW, transmitting on a frequency of 50 kHz, and was capable of transatlantic communication.

6.5 Production and power

By the beginning of the new century in 1900, wireless telegraphy was largely out of its experimental phase, and manufacturers were in a position to produce reliable receivers and transmitters for a willing market. They were also able to supply powerful transmitters with the ability to reach out to receivers in distant parts of the world, reliably and on a commercial basis. Until the outbreak of war in 1914, and supported by legislation passed by their respective governments, manufacturers were busy providing installations large and small to mobile and fixed sites around the world, mainly as a service to shipping needs.

Foremost among these producers was the Marconi Company, based at Chelmsford, England. For a number of years Guglielmo Marconi had been expanding his technical staff at Chelmsford, recruiting able engineers as technical assistants and consultants. Seventeen professional engineers were on site by the year 1900, including William Eccles, James Erskine-Murray, Charles Franklin and Henry Round, with Fleming and Lodge acting as consultants. The inventions and activities of this talented group were a major factor in the commercial dominance of the Marconi Company in the early 1900s. Yet a number of difficulties remained for the company, and these were often of a political nature, and not helped by the attitude of Marconi himself. Austen Chamberlain, then Postmaster General, saw the company as a potential competitor to the government-controlled telegraph industry. The Marconi Company initially laboured under a bureaucratic obstacle imposed by the monopolistic powers conferred on the British Post Office by the Telegraph Acts of 1868–69, which prohibited any private company from setting up a competing inland message-carrying service. The company was not content merely to obtain facilities for wireless telegraphy, which would have been granted under certain conditions by the Postmaster General, but asked for an exclusive right to provide wireless telegraphy throughout mainland Britain, with the Post Office acting as receiving and forwarding agents for their transatlantic services [31]. This the Post Office was not prepared to accept.

From Marconi's point of view, a solution to his production problem lay in the sales of equipment to the shipping companies and the Navy, where the Post Office edict did not run. The Marconi International Marine Company was formed in April 1900 with this in mind. To circumvent the terms of the Act, the company decided that it would enter into agreements with shipping companies not only to install and maintain wireless equipment, but also to provide the services of a Marconi-trained operator, all on a rental basis. The first such contract was to supply an initial order of 32 transmitting/receiving sets for Naval ships, payment for which would be in the form of a royalty paid by the Treasury which also covered training in the equipment for naval personnel provided by the company. The profit to the company from this arrangement was small, and made even smaller by the decision of the Admiralty to take one of the 32 sets and copy it in their workshops (which they were allowed to do under the provision of an 1883 Act which allowed the Crown the use of any patent on whatever terms the Treasury decided when applied to defence equipment). Fifty additional sets were constructed for Navy trials for which no royalties were paid [32]. Nevertheless this was an important order, the first of many orders for equipment for the Fleet, which rapidly increased as the country drew near to the outbreak of the First World War (see Chapter 9).

The first products available from Marconi's Chelmsford Works were spark and arc transmitters and tuned receivers, the latter incorporating Marconi's magnetic detector. Only after about 1906 did thermionic valves make their appearance, first as detectors in sensitive shipboard receivers, and several years later in low-power transmitting equipment. The first order for the installation of wireless telegraphy aboard a British civilian vessel came from the Beaver Line for its ship the *Lake Champlain* in 1901. It was able to establish and maintain contact with Holyhead and Rosslare shore stations on its way out to Canada from Liverpool, and to exchange messages with the *Lucania*, the first Cunard liner to carry wireless. Among the first foreign orders for shipboard equipment were one from France to equip the liner *La Savoie*, and a similar order from Germany for the impressive *Kaiser Wilhelm der Grosse*. Some administrative problems were encountered when ships operating non-Marconi equipment (principally Telefunken apparatus) found that they were unable to establish communication with Marconi shore stations, or with ships fitted with Marconi apparatus, owing to a Marconi commercial edict forbidding such contact. This provoked considerable disquiet amongst shipowners of all nations, and was discussed, following a request from the German Government, at an International Wireless Conference held in Berlin in 1903. The conference recommended that it be made obligatory for all coastal sites to be able to communicate with any vessel, no matter what wireless equipment it carried, despite the opposition to this voiced by the governments of Britain and Italy. The decision was finally ratified by the full International Wireless Conference held in Madrid in 1906, and led to improved safety at sea as well as freeing commercial contracts from unnecessary restrictions on the use of shipboard wireless apparatus.

A number of important stations were being established around the shores of Britain for maritime communication. A significant contract had been awarded by the Lloyd's Corporation for the installation of wireless equipment in ten of Lloyd's signal stations in 1901, with the further prospect of sales of apparatus for ships operated by the line. Operation of these shore stations was taken over by the Post Office in 1909, and thereafter all wireless communication with shipping (apart from Admiralty stations) became the responsibility of the Post Office, with the Marconi Company as the only equipment supplier. Orders for shore stations also came from abroad: a large one from the Hawaiian islands, for example, and supporting orders from the United States. In India, the only major competitor to Marconi was the Lodge-Muirhead Company, which had provided a substantial number of spark installations used to support wireless telegraphy traffic throughout the subcontinent for the Government of India. The Lodge-Muirhead system was originally tried out in communications between the Andaman Islands and Burma by the Indian Government Telegraph Department, and was considered by the government electrician as superior to the current (untuned) Marconi system and to the Slaby-Arco system from Germany, both of which had been tried out in India. The Lodge-Muirhead company was later bought out by Marconi when Sir Oliver Lodge joined the Marconi Company in 1911, and its patents were transfered, leaving Marconi as the sole supplier of coastal telegraphy stations in British India.

6.5.1 The 'Marconi system'

In the first decade of the 20th century the Marconi company had established a number of companies in Britain, Canada and the United States to manufacture a standardised series of transmitting and receiving equipment which became known as the 'Marconi system', used on land stations and at sea, and installed under licence by the governments concerned. In addition a number of high-power stations were established at several key land sites for commercial and administrative traffic.

The Marconi system became available as a series of spark transmitters, the smallest of which was a 0.5 kW set, the type T17, which was installed on many small and medium-sized cargo ships. A synchronous rotary spark gap, which could be operated as a plain spark gap in an emergency, was coupled to the primary of a tuned aerial transformer, as shown in Figure 6.10. The power source was generally a motor alternator providing 100 V a.c. at 500 Hz, giving the transmitted signal a characteristic sound when detected at the receiver. A larger version, the type T18 providing 1.5 kW, was available for use on board larger ocean-going ships and gave the operator the choice of four standard wavelengths, 450, 600, 700 and 800 m. Keying these high-power spark transmitters meant breaking a fairly heavy current and using a wide gap setting for the key, which had substantial electrical contacts. A shipboard wireless operator of the time recounted the difficulties this caused:

I recall the old key which was used to break the heavy current for telegraphic signalling purposes. It was actually pounded and was equipped with giant platinum contacts. The maximum speed we could handle was about ten words a

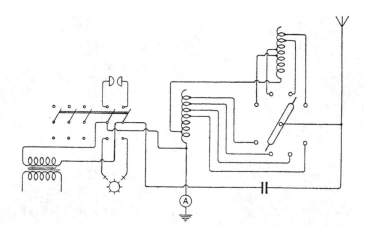

Figure 6.10   Schematic diagram for the Marconi 0.5 kW Type T17 set
Source: Hawkshead, 'Handbook of Technical Instruction for Wireless Telegraphs', 1913

minute. And even at that slow rate of transmission operators would soon tire mostly of strain in the arm, as the key worked like a pump and required considerable expenditure of energy for its operation.

Later additions to the Marconi range were a number of small quenched spark gap transmitters rated at 0.25, 0.25 and 0.5 kW – types T20, T24 and T28 – all of similar design and offering the same wavelengths as the rotary spark gap designs. The advantage of the quenched spark gap sets was the narrower bandwidth occupied by the transmitted signal, a factor of increasing importance as more and more merchant vessels came to be equipped with wireless transmitting equipment. From 1911 onwards some naval vessels and various shore stations were fitted with 3 and 5 kW tuned spark sets based on Marconi's synchronous disc discharger patents of 1907–09. These provided not only a more powerful signal but one that gave out a musical note, as did the earlier rotary spark gap sets, which could be detected over interference. The Marconi Company exploited the possibilities of spark transmission to a greater extent than its rivals in Continental Europe and the United States, and continued to manufacture spark transmitters some time after other companies were turning to CW generators – a factor in their loss of domination in the market after about 1912.

The Marconi receiving stations for installation on shore or on board ships with their spark transmitters consisted of two units: the Marconi short-range multiple tuner, and the 'Maggie' magnetic detector described earlier. The tuner was designed in 1908 as a standard tuning device for receivers, suitable for all wavelengths between 250 and 750 m, and is shown in Figure 6.11. It was a passive device containing two separate LC

Figure 6.11   Marconi short-range multiple tuner, the Mk III
Source: *With acknowledgement to Marconi Ltd*

circuits, one for aerial tuning and the other for detector tuning in accordance with the ideas expressed in Marconi's definitive British Patent No. 7777 and was manufactured in large numbers. The tuner contained several additional features, such as provision for aerial discharge protection (a micrometer spark gap connected between aerial and earth) and a cut-out switch which came into use when its associated transmitter was working [34].

6.5.2 Marconi high-power stations

By 1912 the Marconi Company with its spark transmitters had proved the commercial viability of long-distance communication, and was dominating the transatlantic route. The decision had been taken by the company to build new stations on mainland Britain and in the United States, the latter through its American Marconi Company, which had selected a site close to New York to facilitate this trade. The aim of major companies building large and powerful transmitters at that time was to attempt to achieve CW transmission, or as near to it as was possible with the technology of the time. This would not only reduce the bandwidth occupied by a company's broadcasts, which would improve efficiency, but also permit the transmission of a modulated tone for the Morse signal, which would render it distinctive and easier to distinguish against a noisy background and other stations transmitting on adjacent frequencies. Marconi set out to achieve this in several ways, all present in its flagship installations at Caernarvon in North Wales and its counterpart at Tuckerton, New Jersey. The site at Caernarvon was built on a promontory, some 240 m above sea level and rising a further 400 m along the line of the aerial, which was supported on a set of masts 330 m high. The aerial was directional, arranged to give maximum radiation in the direction of the receiving station in New Brunswick, some 30 km from the Tuckerton transmitting station. This was at a time when long wavelengths were in favour for transatlantic communication. The natural wavelength for this aerial, 5600 m, was increased to 14000 m by including suitable inductances in the aerial circuit.

By 1918 three types of transmitter had been installed at Caernarvon: two synchronous a.c. spark sets; a Marconi 'timed spark' transmitter, used for a commercial service to the United States; and a Poulsen arc transmitter. Only the last of these provided CW transmission, although the others produced relatively undamped repetitive wave trains approaching the ideal CW transmission. For the synchronous a.c. sets, each of 300 kW capacity, power from a hydroelectric power source at Cwm Dyli in North Wales was used to drive a motor alternator connected to the disc discharger. The timed spark transmitter, also of 300 kW capacity, had two distinct primary circuits tuned to the same frequency and coupled to the aerial circuit. A 5 kV d.c. dynamo charged the condensers in the primary circuits, which were then periodically discharged through a spark gap at an appropriate time, when the spark gaps reached their minimum width (see earlier description of this patented Marconi technique). The spark sets were extremely noisy in operation, and the operators were compelled to protect their ears from their incessant cacophony over a range of frequencies well within the audio range. The arc sets, on the other hand, were remarkably quiet and foreshadowed the advent of powerful CW transmitters which were soon to oust the spark transmitters from their position of dominance. The arc set at Caernarvon was rated at 200 kW and driven from an 800 V d.c. dynamo with the field of the arc transmitter separately excited from an auxiliary 110 V d.c. dynamo. Water-cooling was provided for the arc chamber to dissipate the excess heat from the continuous arc. Although primarily intended for a transatlantic service, the spark transmissions were received as far afield as Sydney, Australia, soon after installation in 1918 [35]. In 1921 two Alexanderson high-frequency alternators were purchased from General Electric in America and installed at Caernarvon when the arc transmitter was dismantled.

The Marconi Company's activities abroad were extensive, and it was responsible for providing a number of high-power and other sets to governments and navies in several parts of the world. Typical were its dealings with the Ottoman Empire in the period 1911–14. The company was responsible for the construction of the important Ok Meidan station, standing on a prominence about 4 km from Constantinople. The main transmitter was a synchronous timed spark transmitter similar to the Caernarvon installation, having a working range of at least 1000 km, to communicate with ships of the Turkish Navy, which were themselves all equipped with 1.5 or 3 kW Marconi sets. The Turkish transmitter was powered by a 42 kW d.c. dynamo driving a motor-generator set coupled directly to the rotating disc discharger. The wavelength of the transmission was set by tuning the aerial circuit and coupling to lie between 600 and 2000 m [36].

6.5.3 Duplex working

An important advance in commercial wireless telegraphy took place in about 1913, when duplex working was established from many shore stations. Until then it was the practice to locate the transmitting and receiving aerials on the same site, with the transmissions halted from time to time to permit the receiver to be switched, so that the distant station could acknowledge receipt or ask for repetitions. Marconi was the first to establish the facility for duplex working in his transatlantic service to avoid this inconvenient method of working. To do this he first located the sending and receiving stations several kilometres apart. Two directional aerials were employed for each receiver, one oriented to obtain optimum reception from the distant station, and the other from the local transmitter. The former received the distant signal plus the local interfering signal, while the other received only the local station. When the aerials were coupled in opposition the local signals were balanced out, leaving the distant signal free from interference. A similar idea was patented by H.J. Round in 1911, in which two detector crystals worked in opposition; one was sensitive to the signal, whereas the other was biased so that it operated only when a disturbance exceeded a given value. Again the effect was to balance out the interfering signal so that reception was possible despite the local operation of the transmitter.

6.5.4 Telefunken and Siemens

An early German interest in wireless telegraphy transmission was apparent during Marconi wireless experiments at Lavernock in 1897, referred to earlier, to which the Admiralty was invited to send a representative. William Preece, then Engineer-in-Chief at the British Post Office, also invited Professor Rudolf Slaby of Berlin University to represent the German Government at these tests, despite prior objections by Marconi himself. Slaby was impressed by the trials and delighted at the technical information he acquired during the event. This experience was soon made use of by the Allgemeine Elektrizitäts Gesellschaft, the company founded by Slaby and his colleague, Count Georg von Arco, to exploit the Slaby-Arco system which became a powerful rival to the Marconi Company in Britain [37]. A second German company, also established in 1901, was Siemens & Halske, which cooperated with Professor Ferdinand Braun to produce the Braun–Siemens system of wireless telegraphy and established a number of shore and ship installations at the turn of the century. These were quenched spark transmitters for use at sea; the most important of them was a 0.25 kW set and a larger 1.5 kW set for use in navy and large merchant vessels. Their design was similar to the Marconi T20 series and covered the same range of frequencies. The patents of the two companies soon came into conflict in the German courts, and although a ruling was declared in favour of Braun–Siemens, Kaiser Wilhelm III stepped in and ordered a halt to the rivalry, which was having a deleterious effect on German trade. The two systems were accordingly amalgamated in March 1903 in a new company, the Gesellschaft für Drahtlose Telegraphie, known thereafter as Telefunken GmbH.

The growth of Telefunken was remarkable. Within a short space of time it was able to report at its annual meeting in 1905 that 518 of its wireless installations had been completed in various parts of the world [38]. In 1908 the company began to manufacture a new series of transmitters incorporating a quenched spark gap in a primary circuit, with a tuned aerial circuit coupled to it inductively. Its transmitter was energised by a 500 Hz alternator, designed to produce a sharply peaked waveform rather than a sinusoidal output, a feature which helped to produce a clear audible tone in the receiver, which quickly became recognised as the characteristic 'note' of a Telefunken transmitter. The new design was efficient and was applied to a range of fixed and mobile stations, commencing with low-power stations having an aerial energy of between 4 and 25 kW, and culminating in the construction of high-power stations of 35 kW at Sayville in the United States, and a massive 100 kW installation at Nauen in Germany [39]. A small 1.5 kW set for ship and land stations was produced by Telefunken in large numbers for the German Navy, and was also exported to 30 different countries [40]. These stations provided wavelengths of 300, 450 and 600 m by the simple means of altering the position of taps on the aerial coupling circuit. The standard ship installation may be seen in Figure 6.12, which shows clearly the a.c. dynamo and its driving motor at the bottom, and the quenched spark gap at the top of the picture.

Telefunken was sufficiently confident in 1911 to produce for sale a 'demonstration apparatus for the new quenched spark transmitter and receiver'. This was a smaller version of its wireless telegraph station, intended to be used for experiments in laboratory or classroom, and which could also be used for publicity purposes by Telefunken agents, indicating a measure of confidence in the techniques being developed by the company for civil and military shipping in the period up to 1914. It was marketed at DM 1000 and operated from a 12 V battery, with a vibrator and transformer arranged to provide the high voltage required for the spark generator, and manufactured in quantity at the Berlin factory. The transmitter had a limited range at a wavelength of 25–70 m, and

Figure 6.12　Telefunken quenched spark ship installation
Source: Erskine-Murray, 'Wireless Telegraphy', 1913

sent a signal modulated at the frequency of the vibrator, 350 Hz, so as to be clearly recognised at the receiver, a simple graphite rectifier coupled to the aerial tuned circuit and connected to a pair of headphones [41].

It was, however, the large and powerful wireless telegraphy stations that created an impression of German enterprise in 1914. While the smaller and shipboard stations relied on quenched spark transmitters, not very different from those installed by its rival, the Marconi Company, the larger shore stations designed by Telefunken contained CW transmitters using high-frequency alternators. Several machines were installed in 1911, the largest being a 30 kHz, 40 kW alternator located at Altenburg, near Vienna. The frequency limitations imposed by mechanical construction, discussed earlier, led to the parallel development of a series of frequency-multiplying transformers. These were used in a 100 kW station established at Sayville, New York, to replace the quenched spark transmitter previously installed, and were intended to provide direct communication across the Atlantic with Germany as an alternative to the use of the submarine telegraph cables (which would not be available to Germany in the event of war). With three frequency-multiplying transformers, the station operated at a final aerial frequency of 32 kHz and was working by October 1913, when the first telegram was sent direct to Germany in a daylight transmission; it was called a *Tages-verkehr* (daytime) transmission, indicating a more difficult accomplishment than transmission by night.

The major achievement of the company in the years before the war was, without doubt, the design and installation of the 400 kW high-frequency alternator station at Nauen, near Berlin, in 1914; an improved version of the 100 kW Sayville design. It was an impressive engineering construction, operating at the limits of what was achievable at the time. The rotor was 1.75 m in diameter, weighed 7 tonnes and, rotating at 1500 rpm – 130 m/s^{-1} at the periphery – exerted a tremendous centrifugal force on the rotor. The design of the alternator was due to Georg von Arco, who had been assistant to Professor Slaby at the Berlin University at the time of the Marconi trials. Von Arco later joined the Allgemeine Elektrizitäts Gesellschaft (AEG), manufacturers of the range of alternators used in these pre-war stations. He was responsible for exhibiting his alternator, with its frequency-doubling transformers, at the International Radiotelegraph Congress in London in 1912 just before beginning its installation at Nauen. The alternator generated a comparatively low frequency of 6 kHz with its complement of 240 'teeth' around the rotor plate. Three frequency-doubling transformers were used, giving a 48 kHz output to the aerial. Although a 90 per cent efficiency was claimed for the operation of the doubling transformers, this still meant that over 40 kW of heat needed to be removed by a powerful oil circulation scheme. To achieve a saturating magnetisation of the transformer cores a separate low-voltage generator was used to pass a d.c. current of 300 A through additional windings on the transformer cores (see the earlier description of the working of these transformers). The aerial circuit at Nauen was arranged to give a wide choice of transmitted frequencies for different requirements. By choosing the third harmonic of the resulting waveforms from the transformers, a trebling of the generated frequency was also possible, extending the choice of transmitted frequencies to 6, 12, 18, 24, 36 and 48 kHz. A simplified schematic diagram of the Nauen transmitter is shown in Figure 6.13. To control the 6000 kVA circulating in the aerial circuit at Nauen, a system of several keying relays was used connected in series. These could operate at keying speeds up to 80 wpm and required air-blast cooling

The intention of those who planned and constructed the powerful Sayville and Nauen stations was to erect a chain of similar stations in Togoland, Kiaochow, Yap and Rabaul, Samoa and Dutch Sumatra, but by August 1914 only the Togoland link with the homeland was fully operational, although trials for the Yap station in the Pacific were well under way. The chain was never completed. Apart from the Nauen station which remained in use throughout the conflict, all the other stations were destroyed or taken over by the Allies, as described in Chapter 9.

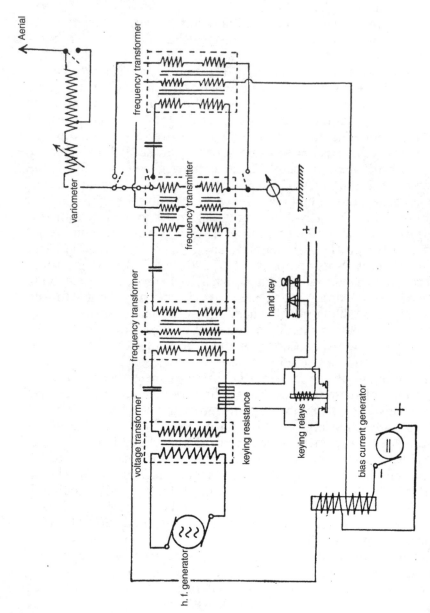

*Figure 6.13   Simplified schematic diagram of the Nauen HF alternator transmitter, 1919
Source: Telefunken Archives, Berlin*

6.5.5 *Keying at high power*

Keying of these high-power transmitters represented a significant design problem, which first arose when spark and arc systems were designed for radiated powers in excess of a few kilowatts and which could not be controlled safely by inserting an operator's Morse key in the power circuit. A solution lay in the use of relays capable of being indirectly controlled from the key. As the power that needed to be controlled in this way increased, several relays were used either in series, where the object was to control a high potential line, such as in the German Nauen transmitter, or in parallel, where a large supply current was to be controlled. The controlled power was often high, and the keying relays required airblast or oil cooling to limit their operating temperature. With many of the high-power transmitters, and particularly with the arc transmitter, keying presented a further difficulty since any sudden change in the circuit conditions could lead to instability.

Two alternative methods were used to enable the arc to be sustained without interruption, and were also applied to valve transmitters for a similar reason. In one of these an auxiliary aerial coil was brought into circuit by means of a relay-controlled Morse key. This had the effect of shifting the transmission frequency by a small amount to a 'spacing frequency', usually about 2 kHz from the 'marking frequency', the latter being easily selected by tuning at the receiver. In the second method the key switched the oscillating circuit from a radiating (aerial) connection to a non-radiating circuit having the same electrical characteristics as the aerial, so that only one frequency was transmitted. In diverting the output away from the aerial to a resistive load it became necessary to apply an air blower or some other means of dissipating the full signalling load, which could be in the region of several hundred kilowatts. In a later development in Germany for the Nauen transmitter, a device incorporating a modification of their frequency doubling-transformer was used (German Patent No. 303094, 1911). Here two magnetised iron cores were inserted in the primary of the machine. By varying the direct current through the magnetising windings, their self inductance could be altered, slightly detuning the radiated frequency. The power required to effect this change was small, and could be keyed without difficulty. For the high-power valve transmitters discussed in the next chapter, keying was often applied by grid control of special control valves instead of slow relays, which enabled faster keying operations to be achieved, reducing the susceptibility to arcing across the key contacts.

6.5.6 *Continuous waves in the United States*

In 1901 the authorities in the United States, and particularly the US Navy, which played a predominant role in technical policy, were still carrying out evaluation tests on a number of wireless transmitting sets from a variety of domestic and European manufacturers in what has been described as 'an unhurried search for radio equipment' [42]. Marconi had visited the United States in 1899 and demonstrated his system by reporting for the press the Americas Cup yacht races, taking place just off New York, and later, with the US Navy, when he installed his equipment aboard the battleship *Massachusetts* and the cruiser *New York* for transmission tests. These tests proved satisfactory with regard to range and reliability, and indeed caused the chief signals officer to the Secretary for War to comment in his report, 'I must accord to Mr. Marconi the credit for having by far the best apparatus, better than any I have seen in America, including my own' [43]. However, it did not escape the notice of Navy officials that interference from other sets working in the vicinity remained a problem with Marconi's current untuned spark technology, and he left the country without having advanced the position of his new American Marconi Company any further. He had under development at Chelmsford new tuned spark transmitters and more selective receivers, but was loath to disclose details before securing full patent rights, despite a request from the US Navy for trials with the new equipment.

Congress continued with its funding of a review of all relevant competitors and authorised the purchase of sets from other European manufacturers, including Ducretet-Popov, Rochefort, Slaby-Arco and Braun-Siemens (soon to become Telefunken), as well as United States companies; the National Electric Signalling Company (NESCO), exploiting Fessenden inventions, the Lee de Forest Wireless Telegraph Company and several other smaller firms. The results were inconclusive. In 1906 the market was shared by the American Marconi Company for both ship and shore communications, and Telefunken for shipboard equipment, together with Shoemaker, a domestic manufacturer, and several other smaller suppliers. According to a report issued by the Secretary to the Navy in 1906, 48.5 per cent of the equipment then in use was of German origin [44]. It is surprising that the highly successful telegraph companies in the United States did not go on to develop wireless equipment. Western Union, the largest of these, was 'apparently more interested in buying up competitors and making protective agreements than in the long range development of communications'. One such agreement was with the newly established Bell Telephone Company, 'to withdraw from the telephone field in 1879 in exchange for Bell's promise to keep out of the [line] telegraph field' [45].

Most of the equipment available from overseas and domestic manufacturers at this time operated with the use of spark, rotary spark or quenched spark technology, with a minority of transmitters using early forms of arc transmitter. Arc transmitters provided a more effective solution to transmission requirements for both civilian and naval use in the United States because of the attraction of CW transmission, for which the arc and the high-frequency alternator were the only serious contenders in the early 1900s. Providers of wireless transmission services were quick to realise this, and spark transmitters were phased out in the United States much earlier than in the rest of the world. Attention was concentrated on CW transmitters and on the engineers and companies who could design and manufacture them.

The two key investigators in the development of wireless telegraphy equipment in the United States in the first decade of the new century were Fessenden and Lee de Forest, and both were to be responsible for major innovations. Professor Reginald A. Fessenden was a long-standing critic of Marconi and his spark technology, maintaining that it was 'essentially and fundamentally incapable of development into a practical system' [46]. He was convinced that only by developing CW transmitters and detectors which could recognise incremental changes in the amplitude of the received signal, rather than just indicate the presence of the signal, as the coherer and magnetic detector did, would a communication system eventually emerge to transmit the human voice – Fessenden's long-term objective. He had already, in 1902, designed a new detector which could operate in this way, which he named a Barretter – a form of electrolytic detector which rectified the signal and provided an output current proportional to the mean level of the a.c. signal [47]. Other rectifying detectors were also becoming available, including the carborundum detector and other crystal diodes which could solve this part of his problem. The lack of an effective generator of continuous waves, however, remained.

In 1900 three methods offered themselves. First there was the use of a spark generator so arranged that the frequency of spark repetition was higher than audio frequency, and where the spark waveform was relatively undamped, producing a noisy but fairly continuous high frequency waveform. Second was the oscillating arc, already used successfully by Duddell. And third was the high-frequency mechanical generator, akin to an a.c. power generator, but operating at a very much higher rate. Fessenden's experiments with a form of rotary spark gap, similar to the highly successful Telefunken quenched spark developed later, gave good results, particularly when applied to Morse telegraphy, and he was later to make use of this in a station at Arlington for the US Navy, manufactured by the company he founded, the National Electric Signalling Company (NESCO). He was less successful with the oscillating arc, which he found difficult to stabilise, but was eventually able to use it in an experimental telephony transmitter, but with poor results [48]. This left the HF alternator as the only remaining possibility for CW transmission. Nikola Tesla had attempted to apply this for wireless transmission in the 1890s with little success. Fessenden was more persistent, and its application became his major interest in the next decade, but first he had to find someone able and willing to build a machine to his specification.

He approached General Electric, and interested Charles Steinmetz of the company in producing the first alternator to approach the kind of frequencies needed. In the event Steinmetz was only able to reach 10 kHz with his machine in 1903, and although Fessenden was able to use it as a high-frequency exciter for a spark machine and thus achieve aerial transmission at 20 kHz, albeit a noisy one, much more development was needed to obtain direct and suitable transmission frequencies. To complete Fessenden's second development order with General Electric (he was meticulous in paying for all development costs, and so remained welcome as a customer, even if no immediate commercial advantage ensued), a gifted engineer was found within the engineering department to carry this out. Ernst Alexanderson was a young Swedish engineer who had recently joined General Electric, and had certain ideas of his own about how to solve the problem of producing high-frequency oscillations by mechanical means. He departed from the usual design of alternators by having a stationary armature on an iron core, located between two rotating steel discs with projecting poles or teeth cut in their circumferences. This was an inductor-type alternator, the frequency of which depended on the speed of rotation and number of poles on the discs, as discussed earlier, and both needed to be very large, with all the mechanical and safety problems that ensued [49]. His machine, producing a frequency of 50 kHz at an aerial transmission power of 50 W, was eventually transferred to Fessenden's experimental station at Brant Rock, Massachusetts, and was successful in transmitting for the first time both speech and music on Christmas Eve 1906 to a receiver in Norfolk, Virginia. Further use of this and later HF alternators in the commercial world was for Morse transmissions, at a time when only the transmission of telegrams brought in the necessary revenue. By 1909 General Electric had designed a 2 kW machine operating at 100 kHz which could be standardised and marketed [50]. Further, it could serve as the basis for larger and more powerful alternators. It was to lead to Alexanderson's finest achievement for General Electric: the construction of the 200 kW alternator discussed earlier in this chapter.

While their purity of waveform and reliability made HF alternators the high-power transmitters of choice, they were difficult and expensive to manufacture, with little saving if a low-power machine was required, and at the high-power end imposing a practical limit of about 200 kW in their maximum capability. The arc transmitter was more flexible in design and could meet a wide range of requirements for both ship and shore stations, including the high-power requirements for long-range worldwide communication. In 1913 several nations were interested in setting up networks of powerful wireless telegraphy systems which would link their scattered colonies and bases abroad. Britain was proposing its Imperial Network scheme; Germany was already going ahead with a number of transmitters in Africa and elsewhere linked to its central station at Nauen; France had plans for communication with its colonies in Africa, the Far East and with America; and Holland was considering links with its East India possessions.

In America, the US Navy was far ahead of any civilian organisation in the construction of powerful long-distance telegraph transmitters. The Navy commissioned its first high-power station in Arlington, Virginia, a 100 kW synchronous rotary spark station installed by NESCO to a design by Fessenden [51]. This was a massive machine consisting of a fibre ring 1.78 m in diameter with an outer circumference fitted with 48 copper vanes, each about 160 mm long. These rotated between stationary spark points at 1250 rpm and were coupled for synchronism with the generator driving shaft. It was intended to be the first and central station in a network of powerful transmitters by which the Navy would be able to maintain communications with its remote bases and ships in any part of the world. Other stations followed in the Canal Zone, on the Californian coast, in Hawaii, on Guam, on Samoa and in the Philippines [52]. Later installations were CW arc transmitters, commencing with the installation of a 100 kW arc transmitter at Arlington in 1913, with a similar 100 kW station ordered for the Canal Zone, situated at Darien, south of the capital at Colón, and ten 30 kW arc transmitters for shipboard use. The Arlington arc set was typical of the simple but effective construction employed at this time. Essentially it consisted of a motor generator, an arc chamber, magnet poles, choke coils and an aerial tuned resonant circuit. Keying was arranged either to alter the resonant frequency by a small amount (typically 1 per cent) or to divert the HF output to a resistive load in order to maintain an uninterrupted arc. The efficiency of an arc is not particularly high, and the arc chamber was generally water-cooled to dissipate the excessive wasted power and supplied with a small flow of alcohol, ignited by the arc to ensure a continuous supply of ionized vapour. Three larger arc stations were later constructed at San Diego (200 kW), Pearl Harbor and Cavite (each 350 kW), and smaller stations (30 kW) at Samoa and Guam. All of these stations were installed by the Federal Telegraph Company of California, whose chief designer was Cyril F. Elwell. Elwell was an Australian engineer responsible for the development of a range of successful synchronous arc transmitters for Federal, the success of which signalled the demise of the spark transmitter in the United States in 1913. He was later to install a 200 kW set to replace the HF alternator at the Sayville station, after the US Navy took over the station from its German owners in 1917; a 500 kW set for Annapolis, Maryland; and a massive 1000 kW transmitter for the Layfayette station established near Bordeaux in France in 1918 to provide secure communications between the Allied governments and the American Expeditionary Force. This last was for a time the most powerful station in existence, and its transmissions could be received all over the world 'with an intensity surpassing that of all other existing stations' [53].

By 1918 the US Navy was equipped with an impressive network of high-power CW transmitters interconnecting with its major bases and the Navy Department in Washington, and was capable of communicating with its fleet in any location in the Atlantic or Pacific Oceans. This was not, however, the only network of arc stations set up by the flourishing Federal Company. The press service in America was clamouring for fast telegraph communication between cities to compete with Postal Telegraph and Western Union over the vast continent. Federal set up a chain of 13 stations over a period of two years to link Los Angeles, San Francisco, El Paso, San Diego, Forth Worth, Kansas City, Chicago and Seattle. Each was equipped with a 12 kW arc transmitter, maintaining a reliable service over distances of 800–1600 km. To achieve higher telegraph rates than the landline competition a high-speed Morse operation was necessary, and was achieved using an off-line Wheatstone perforator system.

All of this installation activity by United States firms during the period 1912–18 was carried out while the country remained neutral as the war raged in France. These companies, taking advantage of the reduced availability of transmitting equipment from Europe, were growing in capability. Further, the position of the American Marconi Company, considerably weakened by its full-time contribution to war work to the detriment of its commercial activities, was in no position to withstand the sale of its holdings to the newly formed Radio Corporation of America (RCA), which came into being with strong Congress backing in 1919. Within a very short time RCA had acquired the rights to fundamental Marconi patents together with those for the Alexanderson alternator, the Westinghouse patents of Fessenden and Armstrong relating to the principles of heterodyne reception and feedback, and finally the de Forest patents on the triode valve – all of which made RCA a major international supplier of communication equipment to Europe and America as the radio world emerged from its wartime preoccupation with military equipment [54].

The second significant figure in the American development of wireless telegraphy equipment at the beginning of the 20th century was Dr Lee de Forest. De Forest was responsible for an outstanding innovation in wireless communication, namely the invention of the triode valve in 1906. The effect of his triode valve, or 'audion' as he preferred to call it, on the design and operation of telegraph wireless communication lies at the heart of the subject matter of the next chapter.

References

1 HERTZ, H.: 'On electromagnetic waves in air and their reflection', *in* 'Electric waves', trans. D.E. Jones (Macmillan, London, 1962; 1st German edn 1893)

2 PREECE, W.H.: 'Aetheric telegraphy', *Journal of the Institution of Electrical Engineers*, 1898, **27**, p. 869

3 BRANLY, E. (1891) 'Variations of electrical conductivity under electrical influence', *The Electrician*, **27**, pp. 221–2 (see also *Comptes Rendus*, 1890, **111**, pp. 785–7)

4 LODGE, O.: 'The work of Hertz and some of his successors', *Proceedings of the Royal Institution*, 1894, **14**, p. 321 (see also *The Electrician*, 1894, **32**)

5 LODGE, O.: 'Past years – an autobiography' (Charles Scribner, New York, 1932)

6 POPOV, A.S.: 'An instrument for detecting and registering electrical oscillations', *Zh. Russ. Fiz.-Khim. Obshchestva (Physics Pt 1)*, 1896, **28**, pp. 1–14

7 HAWKS, E.: 'Pioneers of wireless' (Methuen, London, 1927), pp. 221–2

8 MARCONI, G.: 'Wireless telegraphy', *Journal of the Institution of Electrical Engineers*, 1899, **28**, p. 273

9 POCOCK, R.: 'Marconi's radio trials at Lavernock', IEE PGS7 Summer Meeting, University of Wales, Cardiff, July 1977

10 'Remarks of Captain Jackson on the trial of Professor Braun's system of wireless telegraphy' (1899) Doc. ADM116/563, Public Record Office, Kew, Surrey

11 MARCONI, G.: 'Syntonic wireless telegraphy', *Journal of the Royal Society of Arts*, 1901, **49**, p. 505

12 BAKER, W.J.: 'A history of the Marconi Company' (Methuen, London, 1970), pp. 74–5

13 VYVYAN, R.N.: 'Wireless over 30 years' (Routledge, London, 1933), p. 29

14 ROBINSON, S.S.: 'Robinson's manual of radio telegraphy and telephony' (United States Naval Institute, Annapolis, Md, 1918)

15 ZENNECK, J.: 'Wireless telegraphy' (McGraw Hill, New York, 1915), pp. 210–12

16 BLAKE, C.G.: 'History of radio telegraphy and telephony' (Radio Press, London, 1928), p. 157

17 HOWETH, L.S.: 'History of communication electronics in the U.S. Navy' (Government Printing Office, Washington, DC, 1963), p. 137

18 FESSENDEN, R.A.: 'Wireless telephony', 25th Annual Convention of the American Institute of Electrical Engineers, Atlantic City, 20 June 1908

19 WIEN, M.: 'Über die Intensität der beiden Schwingungen eines gekoppelten Senders', *Physikalische Zeitschrift*, 1906, **7**, (23), pp. 871–2

20 'The Lepel system of wireless telegraphy', *The Electrician*, 14 May 1909, **63**, pp. 174–5 (see also communications by Lepel and Fleming on pp 374–7 of the same issue)

21 HOWE, G.W.O.: 'Historical landmarks in wireless invention', *in* 'Handbook of wireless telegraphy and telephony' (Wireless Press, London, 1921), p. 1153

22 ARCO. Count G. von: 'A new system of wireless telegraphy used by the

Telefunken company', *The Electrician*, 1909, **63**, pp. 89–91; 'The new Telefunken system', *Ibid.*, pp. 370–74

23   Letters in *The Electrician*, 1909, **63**, pp. 142, 228–9, 374–7 (see also under 'Legal intelligence', **66**, pp. 833–4; and Letter in **67**, pp. 25–7)

24   GALLETTI, R.C.: 'Syntony of a quenched spark', *The Electrician*, 20th January 1911, **46**, pp 569–73

25   'Rapport techniques sur les résultats obtenus à Lyon 1911' (1911), Galletti Museum Archives, St Maurice de Rotheren, Savoy, France

26   Proceedings of the International Conference on 'Le origini e lo sviluppo della telegrafia senza Fili', Fermo–Torres San Patrizio, Italy, 4th–5th June 1999

27   TOSI, A.: 'Report by Commandant Alessandro Tosi to Mons. R.C. Danelli, Ingenier' (1923), Galletti Museum Archives, St Maurice de Rotherens, Savoy, France

28   EGLETON, D.: 'L'Ingénier Galletti de Cadilhac', *Le Progrès*, 19th April 1961

29   PEDERSON, P.O.: 'On the Poulsen arc and its theory', *Proceedings of the Institute of Radio Engineers*, 1917, **5**, pp. 255–319

30   ERSKINE-MURRAY, J.: 'A handbook of wireless telegraphy' (Crosby Lockwood, London, 1913), pp. 203–4

31   FLEMING, J.A.: 'Principles of electric wave telegraphy' (Longman Green, London, 1909)

32   MAYER, E.E.: 'The Goldschmidt system of radio telegraphy', *Proceedings of the Institute of Radio Engineers*, 1914, **2**, pp. 69–108

33   Letter from Lamb to Secretary of the Marconi Company, 31st December 1902, Marconi Historical Archives

34   JOLLY, W.P.: 'Marconi' (Constable, London, 1972), p. 91–2

35   Hancock, H.E.: 'Wireless at sea' (Marconi Company, Chelmsford, 1950), p. 46

36   'Modern wireless apparatus – Marconi short-range multiple tuner', *Wireless World*, 1914, **1**, pp. 498–9

37   'Great wireless stations – Caernarvon', *Wireless World*, September 1919, **7**, (78), pp. 301–7

38   'Wireless telegraphy in Turkey', *Wireless World*, April 1915, **2**, pp. 456–8

39   SLABY, R.: 'Fortschritte der Funkentelegraphie', *Zeitschrift der Vereines Ingenieure*, 27th July 1901

40   ARCO, G. von.: 'Die drahtlose Grossstation Nauen', *Elektrotechnische Zeitschrift*, 1919, **40**, Heft 51, pp. 665–7

41   'The Telefunken system of wireless telegraphy', *The Electrician*, 1911, **68**, pp. 213–16

42   DEUTSCHES MUSEUM: 'Demonstrations-Apparat nach dem neuen Telefunkensystem (tönende Löschfunken)' (Gesellschaft für Drahtlose Telegraphie, Berlin, 1911)

43   Discussion: 'The possibilities of wireless telegraphs', *Proceedings of the American Institute of Electrical Engineers*, 1899, 16, pp. 607–88

44   MacLAURIN,W.R.: 'Invention and innovation in the radio industry' (Macmillan, New York, 1949)

45   PIERCE, G.W.: 'Principles of wireless telegraphy' (McGraw-Hill, New York, 1910), p. 202

46 FESSENDEN, R.A.: 'Wireless telephony', *The Electrical Review*, 15th February 1907, **60**, p. 252

47 BRITTAIN, J.E.: 'The Alexanderson alternator: an encounter between radio physics and electrical power engineering', Joint meeting of the Society for the History of Technology, and the History of Science Society, 31st October 1982, p. 5

48 CLARK, G.H.: 'Radio in war and peace' (MIT Press, Cambridge, Mass., 1914), pp. 241–2

49 BULLARD, H.G.: 'Arlington Radio Station and its activities in the general scheme of naval radio communication', *Proceedings of the Institute of Radio Engineers*, October 1916, **4**, pp. 421–6

50 WOODBURY PULSIFER: 'Naval Appropriation Laws from 1883 to 1912', *in* 'Navy year book' (U.S. Government Printing Office, Washington, D.C., 1912)

51 LATOUR, M.: 'The technical progress of wireless telegraphy in France during the year 1920', *in* 'Handbook of wireless telegraphy and telephony' (Wireless Press, London, 1921), p. 1097

Chapter 7

Telegraphy for peace . . .

As the techniques of electrical communications developed in the early 1900s, and telegraph transmission by landline, submarine cable and wireless became widespread, their impact on society began to be felt. Overseas trade prospered, travel on land and sea became safer and a range of new jobs opened up in the industrialised world. For those nations having widely scattered territories abroad, particularly the British Empire, reliable wireless communication was seen as a lifeline and a valuable organising influence in effective management and communication. High-power stations were established in Europe and America, capable of transcontinental communication, to augment the extensive submarine cable networks, which had reached a highly developed state. The simple spark transmitter was rapidly being phased out in favour of the quenched or timed spark and the arc, both of which provided a near approach to the ideal continuous wave transmission, already achieved by the high-frequency alternators installed in some of the high power international transmitting stations.

But the arc and the alternator had reached the peak of their development. Bandwidth at the long wavelengths, deemed necessary for long-distance communication by wireless waves, was at a premium, and a new approach was needed if more transmission channels were to be used without mutual interference, a factor which was beginning to limit the possibilities of this new tool for the community. New avenues were opened up by a number of new technologies which began to emerge after the First World War. Chief among them was the invention and further development of the thermionic valve. This had made its appearance in the first decade of the new century, and although initially confined to receivers, was destined to have a profound effect on all aspects of the new communication methods. In particular it initiated the migration of effective transmission channels to higher frequencies to satisfy the growing demand for more and more channels for business, travel and the military, and eventually for entertainment – although consideration of this last is outside the scope of this book. Before considering the applications of valve developments in any detail, it is necessary to look at how this new technology came about through the fortuitous conjunction of physics and engineering which occurred at the beginning of the twentieth century, initially in the laboratories of Europe and the United States.

7.1 The advent of thermionics

From the time of Hertz's transmissions in 1886–88 to the beginning of the twentieth century, a major problem in the successful reception of wireless waves lay in the choice of a detection device. Resonators, coherers, electrolytic and magnetic detectors, thermal detectors and the more effective crystal rectifier were all used in numerous manifestations and degrees of efficiency, but most fell short in terms of continuous reliability and convenience [1]. Most successful were devices that rectified the alternating signal to produce a direct current, proportional to the average magnitude of the signal, which could accurately convey small variations in the value of the signal as received by headphones, rather than operating directly an indicating device such as a relay or sounder. This would be an essential requirement in the later development of broadcast radio, in which the simple indicators, such as the coherer, had no part to play. The crystal rectifier, discovered by Karl Ferdinand Braun in the 1870s, reached a high degree of popularity in the years between 1906 and 1914 with the use of crystals of galena, copper pyrites or silicon carbide as the rectifying element. These devices offered a much higher resistance to the passage of a current flowing through them in one direction than the other, thus acting as a rectifier of alternating current, but unfortunately varied in sensitivity from sample to sample, and were of uncertain lifetime. They were difficult to adjust and liable to be put out of action by heavy atmospherics, as many owners of the early domestic 'crystal sets' were to discover, and the search for a better and more reliable rectifying detector continued.

7.1.1 Detection

The emission of a stream of electrons from an incandescent filament or cathode contained in an evacuated glass tube had been known from the work of Johan Wilhelm Hittorf and Thomas Edison in 1883. Edison observed this effect in connection with his early light bulbs, and even named it in a patent application as the 'Edison effect', but made no direct use of the discovery [2]. In 1904, during a search for a better detector of electromagnetic waves, John Ambrose Fleming took note of Edison's discovery and, by inserting a metal plate in an evacuated glass tube to collect the electrons emitted from an incandescent filament, was able to show that the quantity of electric charge collected at the plate (which he termed the anode) depended directly on the electrical polarity of the plate with respect to the filament (the cathode) [3]. He later improved the efficiency of the device by placing a cylinder around the cathode in place of a plate, and in this form the diode was applied to wireless telegraphy by the Marconi Company (Figure 7.1a). Fleming had for some time been a consultant to the Edison Electric Light Company, and in 1900 became advisor to the Marconi Wireless Telegraph Company, whilst still retaining his post as the first Professor of Electrical Engineering at University College, London, a position he held from 1885 to 1926. He made many contributions to the development of wireless communication, but his work and patents on the thermionic diode made the greatest impact, and were particularly valuable to the Marconi Company in its commercial exploitation of the growing market for wireless equipment.

Fleming's diode was simple to apply. When the anode was made positive with respect to the cathode, a current flowed; when made negative, the current ceased. He termed his device a 'thermionic valve', controlling the flow of electrons in much the same way as a hydraulic valve controls the flow of water through a pipe. It acted as a detector of electromagnetic oscillations by responding only to the positive excursions of the applied voltage, providing a small but consistent direct current to activate a sensitive device such as a pair of headphones. Following its successful

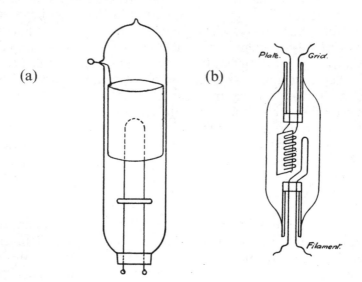

(a)   (b)

Figure 7.1   Early thermionic valves: (a) Fleming's diode valve, (b) De Forest's triode valve
Source: 'The Thermionic Valve', 1924

manufacture it proved reliable and, unlike many earlier devices, needed no troublesome adjustment.

The diode detector, when used in conjunction with a telephone headset for CW telegraphy, had the disadvantage of producing little more than a 'click' heard at the beginning and end of each signal period, resulting from the steady current flowing after rectification of a large number of radio frequency oscillations making up the dot or dash sent by a CW transmitter system. The first solution to this problem was the Pederson Tikker, patented in 1906, which could be used to control a CW transmitter (it was extensively used in the Poulsen arc transmitter, discussed earlier) or within a receiver itself. As applied to a receiver it consisted of a small reed vibrating electromagnetically between two contacts at an audio frequency rate. This interrupted the incoming trains of rectified signals, causing them to repeatedly charge a condenser which was then discharged through the telephones as the reed made contact with the telephone circuit. A tone at the reed frequency could then be heard in the telephones during the dots and dashes of the detected signal, in place of the uninformative 'clicks' heard previously, but had no effect in the quiescent periods between signals. A second solution was a heterodyne method which required a generator of continuous alternating currents at a high frequency to form part of the receiver. This is discussed after we have considered the next major thermionic device to appear, the triode valve.

7.1.2 Amplification and oscillation

Fleming's diode in its role as a detector proved to be the first in a series of what we would now call 'electronic current control' devices which ushered in the thermionic age [4]. A major step forward was made by the introduction of a mesh or grid of wire, between the cathode and anode of a Fleming diode (Figure 7.1b). This provided a simple but effective method of electrical control for the current passing from cathode to anode, and was first applied by Lee De Forest and patented by him in 1907 (US Patent No. 879 532). Initially his new thermionic device proved to be too unstable as a detector for general use, and it was not until after he commenced his association with Federal Telegraph Company in California in 1912 that he discovered that it could also be used as an amplifier and oscillator [5]. His modification to Fleming's invention would have far-reaching consequences for the future development of telegraph receivers and transmitters. He termed his new device an 'audion', but the term never became popular and it became known as a 'triode valve' by Fleming and others in Europe. This new control electrode, termed 'the grid', interposed between the cathode and anode, while allowing electrons to pass through to the anode, gave some measure of control over their passage by means of a potential established between the grid and the cathode. It enabled weak signals applied to the control or grid electrode to be amplified, appearing in enhanced but inverted form at the anode electrode, and thus formed the basis for a highly sensitive receiver of distant transmitted signals. However, De Forest's audion valves were somewhat erratic in their behaviour, because he mistakenly believed that a small quantity of gas within the valve's envelope was essential for a copious electron flow from the cathode.[1] This led to a reluctance to employ the device in apparatus where high reliability was required (e.g. in military equipment) until an improved 'hard valve' version became available following an improvement in evacuation techniques.

De Forest did not fully publish his results on the use of the valve as an amplifier until 1913 [6], and in the meantime Alexander Meissner, working in Germany, utiltized the Lieben–Reisz relay in 1911, the first actual use of this new control method. The Lieben–Riesz relay consists of an evacuated triode valve containing a small amount of mercury within the glass envelope which, when vaporized, provides a plentiful supply of electrons for increasing the conductivity between the electrodes. It was what was later called a 'soft valve', having an imperfect vacuum. Applying a small positive potential to the grid made it possible to initiate an enhanced electron stream, so effectively magnifying the small potential applied at the control grid, which appeared in augmented form in the anode circuit. The device behaved as a very sensitive relay for minute incoming signals, such as a rectified telegraph dot or dash signal, but unlike later applications of the triode its output did not vary with the amplitude of the incoming signal – it only indicated the presence or absence of a signal, and so could not behave as a linear amplifier.

The use of a triode valve as a generator of oscillations was not appreciated until 1912, again by Meissner, then employed by the Telefunken Company, who arranged an inductive coupling between the grid and anode electrodes and so achieved CW oscillation at a frequency determined by the anode–grid coupling circuit. At first the use of the oscillator was confined to low-power applications, again using a 'soft valve', resulting from the inadequate evacuation of the glass bulb, which left residual gases affecting its operation. By 1916 improvements in evacuation techniques at the laboratories of Marius Latour of France, Round and Franklin of the Marconi Company, and Irving Langmuir of the General Electric Company, together with the advent of the tungsten filament, resulted in the production of a much more reliable 'hard valve', which was first used in 1916 by all the armed forces during the First World War for receivers and low-power transmitters (see Chapter 8). Particularly notable was the valve developed by Colonel Ferrié in

---

[1] A few years later, in 1904, Arthur Wehnelt was to show that electron flow would occur in a vacuum, and that an oxide coating on the cathode surface can provide an abundant source of electrons.

France in 1918, which was widely adopted in Europe and the United States because of its superior performance, and known simply as the 'French valve'. It was quickly copied for both the British and German armies in which it was known as the 'R' valve and the 'RE16' valve respectively.

An early use for this new generator of oscillations was as an aid to the reception of telegraphic signals, using the second method of detection referred to earlier: the 'heterodyne method' patented by Fessenden in 1901 [7]. In this technique oscillations from the incoming signal, and those produced by a local generator having a frequency differing slightly from the incoming wireless signal, were mixed together, resulting in a difference or 'beat' frequency in the audio band which may be heard in the telephones. Fessenden used a low-power arc CW oscillator for his local generator, but after its introduction the triode oscillator was used to produce a much more flexible and adjustable beat-frequency oscillator, and the method became wide spread [8]. An important customer for Fessenden's heterodyne method of wireless telegraph reception, using a valve as an oscillator and designated as the 'ultra-audion' by its inventor, was the US Navy which had installed the system as an adjunct to its operational receiver in all its ships by 1917.

7.1.3 Transmission

Despite the improvement in evacuation techniques for thermionic valves, which was adequate for valves used in receiver amplification and low-power oscillator applications, the degree of evacuation achieved was insufficient for the larger and more powerful valves needed for long-range transmitters. With larger electrode systems output powers of 40 to 50 W were achieved, but little more until evacuation procedures could be further improved. Attempts were made to design transmitters using very many low-power valves in parallel. A successful transmitter of this type was designed and built under the supervision of H.J. Round of the Marconi Company for their station at Caernarvon, which already housed a timed spark transmitter, an arc transmitter and a powerful high-frequency alternator for overseas communication (see Chapter 6). The valve transmitter at Caernarvon contained 54 valves connected in parallel, and achieved a transmitting power in the aerial of 100 kW, adequate for communication with Australia, which was achieved in November 1921 [9] (Figure 7.2). As larger valves were developed, the efficiency of the transmitting valve amplifier was such that some portion of the transmitting power appeared as resistive power at the anode electrode, raising its operational temperature, and it proved difficult to remove the heat quickly enough from the valve structure. A limit in valve design of conventional type appeared to have been reached in 1922, when the Marconi-Osram Valve Company produced a glass bulb transmitting

Figure 7.2   The Caernarvon valve transmitter
Source: With acknowledgement to Marconi Ltd

valve of 500 kW rating, several of which were used in Admiralty shore stations.

To achieve higher ratings for long-distance transmitters, two techniques were tried. The first was to replace the valve envelope with silica to enable the valve to run at a higher temperature. With the use of silica an output power in excess of 1 kW became possible, albeit with a device both fragile and expensive to produce. A number of experimental valves of this type were produced for naval purposes by HM Signal School in the period between 1919 and 1921. They were capable of dissipating power at the anode of 6.3, 7.1 and 21.0 kW [10], a remarkable achievement for the time, and were used in several medium-power naval transmitters. A more flexible high-power valve design became possible in 1922 with the introduction of a new form of a copper-to-glass seal, the Housekeeper seal, whereby a special sectional grading of copper was obtained at the joint, decreasing to a very fine edge inside the glass, so that the thermal expansion of the copper matched that of the surrounding glass, despite their differing coefficients of expansion [11]. This enabled the concept of an 'external anode' to be applied, whereby a cylindrical anode could be secured to the lower glass body of the valve, and so permit its anode to become part of the outer valve envelope, and thus directly accessible for external water or oil cooling. These 'external anode' valves allowed very high-power transmitting valves to be designed, a development that was to mark the end of the high-frequency alternator as a generator of continuous waves for telegraphy. The invention also led to the production of a large 'demountable' valve which could be continuously evacuated and its constituent parts made accessible for maintenance and part replacement during its life at the transmitting station.

A major project to use these new high-power transmitter valves was the construction by the Post Office in 1925 of the long-wave telegraph station at Rugby operating at a working frequency of only 16 kHz, whose call sign, GBR, was to become familiar to post office telegraphists at receiving stations all over the globe. Initially the low frequency of the emitted signal, 16 kHz, was controlled by a valve-maintained tuning fork, vibrating at a frequency $\frac{1}{9}$ of the signal frequency; later, quartz crystal control was used [12]. The output of this oscillator stage was fed to a series of five power amplification units, each consisting of eighteen 10 kW water-cooled valves connected in parallel to give a maximum aerial power of 500 kW, with two units held in reserve. Three sets of motor–generator assemblies were installed, each consisting of a three-phase motor driving two d.c. generators. Each generator was capable of developing 250 kW of power at 3 kV and all three could be connected in series to give a maximum of 1500 kW at 18 kV. The aerial system was a complex structure, consisting of twelve steel lattice masts 240 m high, spaced at 400 m intervals and arranged as a figure of eight supporting cage aerials 3.6 m in diameter. The earth system was formed from almost 200 km of copper wire stretching over the entire site, buried in the ground by a wire-laying plough designed originally for use in the Crimean campaign (see Chapter 4). The transmitter was opened for traffic on 1st January 1926, and with its worldwide coverage was an unqualified success. On the outbreak of war in 1939, GBR was used as a telegraph station for the armed services and became of vital importance to the Royal Navy.

7.2 Linking the Empire

The years between 1905 and the outbreak of the Second World War saw the installation of a number of high-power transmitters capable of worldwide telegraphic communications. For those governments whose territories stretched over thousands of kilometres, such as the United States and Holland, the notion of setting up networks of powerful wireless telegraphy systems to link their far-flung colonies and bases and bind them together for organisational, defence and other purposes, was very attractive. Foremost in this development was Germany, already going ahead with a number of transmitters in Africa and elsewhere, linked to its central station at Nauen; France had plans for communication with its colonies in Africa and elsewhere and Holland was considering links with its East India possessions. In America the US Navy commissioned a high-power station at Arlington, Virginia, the first in a network of powerful transmitters by which the Navy would be able to maintain communications with its remote bases and ships in any part of the world.

Britain had plans, proposed by the Marconi Company, for an Imperial Wireless Installation, which had been under discussion by the Treasury, the Committee of Imperial Defence, the Post Office and several manufacturers for a number of years. These meetings culminated in 1913 in a specification and agreement between the Marconi Wireless Telegraphy Company and the Postmaster General, Herbert L. Samuel, on behalf of the government, subsequently ratified by the House of Commons [13]. The detailed specification proposed six major installations to be located in England, Egypt, East Africa, South Africa, India and Singapore, with the last named station expected to communicate directly with Australia as well as India. The proposal was for 'an installation for wireless telegraphy capable of communications over a range of 2000 miles intended primarily for commercial use'.

All stations were to be equipped with high-speed as well as key-speed equipment and capable of duplex working. To facilitate this the transmitter and receiver stations at a given site were to be separated by at least 30 km. The transmitters specified were Marconi's timed spark systems, and the sparking would be synchronised with the rotation of the alternator shaft in order to approximate the generation of continuous waves. This followed the design of the Caernarvon transmitter, already in operation by the Marconi Company and providing a transatlantic service. There was controversy over whether the Poulsen arc or Goldschmidt arc system should be used in place of the Marconi timed spark system to provide a more accurate CW signal. William Duddell FRS, then President of the Institution of Electrical Engineers commented on this in a letter to the Treasury dated 16th July 1913 in which he asked that these systems be looked into, but the specification was not changed. Keying of these high-power transmitters represented a significant design problem. In the proposed Imperial scheme the keying arrangements were specified to include no fewer than three relays in parallel operated by high-tension key switches to divert the output away from the aerial to a resistive load. This latter was fitted with an air blower capable of dissipating the heat arising from the full signalling load, expected to be in the region of several hundred kilowatts. The design of aerials was also specified to be not less than $\frac{1}{12}$ of the wavelength being applied. It was expected that the transmitter wavelengths would be long, in accordance with the currently held view that only with very long wavelengths and high aerials could worldwide communications be secured.

Four meetings of a Sub-Committee of Imperial Defence considered these proposals, and various conflicting views on their acceptance were put forward by the numerous government departments represented at the meetings. The question of defence for any of the overseas stations was discussed at length, and at one point the Treasury expressed its view that the Marconi Company should be required as a condition of its licence to defray the costs of necessary defence arrangements [14]. While no definite conclusions were reached at these meetings, an agreement between the government and Marconi was nevertheless offered and ratified in 1913, but little progress was made in commencing the building of these stations. The difficulties were threefold: political difficulties bordering on corruption in the management of Marconi and in the public offices, a complex situation discussed in detail in Frances Donaldson's book *The Marconi Scandal* [15]; the Admiralty's views on the technical requirements for the proposed transmitting stations; and the reluctance of the Treasury to release funds for the enterprise. The Admiralty was particularly lukewarm towards the project. Its view was that 'Post Office stations are possible sources of leakage of information and communication with the enemy and should be used only if the [submarine] cable breaks down, but if any interference is experienced from them they should be closed altogether. They should be censored when in use . . .' [16], and in a secret document added that 'They cannot of course at this stage commit themselves definitely to the expenditure from the Navy Vote which the erection of such stations would involve.' [17] By August 1914 only two stations, at Leafield in England and one in Egypt, had been begun, and these were to remain unfinished. The materials for the stations, some of which were already in production at the factories, became diverted to a number of other Marconi stations, now required by the Admiralty for wartime control of the fleet [18]. At the end of 1914 the contract with the Marconi Company was cancelled by the Post Office, with no adequate reason given, and resulted in legal proceedings for breach of contract which dragged out over the years and were settled, somewhat inadequately, only in 1919 after the war ended.

The need for efficient and reliable wireless communication with the British Empire remained, and Marconi again put forward new proposals in 1920 to establish direct wireless communications between Britain, South Africa, India and Australia, using powerful transmitters and long wavelengths, with several low-power stations feeding these arteries from local areas. The government's response was to establish a new select committee under the chairmanship of Sir Henry Norman containing members from the dominions and several government departments, but not, however, from the Marconi Company. The Norman Report, announced by Bonar Law in the House of Commons in March 1923, advised against any private company being given a monopoly (indicating perhaps that earlier political events were still having repercussions), and proposed instead that a chain of stations be constructed at intervals of about 3000 km along the route, set up and managed by the Post Office. The spacing of the proposed chain of stations arose from the findings of an Imperial Wireless Telegraphy Committee, appointed in November by the Secretary of State for the Colonies, which formed the opinion that 'no satisfactory commercial wireless service is in operation working over a distance of 2000 miles'. The first two stations to be constructed in response to the Norman Report were to be at Leafield and Cairo, both to be equipped with Poulsen arc transmitters. Objections were raised by the Australian and South African Governments, who were highly critical of the Norman Report, and following an Imperial Conference held in 1923 at which the Empire countries were fully represented, the Norman Report was laid to rest by Winston Churchill, then Colonial Secretary. The entire project was placed in abeyance while further discussions between the Post Office and the Marconi Company took place to consider new ideas for what was now a very much debated topic.

In the meantime the Post Office had received confirmation from the government that 'the Post Office should own and directly operate all stations for communication with the various parts of the British Empire with the exception of Canada, in which connection private enterprise might operate in a competitive basis with the State', and began the construction of its high-power, very long-wavelength station, call sign GBR, at Rugby, described earlier in this chapter. The Marconi Company, on the other hand, had its sights firmly fixed on a private enterprise solution and continued its negotiations with two firms from the Dominions of South Africa and Australia, namely the Wireless Company of South Africa and the Amalgamated Wireless (Australasia) Ltd. The latter had already placed a contract with Marconi for a 1000 kW station to be erected with an elaborate aerial system supported on twenty 240 m steel masts, and preliminary work had also commenced at Klipheuval, near Cape Town, for the South African firm. It was at this moment of disarray in the Imperial Wireless Scheme that a new technical discovery was being made by Marconi which was to have a profound effect on worldwide communication by wireless.

During the war years Guglielmo Marconi had been requested by the Italian Navy to design a system of secure short-range wireless communication for use between units of its fleet. For this proposal he was assisted by C.S. Franklin, who had been with the company since 1899 and was to become a major figure in the communication ventures of the 1920s. Charles Samuel Franklin was born in London in 1879 and received his engineering and scientific education at the Finsbury Technical College under Professor Silvanus Phillips Thompson. After joining the Marconi Company, where he stayed for all of his working life, he spent several years installing wireless equipment for the company in various parts of the world including a spell in Africa, in charge of the equipment used in the Boer War (see Chapter 8). He had already made important contributions in the Marconi research laboratory, resulting in many valuable patents, before he embarked on his major work on long-distance communication in the 1920s, described below [19]. The work for the Italian Navy, which provided the catalyst for the Empire system, involved the use of the previously neglected shorter wavelengths which, because of their expected line-of-sight capability, could meet the secure transmission requirements of the Italian Navy. In asking Franklin to explore the use of short wavelengths, Marconi was conscious of his company's current slavish adherence to longer wavelengths and high powers, then thought essential to achieve reliable long distance communication. 'I could not help feeling', he said, 'that we had perhaps got rather into a rut of confining practically all our researches and tests to waves of some thousands of feet in length' [20]. His concerns proved justified.

The system developed by Marconi and Franklin, with a transmission wavelength of only two metres, used a spark transmitter (no valves were yet available to work at very high frequencies), and the completed system proved highly successful for its purpose. Surprisingly, however, the experimenters found that in some circumstances they achieved much longer transmission distances than the short line-of-sight path lengths. This is now known to be caused by reflection from the ionosphere, but in 1920 little was known of this phenomenon. Marconi and Franklin initiated an impressive series of experiments to investigate this unexpected result. Franklin, based at Caernarvon, established a transmitter system using a wavelength of 15 m between Inchkieth and Portsmouth, radiating an aerial power of only 700 W, and achieved reliable communication across the 150 km separating the two locations. He subsequently operated the equipment between Birmingham and London, and obtained similar results. A more elaborate test was carried out at a site at Poldhu at a wavelength of 97 m, which enabled high-power valves to be used (eight of them in parallel). With an aerial power of only 17 kW, clear reception could be obtained in New York and, with judicious selection of transmitting wavelengths, in Sydney and in Canada. In this way the company gained considerable experience in choosing the correct frequency and time of day to achieve reliable service at these very short wavelengths.

The problem now facing the Marconi Company was how this new information would affect its proposals for an Imperial Wireless Scheme, already in train for South Africa and Australia. Their experiments appeared to show that reliable communication could be achieved at these shorter wavelengths at a much reduced cost for both installation and operating of the stations.[2] The choice lay between continuing to propose and build the long-wave stations for which orders had already been received, while continuing to experiment privately with the new short-wave transmitters, or to take the radical step of offering to design and install a series of short-wave stations, as yet untested in a commercial environment. Marconi, not one to avoid a risk in business ventures, made the decision to inform the authorities in Canada, Australia and South Africa of the possibilities of the new short-wave system. He suggested the abandonment of his earlier proposals to construct long-wave stations on their territories, and instead to design and build new short-wave equipment using highly directional beam aerials, in the process saving them much capital expenditure. This was a bold move for the Marconi Company, which had not had the time to carry out full tests of the capabilities of a beam system and was also fully aware of the short time scale that would be imposed in any joint agreement reached with the British Government and the dominions.

Government support was not long in coming. In a report produced by the Imperial Wireless Telegraph Committee in February 1924 through its chairman Sir Robert Donald, a contract was proposed with the Marconi Company for the establishment of duplex beam circuits between Britain and the dominions. With a number of provisions compelling the company to absorb any financial loss if the performance of the equipment did not match its technical specifications, together with restrictions on the time allowed for experiment and installation (the first station had to be completed within six months of the site becoming available), the British Government, in conjunction with the dominions, finally announced its agreement to the establishment of a World-Wide Imperial Wireless Scheme in July 1924. The proposal was for a group of powerful wireless telegraph stations in England to be established for direct communication with Canada, South Africa, India and Australia, with corresponding stations in these countries constructed for two-way transmission, each station to provide a radiated power of 20 kW, but confined within a directional beam of ±15° from the direction of its communicating station. The effective speed of both sending and receiving simultaneously was to be 100 wpm, and the equipment had to be capable of maintaining a daily average of 18, 11, 12 and 7 hours of transmission respectively with the four named sites. The terms were accepted by the Marconi Company, which established a brilliant design team including Charles Franklin, who had carried out the preliminary design work on the short-wave system, and R.N. Vyvyan, then Marconi's chief engineer, who would be responsible for constructing the stations within the very tight schedule.

---

[2] Figures of $\frac{1}{40}$ power and $\frac{1}{30}$ cost were quoted in Marconi's celebratory book *50 Years of Wireless*, published in 1947.

To achieve the required daily hours of transmission, the following alternative wavelengths were chosen:

To Canada	16.574 and 32.397 m
To South Africa	16.146 and 34.013 m
To India	16.216 and 34.168 m
To Australia	25.906 m
From Canada	16.501 and 32.128 m
From South Africa	16.077 and 33.708 m
From India	16.286 and 34.483 m
From Australia	25.728 m

A complex construction of beam aerials was designed to concentrate the transmitted signal in the required direction, and elevated so that reflection from the ionosphere could be achieved. In England the stations were grouped in two pairs: a site at Bodmin for a two-beam transmitter for Canada and South Africa, with the corresponding receiving stations at Bridgewater, and Grimsby and Sheerness for a similar service to India and Australia. The transmitters were identical and consisted of a carefully screened and stabilised valve master oscillator driving a series of three amplifier stages, each of increasing power capability, culminating in a final stage containing a pair of Marconi CAT2 external anode cooled valves, each rated to take an input of 10 kW at 15 000 anode-volts. Wavelength stability, considered vital, was obtained in the three stages of amplification prior to transmission by using a bridge method of connection whereby the internal capacity between the grids and anodes of the valves was balanced with a second valve acting in push–pull, except for the first stage, where a balancing condenser was used. The keying arrangements were designed to maintain a constant load on the power supply, for which purpose resistive absorbing circuits were provided, controlled by two oil-cooled metal-to-glass sealed valves, type CAM2. Landline signals from the remote keying stations were passed through a relay which provides spacing and marking currents to make the grids of the CAM2 valves conductive, or non-conductive thus affecting the direction of the transmitter power – either to the aerial or to the resistive load [21]. The beam aerials employed at all the stations were supported from a series of steel masts 86 m high, and consisted of a series of vertical wires fed from a simple multi-dimensional feeder with twice that number of parasitic reflector wires spaced about three-quarters of a wavelength behind the driven aerials. The aerials and feeders were balanced through the remarkably simple repeated division technique shown in Figure 7.3, which ensured that the system acted as a pure resistive load on the transmitter, with no reflecting waves along the feeder system. Directional receiving aerials considerably reduced atmospheric interference, as commented on by

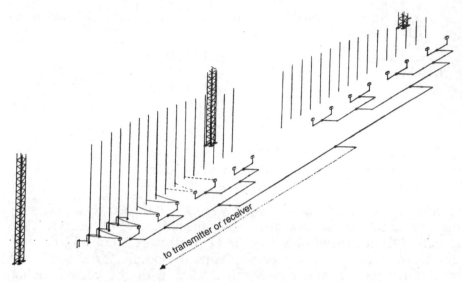

Figure 7.3   Empire beam transmitting aerial
Source: *Post Office Electrical Engineering Journal* **21**, *1928*

Marconi in his address to the American Institution of Radio Engineers in 1928 [22].

In October 1926, the first beam connection with Canada was completed and the other stations in the system followed in rapid succession. The introduction of a short-wave beam system in Britain for wireless transmission over long distances and on a commercial basis proved an outstanding success and for the Atlantic section it quickly provided greater traffic handling capacity than any other long-distance wireless telegraph circuit then in existence. While the profit gained by the Marconi Company was minimal (the actual cost to the company exceeded the amounts it had received for manufacture and construction) it had the satisfaction that wireless organisations everywhere, however reluctant, were now compelled to enter into this new field of development if they wished to remain competitive in the business of commercial wireless communication.

7.2.1 Cable and wireless

With the Empire wireless system in place and providing a reliable link to Britain's overseas dominions, the cost of commercial telegrams fell substantially. This had an immediate effect on the operation of the existing submarine cable network, which had by then developed into a world-wide telegraph system. For the first time the cable companies began to feel the effects of competition. Not only was it cheaper to transmit telegrams over long distances by the use of wireless, but the speed of transmission was greater (100–250 wpm for the beam system, compared with 35–100 wpm over a cable route). Cable communication had the advantage of secrecy, and on certain routes it was possible to collect or drop off messages at intermediate points en route, while the beam system was more economic over long distances. The two systems were in fact complementary, as was being realised in government circles in 1927, despite the migration of nearly half the cable traffic to wireless transmission. Government action was felt necessary to forestall the financial collapse of the cable companies, and at a hastily convened Imperial Wireless and Cable Conference in January 1928, all the cable and wireless interests were invited to 'examine the situation which has arisen as a result of the competition of the beam wireless with the cable services' [23].

The main outcome of the conference was the formation of a new company to take over both the cable and wireless companies. Known as the Imperial and International Communications Ltd, it became incorporated in September 1929. The major companies amalgamated within the new operating authority were: Marconi's Wireless Telegraphy Company, the Pacific Cable Board, the Eastern Telegraph Company, Imperial Atlantic Cables, the Post Office Beam Service, and the British East Africa Broadcast Company, together with 20 smaller companies. At the same time an Imperial Communications Advisory Committee was set up to represent the dominions' interests (it was renamed the Commonwealth Communications Council in 1942). The value of the wireless beam system under this new arrangement was tested in November 1929, when a massive seismic upheaval under the floor of the Atlantic severed 10 of the 21 working submarine cables. The cable traffic was rapidly switched to the beam system without incurring any delay in the transmission of telegrams [24].

One of the anomalies of the merged service lay in the different coding arrangements for the Morse code transmitted by submarine cable and by wireless. Cable code, with its three-conditional form of the Morse code whereby the dot, dash and space are of equal length, is incompatible with the two-conditional code used in wireless transmission. To facilitate automatic interworking between cable and wireless a new two-conditional form of cable code, known as Double Current Cable Code (DCCC), was introduced in about 1930. This was used in conjunction with a direct cable code printer which reproduced the DCCC signals and printed them directly in Roman characters, with an input from either cable or wireless transmission [25].

The Imperial and International Communications Limited was renamed Cable & Wireless Ltd in 1934, and by the outbreak of the Second World War it possessed the most extensive system of world communications ever to exist under the control of a single body. It owned 250 000 km of cable, a cable fleet of 19 ships and a total of 91 wireless stations, all of which were to play a vital part in communications for the British and Allied forces in the war years [26].

7.2.2   Reuters and the news service

Alongside the Cable & Wireless system, Reuters Ltd and other news vendors began in 1923 to broadcast news in Morse code to European countries, renting wireless transmission facilities from the Post Office for the purpose [27]. At first the main transmitter used was the massive wireless telegraphy transmitter, GBR, broadcasting from Rugby on a frequency of 16kHz. Later the service was extended to other geographical areas on HF using two transmitters at the Post Office station at Leafield. After 1933 most of the transmissions made use of a rapid telegraph printing system developed by Siemens & Halske and known as the Hell–Schreiber system. This was essentially a facsimile system in which the letter characters were built up from a series of marks displayed over a rectangular area comprised of seven or twelve vertical lines. This was a departure from normal Morse transmission, although the actual signals sent out were suitable for transmission by a keyed CW transmitter. The signal speed obtained with the Hell–Schreiber system may be defined as several hundred baud, equivalent to a Morse transmission rate of about 100wpm. The Reuters European news service was extended during the war years to cover most of the globe, still using the HF transmitters at Leafield, but with a beamed aerial system requiring a daily schedule of frequency changes to suit the local times and locations of the recipient news distributors.

7.3 Maritime communication

The first two decades of the twentieth century saw the widespread acceptance of wireless installations, carried on board all passenger ships and most large merchant ships in fleets around the world. This was materially assisted by the performance of the new continuous wave or quasi-continuous wave stations, many of them beginning to employ valve receivers and transmitters towards the end of this period, and also to the siting of shore stations around the coasts of all the maritime nations to relay messages between ships and inland locations by wireless, cable or line communications. The Marconi Company established an early lead in the provision of maritime communications with the formation of Marconi Marine Company as early as 1900. It was initially extremely successful in fitting out the coastal stations installed by Lloyd's and the major shipping companies. While nearly all the Empire coast stations installed Marconi equipment, this was not always matched with Marconi sets fitted to communicating ships. The coast stations were staffed by

Marconi-trained operators, and the company's profits were drawn from a charge on the messages conveyed rather than from rental charges for the equipment. The company sought a monopoly in the way in which its equipment was to be used, forbidding its operators to communicate with any other station on board ship or ashore that was not fitted with Marconi equipment. This did not help its commercial activities and both the Telefunken Company in Germany and the De Forest Company in the United States made serious inroads into Marconi's business of establishing new coastal stations in many parts of the world. Further, the ratification of the 1906 International Telegraph Convention forbidding the 'closed shop' activity of the Marconi Company, and the take-over in 1909 of all British coastal telegraph stations by the Post Office, were together responsible for drastically altering the commercial opportunities available for the Marconi Company.

To remedy this situation the company shifted to the supply of equipment rather than services, a process in which Marconi did surprisingly well. Indeed, according to Sturmey in his *The Economic Development of Radio*, the Marconi Company was overstretched at the time and was 'well rid both of the coast stations and the troublesome agreements with the Post Office and with Lloyd's' [28]. Although it opened the way for direct competition with Telefunken and the American companies this shift had no immediate impact on the market for coastal stations in Britain, and Marconi installations continued to dominate until after the First World War. While Marconi was a major supplier for wireless telegraph equipment for the merchant fleet, it was not the only one in Britain. A number of companies were set up in the years before the First World War to provide suitable seagoing apparatus, now being fitted primarily as an aid to commerce and navigation, but also to conform with government regulations now being applied to the merchant fleet. One of the more successful of these was the Helsby Wireless Telegraph Company of London, which produced its Cymophone portable telegraph apparatus 'for ships, yachts and trawlers'. Their spark transmitters used a rotary gap system which was claimed to be 'practically silent in action . . . avoiding the risk of messages being overheard by unauthorised persons' [29]. This says something about the conditions under which the seagoing wireless operator usually worked, often confined in a small cabin containing an extremely noisy spark gap – a situation which was somewhat ameliorated by the development of rotary or quenched spark gaps and the much quieter arc transmitter.

In 1919 a new W/T Merchant Shipping Act made it compulsory for all ships registered in Britain to have on board one or more certified operators. This considerably increased the number of merchant ships fitted with wireless equipment. According to the 1921 edition of *Yearbook of Wireless Telegraphy and Telephony*, there were 725 coast stations and 5821 ship-borne installations then in use, almost all of which were used for public communications. Of these, the coast stations of the British Empire and its dominions amounted to 273 (excluding those established by the Admiralty for naval purposes). The traffic conveyed by these coastal stations was immense. Apart from actual messages – telegrams – ships continually passed their positions, course, speed, and so on to the nearest coastal station, and this information was then passed on to Lloyd's. In the first half of 1921 about a million words were passed and charged for at the full rate, 11d (4½p) per word, with a similar quantity for the press and certain other messages charged at a reduced rate [30]. The immensely important role played by the coastal stations in two world wars is described in Chapter 9. At these times the coastal stations were reinforced with additional equipment and placed under the control of the Admiralty, although they continued to communicate with merchant shipping. In the inter-years the Post Office regained control of the stations, and benefited to the extent of improved capability of their ship-to-shore service as a result of these wartime equipment enhancements.

To provide a wireless telegraphy service over its immense coastline, the United States had 101 non-military shore stations in 1920, including 38 lightships fitted with wireless equipment. The provision of wireless equipment for these shore stations had, in the first decade of the century, been shared between the American Marconi Company and the United Wireless Telegraph Company, but on the bankruptcy of United Wireless in 1912 its assets were acquired by the Marconi Company. This allowed Marconi to gain control of about 500 United States ship installations and 70 shore stations that had belonged to its competitor, giving it 'all the coast stations of importance on the Atlantic and Pacific coasts, besides practically the whole of the American Mercantile Marine at present fitted with wireless installations' [31]. These were all telegraph stations, but in 1921 experiments began to be conducted in radio telephony between Foxhurst coast station and a coastal steamer situated 480 km out in the Atlantic Ocean. By making use of the landline across to the Pacific coast and further radio telephone transmission 50 km into the Pacific, the broadcast was received at the Catiline Islands from the steamer in the Atlantic. This was the forerunner of a general conversion of many of the world's coastal stations from wireless telegraphy to wireless telephony in the remainder of the century [32].

Marconi's control over the United States marine coastal stations was short-lived, for on the entrance of the United States into the First World War in 1916 all the high-power shipboard and shore stations were taken over by the US Navy and the rental system for Marconi's equipment on board ship was abandoned in favour of outright purchase. After the war the American Marconi Company did not resume manufacturing operations for trade within the United States, and was itself absorbed into the Radio Corporation of America. In 1919 the various patents in wireless technology held by the major United States companies and those of the

American Marconi company 'made it difficult for any one company to manufacture any kind of workable apparatus without using practically all the inventions which were then known' (Major Armstrong testifying before the Federal Trade Commission in 1923 [33]). It was Owen D. Young of the General Electric Company, at the suggestion of Franklin D. Roosevelt, then Assistant Secretary of the US Navy, who carried through a plan which broke this bottleneck. Young persuaded General Electric to purchase a controlling interest in the American Marconi Company and to pass this on to a new 'All-American Radio Company' to be called 'The Radio Corporation of America'. The Corporation was formed in 1919 with the support of Congress and was allocated all current and future patents of AT&T and General Electric, including those of the American Marconi Company, all to be available royalty-free for ten years. Westinghouse in the meantime had acquired the patents of the International Radio & Telegraph Company (a successor to Fessenden's NESCO company), giving it access to Fessenden's many patents, and had also purchased Armstrong's regenerative patents. All these were included in the RCA agreements with the result that the Corporation quickly obtained rights to over two thousand issued patents and faced no domestic competition in the radio communications field, which considerably strengthened its position in the manufacture and home installation of maritime wireless equipment for several decades [34].

In the early 1920s a number of technical changes took place in maritime wireless equipment on board ship and at the shore stations. Valves came into general use in transmitting equipment for merchant ships in 1919. After the International Radiotelegraphic Convention, held in Washington in 1927, abolished the use of the spark transmitter for all installations over 300 W, all new equipment was valve operated, although a period of grace was allowed which meant that some spark transmitters were still in use up to the end of 1939. A second major change was the use of wireless equipment on board ship for navigation. Wireless direction-finding was well developed by the 1920s, and had been pioneered as early as 1912 when a Bellini–Tosi direction finder was first installed on an ocean liner, the *Mauretania*, as an aid to navigation. Parallel with this development was the use of a short-wave automatic beacon, the first of which, designed by Franklin at the Marconi Company, was erected at Inchkeith, an island in the Firth of Forth in 1921 [35]. This operated at the short wavelength of 4 m, using a rotating aerial with a parabolic reflector. As it rotated it sent out a different Morse code letter for every two points of the compass. These could be recognised by a suitable receiver carried on board ship, enabling an exact bearing on the beacon to be determined.

7.3.1 Wireless training in the merchant navy

The inclusion of wireless equipment on board commercial ships of all kinds just after the turn of the century brought with it a demand for trained operators to operate the shipboard installations.[3] In Britain most of the applicants came from previous employment with the Post Office, and so had some experience of line telegraph working. Many attended one of the 40 small private training schools which were established to teach the Morse code and the rudiments of wireless telegraphy (Figure 7.4). Foremost among these was the School for Wireless Operators opened by the Marconi Company at Frinton-on-Sea, Essex, in 1902. Morse code was taught in the first week, and it was claimed that the entire course could be completed within one month. The graduate was then considered qualified not only to operate any shipboard installation but also to erect stations and repair apparatus. This was not as presumptuous as might seem, since only those who had had some technical training were accepted for the course, and they would be aware of the construction and operation of much current commercial wireless equipment. 'Hands-on' work for the course included exchanging messages with a Marconi station at North Foreland, some 60 km away and with La Panne in Belgium. There were also short courses, often run on board ship, to train 'watchers'. This new role was created by the Merchant Shipping Act of 1919, which stipulated that certain ships could carry watchers to relieve the trained operator from the task of keeping watch on the reception of distress and safety signals only, but was not expected to be involved in receiving or transmitting messages. This role disappeared as soon as automatic equipment for registering distress signals came to be widely fitted. A thriving trade sprang up in supplying 'Marconi uniforms' (Figure 7.5), which became obligatory for successful graduates of the wireless operator schools, many of which later expanded their syllabus to include full training for mates and merchant marine officers. In the United States the majority of wireless operator seamen originated from the railroads, where they had already received training as telegraphers. The American Morse code was slightly different from the international code in use (Figure 3.3), but after 1912 this was replaced by what was known as Continental Morse code, which became accepted throughout the maritime world.

The changes brought about by the outbreak of the Second World War meant that very many more wireless operators were needed to man the merchant fleet. Ships that formerly went to sea without wireless were now required to carry suitable equipment, and the Admiralty now expected all ships to operate a 24-hour wireless watch. This presented a

[3] In 1921 the need to fit wireless equipment to vessels of 1600 tonnes gross and upwards demanded by the Wireless Telegraphy Regulations of 1912 applied to 3200 British ships alone, excluding those foreign ships which visited British ports and thus also came within the jurisdiction of the regulation.

THE AGE LIMIT IN THE MARCONI COMPANY IS 25

THE

South Wales Wireless Training College, Ltd.

MARKET BUILDINGS

St. Mary's Street, CARDIFF

Phone : 3008 Cardiff

Both Colleges fitted with the MARCONI COMPANIES' installation
throughout

And at CASTLE STREET, SWANSEA

**EVERY STUDENT OF OURS WAS OFFERED A
BERTH THE SAME WEEK THAT HE QUALIFIED**

The above Colleges are the finest in Gt. Britain, and have sent out
over 500 Wireless Students this past 3 years to good berths in
H.M.S. and the Marconi Company.

We have special SHORT Courses for Captains, Mates. and Expert
Telegraphists. WE DO NOT TEACH BY CORRESPON-
DENCE.—" Have no faith in this." We wish to please our Students
and their parents by turning the Students out QUALIFIED operators,
and have done so.

We have 90 per cent. of successes at ALL Exams.

For Terms apply to :

J. R. SCHOFIELD Managing Director

Figure 7.4   Advertisement for telegraphy training, 1919
Source: 'Yearbook of Wireless Telegraphy and Telephony', 1919

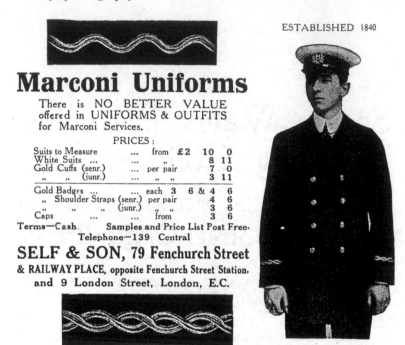

ESTABLISHED 1840

Marconi Uniforms

There is NO BETTER VALUE
offered in UNIFORMS & OUTFITS
for Marconi Services.

PRICES :

Suits to Measure	...	from £2	10	0
White Suits ...	...	„	8	11
Gold Cuffs (senr.)	...	per pair	7	0
„ „ (junr.)	...	„ „	3	11

Gold Badges ...	...	each 3	6 & 4	6
„ Shoulder Straps (senr.)	per pair		4	6
„ „ „ (junr.)	„ „		3	6
Caps	...	...	from	3	6

Terms—Cash.   Samples and Price List Post Free.
Telephone—139 Central

SELF & SON, 79 Fenchurch Street

& RAILWAY PLACE, opposite Fenchurch Street Station,
and 9 London Street, London, E.C.

Figure 7.5   Marconi uniforms available for purchase in 1919
Source: 'Yearbook of Wireless Telegraphy and Telephony', 1919

major training problem, since preparing a radio officer to qualify for the
Postmaster General Certificate in Wireless Telegraphy, as required by
law, took months rather than weeks, and over four thousand additional
trained operators were needed by the merchant fleet during the first year
of the war. The Marconi Company and the many private schools of
telegraphy lengthened their working days and increased their staff in an
effort to reach this goal, and somehow it was achieved, despite the 'competition' for trainees which resulted from the equally urgent requirements
of the armed services (considered in later chapters) [36].

7.3.2 Codes, telegrams and newspapers at sea

A prime task for the telegraphist aboard an ocean liner in the nineteenth
century was the reception and transmission of radiotelegrams for the
passengers. Before the telephone became widely used, the telegram held
an important place in commerce and private life, and, once telegraphic
communication could be maintained by wireless throughout a voyage,
passengers on an ocean liner demanded access to the facilities. The
transmission and reception of radiograms (the generally accepted and
international term for the wireless telegram) was controlled quite closely

by the articles of the International Radio Telegraphic Convention of 1912. As a general principle the telegraphist would transmit the radio-telegram to the nearest coastal station, from which it would be forwarded to its destination through the normal international telegraph network and a charge made accordingly. More distant transmissions could be made entirely by wireless to enable the telegram to reach its destination directly, if the destination was in the country in which the ship was registered. In order to reduce the length – and, of course, the cost – of a telegram for transmission, a number of codes were used. Some were industrial codes, similar to those discussed in Chapter 3, and were specific to the sender of the telegram; others were operator codes, of which the most widely used was the international Q code established by the 1912 International Radio Telegraph Convention and given in Table 7.1. This was in general use for ship-to-ship and ship-to-shore communication, and versions of it were widely adopted for the services, civil aviation and radio amateur use. Two of these were the Z code, originated as a company code by Cable & Wireless and limited mainly to high-speed machine Morse operation, and the X code used by the European military services as a W/T code. This latter consisted of the letter X followed by a number, for example X100, 'Affirmative', or X279, 'What is the strength of your signal?' These were widely used until 1942 when, at the request of the US military, they were replaced by elements of the international Q code, now extended to cover newer military needs.

The ocean newspaper was initiated in November 1899, when Marconi and two of his engineers were on board the American liner *St Paul*, returning home from a successful attempt to report on the Americas Cup races in New York Harbor for Associated Press. When they reached the point at which they could expect to be in touch with the Needles wireless station and its telegraphic news, and with several days sailing yet to come, they conceived of the idea of publishing a four-page newspaper on board to be sold to passengers at a dollar a copy for the benefit of the Seaman's Fund. *The North Atlantic Times* was a great success, and was followed by similar newspapers issued on board a number of the large ocean liners. In 1902 telegraphic communication was established and maintained throughout the entire voyage of the RMS *Lucania*, giving the captain the opportunity to publish a daily bulletin of news for the benefit of his passengers. Thereafter the Cunard Company began publishing its own daily bulletin on all its Atlantic crossings (Figure 7.6). Other lines followed suit, including the French Compagnie Général Trans-Atlantique, the Koninklijke Hollandsche and the Scandinavian-American line with its own *Atlantic Daily News*, and in other waters, the Belgian Congo Line, and the Union Castle Steamship Company plying between Britain and the continent of Africa. Many of the lines relied on messages transmitted by wireless telegraphy from the Marconi station at Poldhu in Cornwall. Their bulletins, at that time prepared at Marconi

Table 7.1 The international code as defined by the International Radiotelegraphic Convention, London, 1912

QRA	What is the name of your station?	QSF	Are the radiotelegrams to be transmitted alternately or in series?
QRB	How far are you from my station?	QSG	Transmission will be in series of 5 telegrams.
QRC	What are your true bearings?	QSH	Transmission will be in series of 10 telegrams.
QRD	Where are you bound?		
QRF	Where are you coming from?	QSJ	What is the charge to collect for?
QRG	To what company or line of navigation do you belong?	QSK	Is the last radiotelegram cancelled?
		QSL	Have you got the receipt?
QRH	What is your wavelength?	QSM	What is your true course?
QRJ	How many words have you to transmit?	QSN	Are you communicating with land?
		QSO	Are you in communication with another station?
QRK	How are you receiving?		
QRL	Are you receiving badly?	QSP	Shall I signal——that you are calling him?
QRM	Are you disturbed?		
QRN	Are the atmospherics very strong?	QSQ	Am I being called by——?
		QST	Have you received a general call?
QRO	Shall I increase my power?	QSU	Please call me when you have finished.
QRP	Shall I decrease my power?		
QRQ	Shall I transmit faster?	QSV	Is public correspondence engaged?
QRS	Shall I transmit more slowly?		
QRT	Shall I stop transmitting?	QSW	Must I increase the frequency of my spark?
QRU	I have nothing to transmit.		
QRV	Are you ready?	QSX	Must I decrease the frequency of my spark?
QRW	Are you busy		
QRY	Are my signals weak?	QSY	Shall I transmit with a wavelength of——metres?
QSA	Are my signals strong?		
QSB	Is my tone bad?	QSZ	Transmit each word twice.
QSC	Is the spacing bad?	QTA	Transmit each radiotelegram twice.
QSD	What is your time?	QTC	Have you anything to transmit?

House, London, were transmitted on long wave and extended to five or six hundred words, being particularly appreciated by passengers for the war news they gave during the first few years of the First World War [37].

7.4 Life-saving at sea

One of the most manifest benefits of marine telegraphy is undoubtedly its value in the saving of life at sea. A special distress signal was first suggested at the International Conference on Marine Telegraphy held in Berlin in 1903, when the letter sequence 'SSSDDD' was proposed but

Figures and text within the newspaper image:

The NORTH ATLANTIC TIMES

...ul
defend
...eption can
the part of the

.N BELFAST.

.LTIES IN STREETS
, SHIPYARDS.

...ersons were killed and eighty including forty seriously, during ghting between the Unionists and .einers in the shipyards and streets of yesterday.—Reuter.

Reuter message at six p.m. this evening .ttes that rioting was renewed in Belfast .oday The Post Office was wrecked. More soldiers and several civilians were wounded as the results of firing.

It was stated in the House of Commons this afternoon that considerable military forces were in the city One of our ablest Generals was on the spot, and he would endeavour to keep order regardless of the political or religious views of any person concerned.

TRAINLESS DUBLIN.

With the exception of the Killarney express, not a single train left Dublin again today because police were aboard. Crowds for the Curragh races were again held up — Reuter.

WIRELESS RATES TO SHORE.

Radio-telegrams may be transmitted from the *Victorian* through its special Marconi long-distance apparatus. The charge to the United Kingdom is 10½d. per word. The Postmaster-General has agreed to a special Press rate of 5½d. per word for Press messages sent, during this voyage only, to newspapers within the United Kingdom All radio-telegrams should be handed to the Purser.

TO-MORROW'S SPORT.

CRICKET
SURREY v. LANCASHIRE.
WARWICKSHIRE v. HAMPSHIRE.
ESSEX v SOMERSETSHIRE.
KENT v NORTHAMPTONSHIRE
YORKSHIRE v GLOUCESTERSHIRE.
ROYAL ARTILLERY v ROYAL ENGINEERS, at Lord's.

RACING
LIVERPOOL MEETING (third day).
HURST PARK Henry VIII Plate T Y.O

.eter, Faulkner,
C. today in their
Engineers.--Reuter

.CING RESULTS.

.Molyneux Plate, 1st Racket, rd Tut tut, ten ran Betting to 4, 4 to 1

.erpool Cup – Probable starters and .ockeys, Targiers (Hulme); Rich Gift, .(Slade), Midshipmite (Whalley); Perion ...sed against O Bullock), African Star (L. B. Balding), .espite the A L Ordre (Blank), Alasnam (Fox); ...us the German Devizes (Garnett), Red Head (Weston); .. landing. Reuter Cumberland (T Morgan)

.ITUARY.
is announced of Mrs. Corn-
. est.—Reuter

LOOKING BACKWARDS

JULY 22—IN OTHER YEARS.
1298—Battle of Falkirk : Wallace defeated.
1562—Birth of Robert Southwell, Jesuit poet who was hanged at Tyburn, February 22nd, 1595
1914—Austrian ultimatum to Serbia shown to King of Rumania, 1914.

MESOPOTAMIAN ENGAGEMENT

Mr. Churchill in the House of Commons today stated that the relief column advancing in Mesopotamia is engaged with 2000 Arabs near Rumeita. The situation in the Samara area has improved. The tribesmen are suffering heavily and there are signs of dispersal – Reuter

---

Figure 7.6   The North Atlantic Times – the first ship's newspaper
Source: 'Yearbook of Wireless Telegraphy and Telephony', 1914

not generally adopted. The Marconi Company, which since the inception of marine telegraphy had been using the general call 'CQ' to alert all stations to the reception of a transmitted message, proposed that the call 'CQD' (D for 'distress') be used in their installations from February 1904. This was subsequently adopted by Britain and the United States, until the matter was raised again at a further Radio Telegraph Convention in Berlin in 1906. For some time Germany had been using the general call 'SOE' for its maritime fleet, equivalent to the British and American 'CQ'. At the 1906 Convention it was suggested that the last letter of this code be changed to 'S' to give the now familiar 'SOS',[4] which was universally adopted at the Radio Telegraph Convention held two years later in 1908, although the call 'CQD' continued to be used for a few years, particularly by the British.

The first notable occasion of its use was in July 1909 when the SS *Republic*, an ocean liner, collided with the SS *Florida* in thick fog [38]. Jack Binns, the wireless operator on board the *Republic*, had little opportunity to radiate on full power since the ships generators were immediately disabled by the force of the impact, and he was compelled to switch to his emergency batteries which reduced the range of the spark transmitter to about 80 km. However, the SOS transmission giving the location of the sinking ship was picked up by a shore station in Siasconset, Massachusetts, which relayed the call to the SS *Baltic*, the SS *Lorraine* and other vessels in the vicinity, an action which successfully saved all hands. The most spectacular but less than successful emergency call was the SOS call from the White Star liner *Titanic*, which collided with an iceberg on its passage to New York in 1912 during its maiden voyage. Some eight hundred passengers were saved out of the two thousand carried on the vessel as a result of a wireless message received by the USS *Carpathia*, USS *Mount Temple* and other ships [39].[5]

The period from the spring of 1912 to the outbreak of the First World War saw an extraordinary number of sea casualties. Ships were lost on widely scattered oceans: off the coast of Australia, in the Sea of Marmara, in Alaskan waters, off Tierra del Fuego, in the St. Lawrence, around Japan, in the Black Sea. In most of these events wireless telegraphy played a major role in bringing relief to the stricken ship, and many lives were saved. Even before 1914, when the outbreak of war added hostility to accident, Ambrose Fleming estimated that over five thousand people were saved in peacetime by timely use of wireless telegraph SOS rescue procedures [40]. 'Listening out' for SOS calls was introduced by the 1912 International Radio Telegraph Convention, and was universally observed. Every half-hour while a ship is at sea, the receiver is tuned to 600 m (500 kHz) at 15 to 18 minutes and 45 to 48 minutes past each hour, and all transmission must be stopped. These two 3-minute silent intervals are internationally observed to protect possible

[4] The derivation of 'SOS' as standing for 'save our souls' is a folk etymology. The letter combination was chosen simply for ease of transmission when using Morse.
[5] Had wireless apparatus been carried on a freighter, without wireless, known to have been within 40 km of the stricken ship, and if the wireless equipment carried on board the USS *California* (which was even closer) had been operational, then the death toll would almost certainly have been considerably less.

SOS calls which, through weak transmitters or heavy interference, or because of great distances, might otherwise be lost and unheard. In modern ships the process is automatic and supplemented by other emergency transmission procedures, but in the early part of the twentieth century this was a vital part of a wireless operator's duties.

The notable successes of these life-saving operations led to more international cooperation on ways to improve the safe passage of ships at sea by the widespread use of wireless. The first such joint safety conference was proposed by the German Emperor, probably mindful of an earlier action by Germany in 1903 on behalf of the international community in making the ability to communicate with ships of all nations obligatory, independent of the manufacturer of the wireless equipment available on board ship (see Chapter 6). However, it was not in Berlin but in London, in 1914, that the first International Conference on the Safety of Life at Sea was held, attended by representatives of 17 maritime nations, including the United States, Japan and all the dominions of the British Empire [41]. It was to lead to a number of such conferences over the next fifty years which were responsible for considerable changes in the practice of marine wireless communication, to both the equipment carried and the procedures employed.

Lord Mersey was appointed chairman of this first conference. He had previously presided over the court of enquiry into the loss of the *Titanic* and was well aware of the need for promoting safety at sea. The range of enquiry and the eventual recommendations of the conference were extensive, contained in 74 articles, many of which are extremely detailed. If all had been universally enforced they would have led to an enormous improvement in safety for passengers, crew and ships. Initially, however, the recommendations were applied only to merchant ships which were mechanically propelled, carried more than 12 passengers and were proceeding to or from a port situated within the domain of the contracting parties. Several exceptions were also agreed, including a major one permitting ships which, in the course of their voyage, did not go more than 150 nautical miles (276 km) from the nearest coast. Despite these loopholes the new regulations considerably improved the safety of sea travel, not least because all large and many medium-sized vessels on which passengers were carried were obliged to have wireless equipment. Merchant ships carrying 50 or more persons (including crew) had to be fitted with wireless telegraphy apparatus capable of a minimum range of 100 nautical miles (184 km) by day, and carry at least one suitably qualified wireless operator A comprehensive list of transmission codes was stipulated for use in broadcasting information on ice, derelicts, weather, location, nationality and safety signals, with definitive procedures for conveying this information using the Q code. The transmission of information on ice or the location of derelicts endangering traffic was made obligatory for the captain of a ship sighting such obstacles to navigation.

A requirement for a continuous wireless telegraphy watch during navigation, for vessels carrying more than 50 persons, meant that one or more operators had to be employed, each holding appropriate certificates, together with supporting watchers holding lower qualifications. In Britain this initiated the boom in the provision of training courses referred to earlier, provided principally by the Marconi Company. The 1914 conference defined the range and power of the wireless equipment considered necessary for various classes of vessels, and stipulated a separate emergency installation having a range of at least 80 nautical miles (147 km), together with an independent source of power, capable of working the equipment for at least six hours.

Although they were generally respected by the nations participating in the 1914 conference, the recommendations were never ratified, being overtaken by the outbreak of the First World War. The proposals were revived at a Second International Conference on the Safety of Life at Sea, held in London in April 1929 [42]. The stipulation that all ships having more than 50 people on board should be fitted with wireless apparatus was extended to all passenger ships of any size and all merchant ships of 1600 tonnes gross or more. New regulations were proposed governing the provision of wireless telegraph equipment in lifeboats. For a ship with more than 10 lifeboats, it was agreed that one of them should be fitted with wireless, and if more than 13 then this should also be motorised in order to automatically transmit SOS. On ships requiring continuous watch the qualification of the watchers was raised to almost operator standard, requiring the ability to read Morse code at 16 wpm.

Two new requirements were included. For the first time it was stipulated that within two years every passenger ship of 5000 tonnes gross or more should be fitted with an approved direction-finding apparatus or radio compass; auto-alarm devices had to be used at all times when a ship was at sea. It is generally considered that the tragic loss of the *Titanic* initiated the move to design and install auto-alarm devices on almost all seagoing vessels. A more prosaic reason was the cost to shipowners in supporting the extra wireless staff needed under the Merchant Shipping (W/T) Act of 1919. The Act required any seagoing ship carrying 200 or more persons to carry three radio officers on voyages lasting from 8 to 48 hours, one of whom was to stand by in the wireless room as a 'watcher' in case a distress signal should be received while one of the qualified operators was off duty. Other ships were subject to similar provisions depending on the number of passengers carried and the duration of the voyage [43].[6] Experiments to design a suitable set were carried out by the Marconi Marine Company in the early 1920s, and by

---

[6] Similar provisions were made in the United States under the US Radio Law (Wireless Ship Act) of 1910.

1927 the British Government decided to make the fitting of an auto-alarm device compulsory, using the Marconi system. This consisted of a receiver tuned to the distress frequency of 500 kHz and responsive to a series of 4 s dashes which triggered an alarm circuit containing a bell or siren, the circuit for which is shown in Figure 7.7. In the United States a similar requirement led to the production and widespread use of an auto-alarm system manufactured by the Mackay Radio & Telegraph Company which operated in the same way [44]. Watch periods were not removed completely with the use of this device, but were now confined to four periods of half an hour each: 08:00–08:30, 12:00–12:30, 16:00–16:30 and 20:00–20:30. At all other times the auto-alarm was in operation.

The Third Conference on the Safety of Life at Sea was held in London in 1948 and took another step forward by specifying that all passenger ships and all merchant ships of greater than 1600 tonnes gross should maintain a continuous watch on the distress frequency of 500 kHz with an operator/watcher assisted by an auto-alarm. Further, all merchant ships between 500 and 1600 tonnes gross should fit wireless equipment and carry a suitably qualified operator. It also became obligatory to carry radio telegraphy sets in all lifeboats, and suitable small sets were produced which could be used by an inexperienced operator (described in Chapter 10 for air–sea rescue). Their installation in all classes of ship was

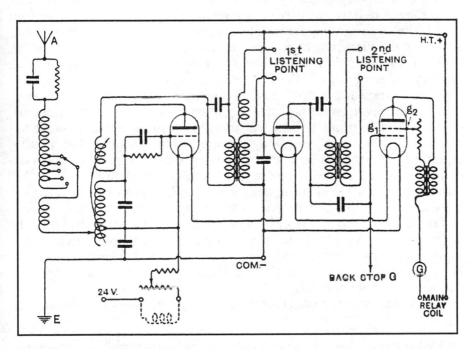

Figure 7.7   Schematic diagram of the Marconi auto-alarm receiver
Source: Dowsett 'Handbook of Technical Instruction for Wireless Telegraphists', 1930

accelerated by the well publicised case of the Hain Steamship Company. One of its vessels, the *Trevesso*, was sinking on a voyage from Australia to England and launched its lifeboat, which was not fitted with wireless. The survivors then made a spectacular voyage of over 2700 km in the open boat, lasting for 22 days before they reached land. Had wireless been carried it was generally agreed that rescue would have been possible within a few days [45]. Other conferences on the safety of life at sea (SOLAS), now under the auspices of the International Marine Organisation, have been held in 1960, 1974 and, most recently, 1988, and continue to expand and refine the instruments for safe travel at sea, now however with other systems than W/T [46].

7.5 International agreements

The SOLAS conferences discussed above form part of a set of international agreements which have developed over a long period, almost as long as the history of telegraphy itself. In 1868 the Bureau of the International Telegraph Union was established at Berne, to act as a central agency for matters concerning the regulation of international telegraphy, then concerned with line transmission between countries, and was extended at the beginning of the twentieth century to include wireless telegraphy as well. It was under the bureau's auspices that the increasing use of wireless telegraphy for maritime purposes prompted the formation in 1903 of the First International Convention on Telegraphy. This met in Berlin at the invitation of the German Emperor, and was attended by representatives of most of the European governments. The most important decision reached was the abolition of the isolation principle carried out by some organisations (principally Marconi Ltd) in which communication was refused unless both parties to it operated the same make of equipment.[7] At the convention, wireless communication was jointly agreed by all parties to be a universal as opposed to a private means of communication, and all requests for wireless connection were to be answered irrespective of the system of telegraphy employed (Article 3 of the convention). A valid licence to transmit was still needed, however, and in Britain a Wireless Telegraphy Act was passed in 1904, and periodically renewed, to prohibit the installation and working of wireless telegraphy apparatus in the United Kingdom or on board ships except under licence from the Postmaster General. Similar restrictions were applied nationally in other countries.

The Second International Convention on Telegraphy was held, also in

---

[7] At least one Marconi station was closed down by government intervention because it would not accept non-Marconi messages. This was the Marconi station at Nantucket lightship, off the coast of Massachusetts.

Berlin, in 1906. This was to a large extent concerned with the operation of a telegram service, and had six primary objectives:

1 acceptance and transmission of telegrams,
2 adoption of international rules of working,
3 provision of means of collecting charges,
4 publication of information on inter-communication,
5 rules to prevent wireless interference, and
6 provision that inter-communication (by wireless) may not be refused on the grounds of different equipment design.

Documents relating to these objectives were revised and agreed internationally at the next convention, the London International Radiotelegraphic Convention, which took place in 1912. At this convention a detailed agreement on telegraphic matters was reached between nations including the United States, Russia, Japan and China. It was a far-reaching agreement which, besides covering the objectives of the 1906 meeting in its 23 detailed articles [47], defined such terms as 'coast station', 'ship station' and 'charge'. An important regulation was the necessity for every ship station to be able to use wavelengths of 600 and 300 m, reserved for emergencies at sea (later a wavelength of 900 m was also agreed for emergencies associated with civil aviation). Some exceptions for certain countries relating to several articles were incorporated in the final protocol, in particular concerning tariffs imposed by the United States and Canadian authorities. A separate set of service regulations was issued, covering minimum equipment configuration for coastal and other stations, the wavelengths and procedures to be used, a definition of emergency equipment to be carried, certification of operators, compulsory conditions for the stations, including power rating of the transmitting equipment, hours of service, charges, form of telegrams sent, codes, calling signals and many other matters. The abbreviated codes used for the transmissions had an important bearing on the economics of radio telegrams, and the convention laid down those to be used and the method of using them. The list for the Q code which formed a major part of this regulation as it appeared in 1912, has been given in Table 7.1.

It was the intention to hold an international convention every five years, but this did not always prove possible, and although a limited Inter-allied Convention on Radio Telegraphy took place in Paris in 1921, the next full international convention, the third to be held, was not until 1927 [48]. This conference was long overdue, its main task to revise the International Radiotelegraphic Convention Regulations which had been agreed in London in 1912. The convention took place in Washington, and was concerned with allocating frequencies to wireless and radio services as well as revising the 1912 agreements. It was at this convention that the decision was made to specify transmissions by their frequency in kilocycles per second (now kilohertz), rather than their wavelength in metres, in all future international documents. While the earlier international conventions in telegraphy had been concerned specifically with communication by wire and wireless, the 1927 convention was the first in which frequency allocation for public broadcasting was also considered. This was later to form a major part of discussions at international level, but its consideration falls outside the scope of this book [49]. The regulations had applied exclusively to ships and ship-to-shore communication in 1912, but in 1927 they were applied also to mobile services, including aircraft, and provision was made for distress calls and messages for these services. In addition to the well-established use of the Morse SOS signal on the 500 kHz international distress frequency, a R/T equivalent, 'mayday', was proposed and accepted. Provision was made for the use of an alarm signal, a series of twelve dashes sent in one minute, to actuate automatic receiving equipment which was now being fitted to many oceangoing ships. A major topic of discussion and a source of dissension at this conference was the continued use of spark transmitters in the maritime service. It was recognised that heavy investment in wireless equipment by shipowners and coastal stations made it difficult to replace all of this interference-producing equipment at once, and only modest curbs were placed on its use, namely its elimination from the long wavebands below 375 kHz (except for transmitters of less than 300 W) and no new spark equipment to be installed. Compete suppression of spark transmitters did not occur until 1940, although for many years spark transmission was allowed for emergency use on the distress frequency of 500 kHz.

The Fourth International Radiotelegraphic Conference was held in Madrid in 1932 and covered both line, wireless and radio systems [50]. It was in effect a merger of the International Convention on Telegraphy and the International Radiotelegraphic Conference to form a single body, to be known thereafter as the International Telecommunications Union (ITU). Further reallocation of frequencies was made, and one effect of this was to restrict to very narrow frequency bands all ship transmissions, a response to the ongoing problem of crowded communication bands. Spark transmitters were further confined to a small number of frequencies in the long-wave band, including the distress frequency of 500 kHz. By the time of the next international conference (the Fifth), in Cairo in 1938, now designated as an ITU Conference, little further change had taken place in wireless telegraphy or line telegraphy requirements, although a large number of new regulations were considered for radio entertainment, including television transmission. By this time all regulations regarding frequency allocation had been taken over by the ITU, which ran its own series of international meetings at Geneva specifically for this purpose, now held under the auspices of the United Nations [51].

7.6 Civil aviation

The first attempts at wireless transmission in aviation are inseparably linked with military use, and will be considered in detail in Chapter 10. Even at the time of these early experiments, which were made with spark telegraphy transmitters, it was appreciated that as soon as aircraft became dependable enough to carry fee-paying passengers, then wireless communication involving telephony would enter the picture. The initial wireless experiments showed that aircraft could maintain continuous contact with the ground station by the use of Morse transmissions, but this would not be ideal for civilian aviation traffic. Not only would a Morse telegraph operator need to be carried in addition to the pilot, who would not be in a position to operate a Morse key in flight, but a more immediate contact would be required between the pilot and ground and this could only be provided by telephony transmission. The first civil aircraft telephony installation to go into operational service in Europe was used on the London–Paris route by Imperial Airways. This was a Marconi type AD 1/S set fitted initially in a Handley Page machine (a converted wartime bomber). The pilot was able to maintain contact with both sides of the English Channel up to a distance of 200 km, and this success initiated a regular service between the two capitals from 1919. At about the same time experiments were being made with both R/T and W/T from a fleet of civil airships, the R33, R34, R36, R38 and R40, with the intention of instigating regular commercial flights between Britain and the Continent, later to be extended to cover the Mediterranean and Egypt. However, with the tragic loss of both the R34 and R38 in 1921, and the R100 a few years later, this enterprise was abandoned, at least in Britain, and civil aviation was confined entirely to heavier-than-air craft [52].

   The Marconi equipment employed in regular service by 1921 consisted of a transmitter/receiver unit, type AD2, connected to a remote control unit which could be placed near at hand for the pilot or navigator. This also switched the aerial between the transmitter and receiver as required, as well as making other changes required to permit either telephony or telegraphy transmission. The transmitter consisted of a single-valve master oscillator with a modulator valve accepting input from a microphone or a keyed buzzer, with the transmission switched from CW to ICW as required. The receiver for the Marconi installation was a five-valve tuned radio frequency (TRF) receiver containing three tuned HF stages, and the whole thoroughly screened from aircraft ignition interference [53]. The AD2 was well received and used by several other European airlines coming into existence at that time, including KLM and SNETA (the forerunner of Sabena, the Belgian airline). A Marconi ground station was installed at Croydon, London's first regular civil airport, consisting of a 100 W CW/ICW telegraph and telephone transmitter and a Type 55

Bellini–Tosi direction-finding receiver installation. The transmitter was afterwards replaced by one of 1.5 kW, designed by P.P. Eckersley, a leading Marconi engineer. Telegraphy transmission using Morse code, and the provision of a wireless operator (often the co-pilot), was still required on long-distance routes, and even on shorter flights it became necessary for a number of years to carry a Morse key and the ability to provide a telegraph service on the 600 and 900 m distress wavelengths in the event of emergency, in order to comply with international regulations. The increase in sensitivity of the valve receivers carried in the aircraft opened up the possibility of improved direction-finding, and considerable effort was expended on achieving this by using a loop aerial within the aircraft. Earlier direction-finding work had been confined to ground stations which, after taking a bearing on the transmitting aircraft in the air, were able to transmit the results to the aircraft. The more accurate and convenient technique of taking a bearing from the aircraft itself was pioneered in Britain and Germany, and was soon to become a necessary part of civil aircraft wireless practice.

In Continental Europe the use of Morse code for long-distance and emergency operation with civil aircraft was retained up to the outbreak of war in 1939. Germany had established a civil aviation presence in 1926 with the formation of Deutsche Lufthansa (DLH), and Telefunken, drawing on its experience from the First World War had produced two suitable airborne wireless sets, Spez 205F and Stat 257F, each of 70 W aerial power and between them covering a frequency range of 230–1000 kHz. The sets could be used for telegraphy CW, telegraphy ICW and telephony. Shortly afterwards, Lorenz AG brought out an airborne civil aviation set, the SERFOIV 28, providing a larger output power of 100–120 W and a frequency range of 222–546 kHz. The receivers for both manufacturers were TRF sets containing two tuned HF stages, a reactive detector and two LF stages. Both sets were produced in large numbers for the bigger aircraft, such as the Albatros L73, the Dornier Merkur and the Dornier flying boats. Their weight (54 kgm) was too great for the smaller aircraft, and for these a smaller set with a lower aerial power of 20 W was produced. For internal flights DLH used telephony to keep in touch with the aircraft control stations, but used telegraphy for international flights, not only for longer-distance transmission, for which telephony was not yet suited, but also to obviate language difficulties by the use of international codes.

Initially all communication with aircraft was carried out on MF and HF. The development of SW transmissions, as in other European countries and the United States, stemmed from the experiments of radio amateurs, who in Germany had been using VHF for several years in the range 3–30 MHz. In 1928 several firms, Telefunken, Lorenz and DVG, took notice of these developments and began to work on equipment for these frequencies, carrying out air-to-ground tests with apparatus installed in a

Dornier flying boat. Lorenz produced a crystal-controlled VHF transmitter, model SERKTII282, with an aerial power of 1.5 kW operating over a frequency range of 2.9–7.5 MHz; Telefunken produced similar sets in their Spez series, Spez 327F and 366F. In the meantime improvements in valve technology allowed UHF sets to be constructed, and a short and better matched aerial to be applied (long-wave sets did not reach their maximum efficiency with short fixed aerials, and long trailing aerials were beginning to be incompatible with the performance of the aircraft). Both companies produced a small UHF set working on the 100 MHz band and showed that air-to-ground transmission was possible at these frequencies. Although it was not immediately taken up by DLH, this experience proved invaluable for the development of German military aircraft equipment in the next decade (see Chapter 10).

In January 1930 the International Commission for Aerial Navigation agreed that wireless equipment should be carried on all aircraft carrying at least ten passengers. Improved sets had been developed for civil aviation and were installed in all aircraft of the British and German civil airlines operating over Europe. In Germany, the re-equipping of DLH's fleet in 1931/2 coincided with a significant change in construction methods practised by its principal suppliers of wireless equipment. Instead of using front plates of mahogany and pressboard to mount the equipment, a new technique of rigid aluminium die-casting was used. This gave improved mechanical stability and good electrical shielding between units, and was to become the standard constructional technique for wireless equipment in all civil and military applications from the 1930s onwards. Improved valve design made possible the superheterodyne ('superhet') receiver, which began to be tested for aircraft use, making its appearance most notably in the military airborne equipment soon to be produced by Telefunken and other manufacturers, as the emphasis shifted in Germany towards military production.

Morse telegraphy was still being used to obtain the greatest range with the co-pilot acting as the wireless operator [54] as on the long-range routes now being established by Imperial Airways. The equipment finally developed by Marconi for this work consisted of the AD37a and AD38a, operating on the dual wavebands of 40–89 m and 500–1000 m and capable of communication by R/T over a maximum distance of 2000 km, extended to 8000 km for W/T. By this time an essential shortening and standardisation of telegraph traffic for civil aviation had taken place, with an international agreement to use the Q code in 1931.

The war years of 1939–45 saw little development in commercial aviation, the major effort by manufacturers being switched to the production of equipment for the armed services. The requirements for radio communications for heavy aircraft in wartime and civilian aviation is not, however, very different, and shortly after the end of the war the wartime equipment became available for the various types of civilian aircraft now becoming operational on international routes. An early type of airborne wireless installation used by British and other airlines was the Marconi AD87 transmitter and AD8882 receiver, which were slightly modified versions of the RAF T1154/R1155 transmitter/receiver equipment to be described in detail in Chapter 10. These wartime modifications were replaced by Marconi a few years later with a lightweight MF/HF transmitter type AD97 and receiver AD107A/B. The transmitter contained eight crystal-controlled channels and radiated a power of 20 W on the civil frequency bands of 2.5–9.5 MHz and 320–520 kHz. Alternative equipment used for civil aviation in the 1950s were the Standard Telephone and Cable STR18B/2 transmitter, the American Collins 618S transmitter and the Bendix TA-20A VHF transmitter. All these later designs were for telephony transmission only, and Morse code ceased to be used in regular airline work [55].

7.7 The role of amateurs

The contribution made by non-academic experimenters to the development of wireless telegraphy in the 1920s and 1930s is significant, particularly in the field of military communications and civil aviation. This became apparent following the shunting of their operations to the higher-frequency bands in 1923, considered at the time to be a 'harmless' part of the spectrum, posing no threat to commercial or military communications in terms of interfering signals, since it was the lower frequencies that were considered essential to secure long-distance commercial wireless communication. Frequency allocation for radio amateurs became formalised on a worldwide basis at the International Radio Telegraphic Conference held in Washington in 1927. A number of higher frequencies were allocated for amateur use; 1715–2000, 3500–4000, 7000–7300, 14000–14400, 28000–30000, and 56000–60000 kHz, and it was within these frequency bands that the major amateur contributions were made. In Great Britain the amateurs were supported by the Radio Society of Great Britain (RSGB), which grew out of the London Wireless Society, formed in 1913. At its peak the RSGB had more than ten thousand members – a sizeable body of informed laymen well versed in the techniques of wireless design and operation. This large membership made possible fairly wide ranging group experiments in long-distance high-frequency transmission. One such exercise was initiated by Sir Henry Jackson, a pioneer from the early days of wireless telegraphy, who by 1925 had become Admiral of the Fleet. He invited the amateurs to assist in an extensive programme of ionospheric research, then being carried out by the Radio Research Board, by reporting on the propagation of long-distance short-wave signals transmitted via the upper atmosphere. This was taken up enthusiastically by all the members. One notable report from an amateur taking part in the tests, described the reception of some 50 United States stations transmitting at a wavelength of 45 m and a further 25 stations on 23 m. These investigations, coupled with the direct assistance of radio amateurs in the development of short-wave communications for the RAF (described in Chapter 10), were an important factor in changing the direction of commercial and military communications in the 1920s, which until then had been confined to the longer wavelengths. The work of the amateurs had shown that, by suitable choice of frequencies and times of transmission to conform with the condition of the ionosphere, enormous distances could be covered with extremely low powers. A typical 'contact' at the time was a two-way transmission between the United Kingdom and an amateur in New Zealand, transmitting on a wavelength of 45 m using an aerial power of a little over 11 W [56].

A second contribution by the radio amateur lay in the techniques developed for reporting the results of their numerous transmission experiments on the shorter wavelengths. While most amateurs were familiar with the abbreviated Q code established by the 1912 International Radio Telegraph Conference, this allowed rapid exchange of operating procedure, but not detailed reporting on the actual characteristics of a telegraphy signal. In 1934 an American amateur, Arthur Braaten, introduced what became known as The Braaten code, which has since been used as a standard reporting code by bodies both inside and outside amateur circles. The code is based on the three important characteristics of every telegraphy signal, namely its readability, signal strength and tone. Each of the 'RST' characteristics is rated on a scale of 1 to 9, so that S1, for example, indicates 'very weak signals', and S9 'extremely strong signals'. In this way the essential characteristics of a W/T transmission could be accurately and quickly transmitted to the sender, much facilitating the process of information exchange.

Finally, the special skills of the amateur fraternity, their familiarity with transmission equipment and use of the Morse code, became especially relevant in both Britain and the United States during the late 1930s when trained reserves were sought in all of the armed services for enrolment in the event of war. This did not, however, happen in Germany. In the 1920s and 1930s the German authorities viewed the role of radio amateurs with considerable distrust and, while not actually banning their activities, made the process of obtaining a permit for overseas transmission a lengthy and difficult operation, adding to the problems the amateurs experienced in obtaining wireless apparatus with which to build suitable equipment. As a consequence the German armed forces were not able to benefit from a skilled and knowledgeable pool of wireless enthusiasts as became available for the Allies. In Britain the first service to seek such assistance was the Navy, which established the Royal Naval Wireless Auxiliary Reserve in 1932, later to be known as the Royal

Naval Volunteer (Wireless) Reserve (RNV(W)R). This consisted of a reserve of licensed radio amateurs who would be prepared to exercise with Admiralty personnel and be available for immediate enrolment in time of war. The exercises took the form of transmissions on the long wavelength of the Admiralty station at Cleethorpes (3325 m), to familiarise the recruited amateurs with Admiralty procedure for the interchange of service messages. A similar organisation was set up by the Air Ministry in 1938, known as the RAF Civilian Wireless Reserve, members of which served early in the war as fitting parties, others becoming attached to the Wireless Intelligence Screen in France and Belgium. Both these organisations benefited considerably by the National Field Days (NFD) which had been arranged as an annual event by the RSGB since 1933. The procedure for an NFD was to set up, at various locations in the countryside (usually on a hilltop or on top of a tall building), a temporary series of low-power amateur stations which would communicate with one another, thus forming a network for the transmission of information using an agreed operational procedure. The results of these exercises showed 'that if the necessity arose the Amateur Radio movement in the UK could place into operation an emergency network of stations at short notice'.

With these and other exercises the amateur radio movement was able to make available to the armed forces a large number of potential recruits, already knowledgeable and trained in wireless communication techniques, a factor which had an important bearing on the setting up and manning not only of the armed forces, but also of the wartime Post Office and Cable & Wireless communication organisations. Another, and unexpected outcome was that an RSGB publication, well-known amongst members in 1938, *The Amateur Radio Handbook*, found a new role as a standard instructional text for signals recruits in all branches of the armed forces during the war. By the end of the conflict in 1945, more than 180 000 copies had been printed and distributed among military and civilian personnel.

References

1   PHILLIPS, V.J.: 'Early radio wave detectors' (Peter Peregrinus, London, 1980)
2   DYER, F.L., and MARTIN, T.C.: 'Edison; his life and inventions' (Harper Bros, London, 1910)
3   FLEMING, J.A.: 'Improvements in instruments for detecting and measuring alternating electric currents', British Patent No. 24 850 (16th November 1904)
4   FLEMING, J.A.: 'The thermionic valve' (Iliffe & Sons, London, 1924)
5   DE FOREST, L.: 'Wireless telegraphy and telephone system', US Patent No. 1 507 017 (20th March 1914)

6 DE FOREST, L.: 'Method of and apparatus for amplifying and reproducing sounds', British Patent No. 2059 (24th June 1913)

7 FESSENDEN, R.: 'Heterodyne methods of reception', US Patent No. 706740 (1901)

8 CLARK, G.H.: 'Clark radio collection', Smithsonian Institution, C15, Box 67, 3rd April 1913

9 BAKER, W.J.: 'A history of the Marconi Company' (Methuen, London, 1970), p. 207

10 SHEERING, G.: 'Naval wireless telegraph transmitter employing rectified alternating current', *Journal of the Institution of Electrical Engineers*, 1925, **63**, pp. 309

11 HOUSEKEEPER, W.G.: 'The art of sealing base metals through glass', *Journal of the American Institute of Electrical Engineers*, 1923, **41**, p. 870

12 WALMSLEY, T.: 'Rugby Radio Station', *Post Office Electrical Engineers' Journal*, 1925, **1**, (1), pp. 37–57

13 TREASURY OFFICE: 'Marconi's Wireless Telegraph Company Ltd, general specification for imperial wireless installation' (HMSO, London, 1913)

14 Report of the Committee on Imperial Defence (1914), Doc. CAB16.5.189, Public Record Office, Kew, Surrey

15 DONALDSON, F.: 'The Marconi scandal' (Rupert Hart-Davis, London, 1962)

16 'The Empire's wireless telegraph communication' (1914), Doc. ADM1/ 8380/151, Public Record Office, Kew, Surrey

17 Letter to the Secretary of the Committee of Imperial Defence, 25th June 1914, Doc. ADM1/8380/151, Public Record Office, Kew, Surrey

18 MARCONI, G.: 'Radio telegraphy', *Proceedings of the Institute of Radio Engineers*, August 1922, **10**, (4), p. 215

19 'The "Empiradio" beam stations', *Post Office Electrical Engineers' Journal*, 1928, **21**, pp. 55–65

20 MARCONI, G.: 'Radio communications', *Proceedings of the Institute of Radio Engineers*, 1928, 16, pp. 40–69

21 'Imperial Wireless & Cable Conference 1928', *Telegraph and Telephone Journal*, 1928

22 COGGESHALL, I.V.: 'An annotated history of submarine cables' (University of East Anglia, Norwich, 1993)

23 WOOD, K.L.: 'Empire telegraph communications', *Journal of the Institution of Electrical Engineers*, 1939, **84**, p. 638

24 GRAVES, C.: 'The thin red line' (Standard Art Book Co., London, 1946)

25 WEST, W.: 'Reuter's wireless services', *Post Office Electrical Engineers' Journal*, 1946, **39**, pp. 48–52

26 STURMEY, S.G.: 'The economic development of radio' (Duckworth, London, 1958)

27 'The Helsby wireless system', *The Electrician*, 1910, **65**, pp. 142–4

28 CRAWLEY, C.G.G.: 'Wireless communications', *Telegraph and Telephone Journal*, 1921, **7**, pp. 92–3

29 'Annual report of the American Marconi Company for the year ending January 31st 1913', p. 5

30 GOLDSMITH, A.N.: 'Progress in radio communication in the United

States, 1920', *in* 'Yearbook of wireless telegraphy and telephony, 1921' (The Wireless Press, London, 1921), pp. 1115–20

31   From 'The radio industry Federal Trade Commission report' (Government Printing Office, Washington, D.C., 1923)

32   MACLAURIN, W.R.: 'Invention and innovation in the radio industry' (Macmillan, New York, 1949)

33   'JIGGER': 'British Empire Exhibition', *The Electrical Times*, 29 May 1924, **65**, pp. 633–4

34   GODWIN, G.: 'Marconi 1939–1945 — a war record' (Chatto & Windus, London, 1946)

35   'Wireless newspapers at sea', *in* 'Yearbook of wireless telegraphy and telephony' (The Wireless Press, London, 1914), pp. 614–16

36   'Heroism at sea', *in* 'Yearbook of wireless telegraphy and telephony' (Wireless Press, London, 1918), pp. 854–7

37   BAARSLAG, K.: 'SOS — radio rescues at sea' (Methuen, London, 1937), pp. 63–100

38   FLEMING, J.A.: 'Fifty years of electricity' (Wireless Press, London, 1921), p. 325

39   'International conference on the safety of life at sea' (Washington, DC, and HMSO, London, 1914)

40   CRAWLEY, C.: 'Safety of life at sea', *Electrical Review*, 19 July 1929, **105**, (2695), pp. 106–7

41   'Merchant Shipping (W/T) Act 1919' (HMSO, London, 1919)

42   STERLING, G.E.: 'The radio manual for radio engineers' (Van Nostrand, New York, 1938), Chap. 15

43   HANCOCK, H.E.: 'Wireless at sea — the first fifty years' (Chelmsford, 1950)

44   GODWIN, W.D.: '100 years of maritime radio' (Brown, Glasgow, 1995)

45   'International Radiotelegraphic Convention 1912', *in* 'Yearbook of wireless telegraphy and telephony' (Wireless Press, London, 1921) pp. 48–63

46   TERRELL, W.D.: 'The International Radio Telegraphic Conference at Washington 1927', *Proceedings of the Institute of Radio Engineers*, 1928

47   BRIGGS, A.: 'The birth of broadcasting', vol. 1 (Oxford University Press, Oxford, 1995)

48   'Sidelights on the Madrid Conference', *Telegraph and Telephone Journal*, 1933, **19**, (217), pp. 152–4; (218), 176–8

49   CODDING, G.A., and RUTKOWSKI, A.M.: 'The International Telecommunications Union in a changing world' (Artech House, Dedham, Mass., 1982)

50   SINCLAIR, D.: 'Civil airship wireless in 1921', *Wireless World*, 1922, **11**, pp. 436–42

51   CHILDS, H.B.T.: 'Wireless telegraphy as applied to commercial aviation', *in* 'Yearbook of wireless telegraphy and telephony' (Wireless Press, London, 1921), pp. 1196–1208

52   TRENKLE, F.: 'Bordfunkgeräte vom Funkensender zum Bordradar' (Bernard u. Graefe Verlag, Koblenz, 1986)

53   GROVER, J.H.H.: 'Aircraft communication' (Heywood, London, 1958)

54   CLARRICOATS, J.: 'World at their fingertips' (Radio Society of Great Britain, Potters Bar, Hertfordshire, 1967), p. 112

55 Deutscher Amateur-Radio Club e.V: http://www.darc.de/vorstand/sdar/index.html (in German)
56 'Results of the RSGB National Field Day', *RSGB, T & R Bulletin*, 1933, **9**, (2)

Chapter 8

... and at war

The gradual development of wireless telegraphy into a reliable communications service in wartime was perceived quite differently in the various branches of the British armed services. In the Navy it was seen from the first as a vital adjunct to operations, particularly with the development of long-range guns and a war in which close engagement was the exception rather than the rule. Apart from its undoubted value in reporting events, Nelson would have had little use for the technology at Trafalgar, where close contact with the enemy was paramount.[1] The British army had a long history of terrestrial telegraphy before taking its first tentative steps towards wireless telegraphy as a tactical facility in the Boer War, but made no great use of wireless thereafter until the closing years of the First World War, when technical developments became rapid and new sophisticated equipment assumed a central role in all branches of the service. The Air Force, emerging as a formidable fighting arm by 1918, made considerable use of wireless telegraphy and wireless telephony as soon as lightweight sets became available. Navigation and air-to-ground communication equipment were the cornerstones of wireless telegraphy development in the inter-war years for the RAF, leading to their dominance in operations by the opening of the 1939–45 conflict.

In this chapter the use and development of wireless communication for the ground forces will be considered in some detail. The equipment and operations problems of both the navy and the RAF deserve separate chapters and will be considered later. Here the tactical and operational developments for the Army are described, paying particular attention to

---

[1] A first major use of long-range wireless contact in a naval context occurred during the Russo-Japanese war in 1905 at Tsushima Strait. In its lengthy passage from its home base, around the Cape of Good Hope and through the South China Sea to the Sea of Japan, the Russian Baltic fleet was able to keep in touch with St Petersburg, using a combination of wireless telegraph transmission from its warships to the nearest land station and onward travel by a commercial telegraph overland route. The Japanese fleet also benefited from wireless communication. (See Chapter 9.)

the new techniques and problems which emerged for ground communications, for which no counterpart existed in civilian life.

8.1 Army wireless before 1914

The British Army had experimented with wireless equipment in the Boer War. Accompanying the relief troops sailing from Southampton in October 1899 were a number of sappers under the command of Captain Kennedy and six engineers from the Marconi Company bringing with them five newly developed 'portable wireless stations'. On arrival the sets were quickly transferred with their civilian operators by rail to the front – at that time along the Orange and Modder rivers adjoining the Orange Free State. The intention was to establish wireless telegraph communication between the British forces encamped in this region, about 80 and 112 km apart, but despite the opportunity to fly kite aerials for reception up to 152 km in height, as well as the use of bamboo masts of 18 m or more for the transmitters, this was not a success. Lightning-induced charges affected the performance of the coherers used, and dust storms at the chosen sites did not help matters. By February 1900 the Director of Army Telegraphs abandoned the project. However, the equipment itself was later successful, for the discarded transmitters were requisitioned by the Royal Navy and installed aboard several of their ships carrying out a blockade in Delagoa Bay (now Maputo Bay; Figure 4.8). The equipment enabled the ships to cover a wider search area while still maintaining contact with one another, and by using ships in a relay mode, communication was maintained with headquarters in Simonstown, some 1600 km away, and from there along the land-based telegraph network to Cape Town [1].

The equipment in use at the time consisted of an induction coil producing a 10-inch (250 mm) spark between the spheres of a spark gap, when keyed from a primary circuit including a d.c. supply. Tuning was fixed and relied entirely on the natural resonance of the aerial. Similarly, the receiving aerial was untuned and connected to a Marconi coherer fitted with an automatic 'tapper' device to restore the coherer to its nonconducting state after the receipt of a signal. This was essentially the standard Marconi 'plain aerial' transmitter and receiver described in Chapter 6, but with the addition of an inking printer, operating by clockwork, to make a record of the received signal. The difference in performance between the army and naval installations using the same equipment has led to the suggestion that 'plain aerial working', with the predominant frequency at which radiation occurs influenced mainly by the aerial length and its ground connection, the particular circumstances of the South African terrain, with its poor ground conduction, severely limited the energy transmitted [2]. The practical outcome of these experiments for the Army was to seriously compromise the build-up of army technical equipment during the military preparations for the First World War, and it was several years before wireless became a dominant factor in the land campaigns.

Elsewhere the Marconi Company was more successful, and similar apparatus sent to Russia for trials at the commencement of the Russo-Japanese war of 1904–5 was accepted by the Russians, who ordered a number of units that were used during the war, in Manchuria. The spark transmitter was carried in covered wheeled transport, and there was a dynamo driven from a cycle pedal gear. The receiver was a Marconi magnetic detector, which had replaced the coherer as a much more reliable device. The range of the equipment was only 30 to 40 km, being limited by the power which could be taken from the pedal generator. Later development of a similar portable set for the military by Marconi resulted in sets being used by the Italian Army and for demonstration in Switzerland, where a range of about 60 km was achieved in mountainous terrain. Some larger sets were produced for the British army in 1913. These were transported by motor-car and had an output of 1.5 kW with a range up to 250 km, using a 21m mast aerial [3]. A second attempt to interest the British Army in the advantages of spark telegraphy was made in 1909, through the London Wireless Company of Territorials attached to the Royal Engineers. The equipment used was the Lepel quenched spark system, described in Chapter 6, and stations were established at Slough and Twickenham. These were used on manoevres in cooperation with several units of the Territorial Force, which had enthusiastically created no fewer than five wireless companies – one for each command – although little actual transmitting equipment reached the Force before the onset of the First World War.

8.2 War on the ground, 1914–18

The Army's intercommunication requirements could for the most part be met very economically by a system of point-to-point telegraph links, electrical or visual, augmented by a signal despatch and unhampered by the numerous complications which were soon to develop. Telephones were few and their use, apart from artillery fire-control, was confined to the rear areas. The small amount of rather unreliable wireless equipment was exclusively allocated to the cavalry.[4]

That was one later assessment of the status of wireless telegraphy in the British Army at the outbreak of war. It was considered most convenient to associate wireless with the mobile cavalry corps because of the difficulty of providing rapid communications by other means. As a consequence, wireless telegraphy in 1914 was organised and equipped entirely on this basis, which had long-term effects on the adoption of wireless by the army [5]. The low-level status of military communication by wireless that followed this decision was to change dramatically as the war progressed.

The army had little field experience of wireless communications in 1914, and few suitably trained signal personnel. From the commencement of the campaign in France, army recruiting offices called upon volunteers from the Post Office to meet the growing signalling requirements of Kitchener's army, for both line and wireless. Signal troops trained on Marconi wireless equipment were attached to the cavalry brigades during most of the war. Training in the operation and maintenance of service wireless equipment was badly needed during the early part of the war, before the armed services had built up their own establishments. The Marconi Company undertook to train large numbers of wireless operators to use its equipment, and at the commencement of hostilities lecture rooms at King's College and Birkbeck College were taken over to augment the training facilities already available at Marconi House, London [6]. For the army, the chief of the Marconi school at Broomfield was seconded to the War Office, and gradually established a further training school at the Crystal Palace, which was effective in producing large numbers of trained telegraphists – over ten thousand in 1916 alone [7].

8.2.1 Wireless direction-finding

At the start of the war a typical wireless troop in the field carried three wagons fitted with Marconi pack sets, an improved version of the mobile sets demonstrated to the Swiss Army, together with some cable and ground telegraph equipment. Only about a dozen of these were brought over to France with the original Expeditionary Force in August 1914. The point-to-point facilities offered by these pack sets were far from being fully appreciated by the senior officers in the field. The wireless receiving equipment was used almost exclusively for intercepting enemy transmissions and, rather surprisingly, for direction-finding, in which it achieved a notable and early success. The direction-finding equipment, which used soft 'C' valves and a modified Bellini–Tosi directional system [8], had been developed by H.J. Round of the Marconi Company just before the war.

Henry Joseph Round was born in 1881 and, following a scientific education at the Royal College of Science, joined the Marconi Company in 1902 to become a leader in the development and application of thermionic devices. In his lifetime he filed 117 patents for the company, many of them concerned with direction-finding, for which his ideas were outstanding. His was a vital contribution to the war effort. The Marconi company had acquired the Bellini–Tosi patents in 1912, and Round had developed them into a practical scheme for wireless direction-finding which was already in an advanced state of readiness when war broke out. The equipment he transported to northern France in 1916, as Captain

Round seconded to Intelligence, consisted of 21 m demountable masts, the Bellini–Tosi aerial with its goniometer (Figure 8.1), a Type 16 Marconi receiver, which incorporated a crystal detector and a soft-valve audio frequency amplifier. This was a long-wave system operating over the 400–2000 m band. After a number of tests at different locations, Captain Round's two first stations formed a baseline running from Amiens to Calais, which enabled the requisite cross-bearings on a transmission to be taken over a wide area covering northern France and part of the North Sea. Following this success, a large network of similar stations was quickly established covering the entire Western Front. By 1916 chains of these direction-finding stations were in operation, not only in France but also, operated by the Admiralty, in England [6]. A major function of the stations was to track the movements of German warships and of the German Zeppelins, then raiding over the east coast of England. An operations map of the latter activity is shown in Figure 8.2. For the Zeppelins this was made easier not only by their slow speed, but because the German craft also made use of direction-finding techniques by

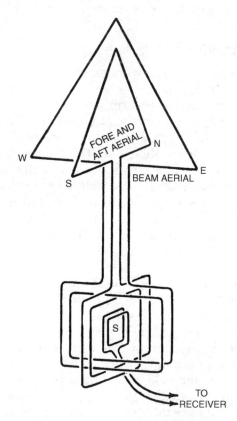

Figure 8.1   Bellini–Tosi directional system: aerial and crossed coil connections

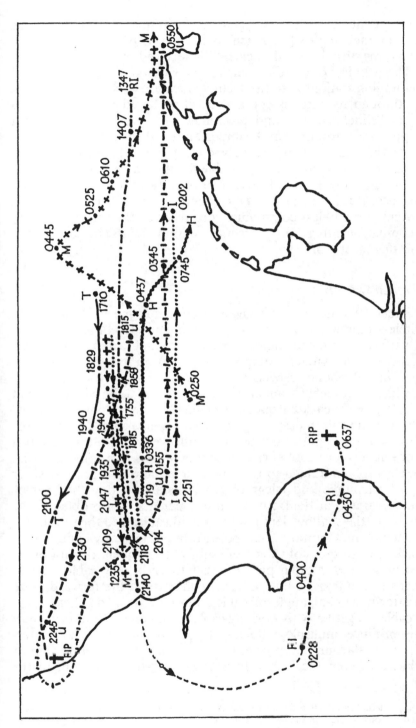

Figure 8.2   Routes of German Zeppelins raiding over the east coast of England in 1916
Source: Baker 'History of the Marconi Company', 1970

transmitting continuous 'fixing' signals to Continental ground stations and receiving their replies [9], which could then be plotted by the operator aboard the dirigible (and, incidentally received by the British operators).[2] The Admiral of the Fleet, Sir Henry Jackson, credits the Round direction-finding stations with the accurate location of the German Fleet in 1916, through its wireless contacts, as it prepared to leave the safe harbour in Wilhelmshaven, and precipitated his order for the British Fleet to put to sea, resulting in the important Battle of Jutland [10].

Later in the war, and increasingly during the 1917–18 campaigns, wireless equipment became indispensable for the artillery, for line communication and, towards the last year of operations, for forward troop positions also. But it was in the static conditions of trench warfare that both ground and wireless communications received their initial trials in actual combat, leading to spectacular developments in techniques, unique to this theatre of war.

8.2.2 Trench warfare

By December 1914 most of the cavalry were employed in the line, and their wireless equipment lost its vital role in mobile communications. Trench warfare dominated military operations, and a minor static role was allocated to the wireless equipment (apart from its use in direction-finding). The telephone, especially the civilian magneto-electric telephones, which were easily available in France, began to be widely used at the front, despite the earlier negative views held by high command on the use of this new invention (see Chapter 4). This was brought about by the lack of trained telegraphists through casualties and expansion, as much as the convenience of telephones in local operations. For long distances the line telegraph continued to be used in large numbers, and the sheer quantity of telephone and telegraph wires installed along the lines of communication and at the front began to create its own problems. The difficulty lay in the shallow depth of the buried cables and the precarious state of the air lines supported on poles, often along busy communication trenches. Both types of cable were subject to frequent disruption by enemy shelling, and it often proved easier to lay a new cable than to repair gaps in a damaged line. In a very short time the forward areas became strewn with a tangle of cables, most of them derelict. These disused cables, together with working cables using an earth return, combined to produce multiple inductive loops so that cross-talk interference rose to alarming proportions. Telephoning was rendered almost impossible, and even the Morse lines began to strain the skill of the

---

[2] The German Telefunken direction-finding system also made use of a Bellini–Tosi aerial with two large ground stations situated on the north German coast, and 'gave good service in guiding our aircraft in their voyages over England' [9].

telegraphist in deciphering the wanted signal from a multitude of interfering signals.

The problem was not limited to electrical interference on the lines, but was also a source of leakage of operational intelligence, particularly from telephone transmissions which could not be coded directly. 'Overhearing' enemy telephone and Morse transmissions was, in fact, a major component of signal operations for both sides. For this technique the minute earth currents initiated by induction from operational telegraph lines could be detected by suitable 'wireless' equipment and amplified by a sensitive valve amplifier. With the widespread use of these techniques, breaches of security were common until officers learned to limit their telephone calls and ensure that all important signals were encoded and transmitted by telegraph. The unease about the military use of the telephone at the front was enhanced by the discovery in a later phase of the Battle of Mons that it was possible to speak to many towns far behind German lines via the intact portions of the civilian magneto-telephone service. In addition, the British military buzzer signals could be heard by anyone on any of the telephone lines near the battle area. It proved very difficult to impose effective security measures for telephone usage, and by 1916 the use of the telephone within a couple of kilometres of the German lines was forbidden [11]. The German Army was at that time ahead of the British in the detection of earth currents, for which they made use of the effective amplification provided by a valve amplifier [5]. They also applied earth telegraphy with a 600 Hz *Pendelunterbrecher* vibrator, which contributed much interference to telephone lines in the neighbourhood, and which the German High Command eventually decreed should be used only in an emergency.

As the second year of static trench warfare began, on the British side a more organised approach to signal communication was taken. The scattered disused cables were recovered; better construction, maintenance and labelling of working cables was adopted; and a significant improvement was achieved by using twisted line pairs, thus avoiding an earth return. To minimise inductive pick-up and to strengthen the cables, steel-armoured and brass-sheathed cables were laid wherever a semi-permanent installation could be justified. Where an earthed plate was considered necessary, it was sited several hundred metres behind the front line – in some cases as much as 1600 m.

By 1915 the use of visual indication of the received Morse signals, such as magnetic needles, had already been abandoned in favour of aural devices, whereby a sensitive head set could be used to 'listen to' the telegraphic information being transmitted. This had the advantage of leaving the operator free to write down the signal without the assistance of a 'reader' to interpret the moving needle, and also provided a system less susceptible to electric or audible interference. The sounder, one of the earliest audible reading devices, found use in quieter areas of the front, but was overwhelmed by the noise of continual bombardment, so an electric buzzer connected to headphones was the instrument of choice for the forward areas.

In that year too a new technique and use for simple spark transmitters was being devised by both sides as part of the development of earth induction telegraphy (the German *Pendelunterbrecher* was the earliest of these devices to appear). The apparatus used consisted of an induction coil and interrupter, powered by a hand-driven alternator which conveyed a current into the earth via a pair of buried copper plates placed several metres apart. A similar pair of plates constituted the receiving 'aerial'. A high-pitched buzz was heard in a listening set, often established in a deep trench, the set consisting of a crystal detector followed by a three-valve amplifier and a pair of headphones for increased sensitivity. The normal range of such a 'power buzzer' was about 2 km. An operational disadvantage of the sets was their use of plain language transmission, with consequent lack of security, although this could be overcome with some difficulty by introducing message coding. A more severe problem in their use, which eventually led to the withdrawal of this form of ground transmission, was the interference they created in any grounded receiving set nearby. Nevertheless the system proved invaluable to the Allies when used to connect advanced sections at the front, ahead of the buried cable positions, and was used successfully for infantry communications.

The use of the vibrator and buzzer in trench communications was however, a major source of security leakage, owing to the comparatively high potential used and the rapid alternations in the currents sent down the line, which could be detected by inductive or earth current methods. This disadvantage led to the development of a unique form of communication telegraphy in the British Army, known as the 'Fullerphone' after its inventor, Algernon Clement Fuller, then a lieutenant in the Royal Engineers and later to become a major-general. He hit on the idea of substituting the alternating line current sent out by buzzer or vibrator with a very small direct current which would be broken up in the receiving instrument itself at an audio frequency [12]. The Morse key simply acted as a switch to permit the direct current to pass down the line in accordance with the dots and dashes of the code, while the vibrator, now located within the receiver, acted on the received current to interrupt it at an audio rate which could then be heard in the receiving headphones.[3] The direct current in the line caused extremely little induction, and the danger of overhearing was reduced. Incorporated in the instrument was a hand telephone set (Figure 8.3) which could use the same lines, but

[3] This was not a new idea, and had been embodied in several patents by Tesla, Poulsen and Pederson in the years 1901 to 1907, and used by von Arco who referred to the device in 1907 as a *Ticker-schaltung* – a 'ticking contact'. Fuller, however, was the first to apply the technique for the purpose of military security in which it was pre-eminent.

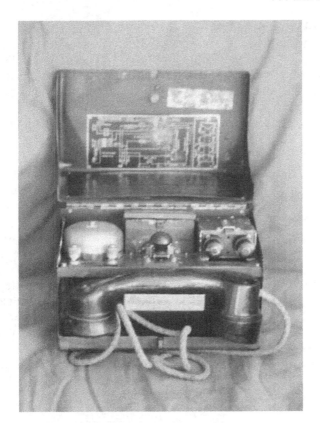

Figure 8.3   The Fullerphone line instrument
Source: *The Air Defence Radar Museum, RAF Neatishead*

there was a warning notice affixed to the case; 'Speaking is only to be resorted to when specially permitted; it is not immune from overhearing in the same way as signals sent with a Morse key.' Although the vibrator was somewhat difficult to maintain in adjustment, the Fullerphone proved very successful and was being manufactured in large numbers by 1916 [13]. It was still in use in the 1939–45 conflict as a main line communications device and for field operations in the Far East.

The major problem of enemy disruption remained, and grew worse as more powerful explosives began to be used, necessitating very deep burial of the cables to avoid destruction. Eventually by 1916 nothing less than a 2 m trench was required, demanding a considerable amount of labour for its construction, making the cable routes not easy to change. Both sides were well aware of this, and would specifically target cable routes and exchanges for their bombardment. During the fighting on the Somme a new technique of laying a standardised grid for front-line cables began to be used by the British whereby each division would have a central cable running from front to rear, with side branches at the level of batteries, brigade headquarters and divisional headquarters, as shown in Figure 8.4. In the event of a break it was thus possible to pass a message laterally to an adjacent division, which could then transmit via its control cable to the next lateral connection. These lines were laid deeply and contained armoured test boxes at regular intervals. If the front line advanced, new side branches would be added, while lateral movement of a division meant taking over the cables of an adjacent formation. This arrangement was found to be labour-saving and effective, and was used up to the end of the war by all the Allied signal services.

8.2.3 Wireless at the front

The German Army became aware of the potential of wireless transmission in trench warfare a little earlier than the British [9]. The Telefunken organisation, the main suppliers of telegraph equipment for the German forces throughout the war, had oriented its technological development towards a mobile war and had produced a military station carried in a horse-drawn cart as early as 1911. This contained a 1 kW quenched spark transmitter giving a range of up to 300 km, with a sectional steel mast 25 m high and petrol generator carried in another cart [14]. The telegraph corps responsible for the transport and operation of this system consisted of an officer and six telegraphists, in addition to the drivers of the six horses limbered to the two iron-wheeled carts used to carry the

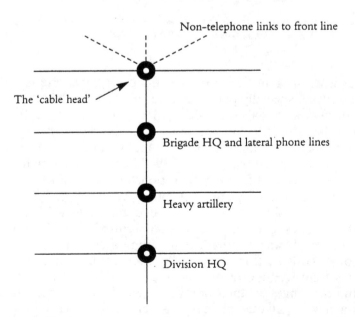

Figure 8.4   Standardised grid for the front line cables on the Western Front, 1917

equipment (Figure 8.5) [15]. When the war began, Telefunken had nothing available in the quantity required for close-contact trench warfare, but had made some effort to produce a short-range wireless communication set using thermionic valves, but this was still at the development stage. Early in 1915 a spark transmitter was hurriedly produced by Telefunken, and made available for trench use and distributed over much of the front. This was the *Funkensendern*, a small quenched spark set providing 100–200 W of radiated power at the low frequency of 500 Hz. It had a range of 300–800 metres and was powered by a bicycle generator or petrol dynamo. This completely replaced the *Pendelunterbrecher* earth telegraphy system which had proved so unsatisfactory. The aerial used was a 20–30 m cable arranged as a V-shaped line, only 1–3 m high, and so did not attract the attention of the enemy. The importance of this development can be seen from its use along a typical front of 40 km, where 150 to 250 of these stations would be distributed, enabling the *Funkentelegraphie* development to provide a reliable front-line communication system.

The Telefunken set was followed by a more flexible 'trench set', manufactured by Siemens & Halske, which incorporated a number of advanced features. This was also a quenched short spark set with a multiple gap assembly arranged so that the number of segments used could be modified to suit the transmitter power required. The set contained both a spark tuned circuit and an aerial tuned circuit with adjustable coupling between them, so that a narrow bandwidth could be maintained which would assist discrimination at the receiver. Six fixed wavelengths were possible between 150 and 470 m (2000 and 638 kHz). A vibrator power supply was interspersed between the 24 V d.c. battery input and the a.c. output of the vibrator applied to the spark unit [16]. At headquarters the German telegraph organisation was equipped with Hughes printers and a Siemens rapid paper tape system, and was in direct contact with Berlin. Communication speeds using punched tape input were as high as 120–140 wpm enabling main line connections to the front in France and Belgium and to their allies in Bulgaria and Turkey to be maintained [17].

At home, to support its valve developments, Telefunken had produced a gas-filled triode in 1913, and later a high-vacuum amplifying valve. The latter was put into use as an amplifier to enhance the reception of overheard enemy telephone conversations by earth current induction before being applied to a lightweight CW transmitter, used by both army and air force in 1917. The transmitter operated on three wavelengths, 150, 200 and 250 m, with a power output of 500 W. Later a 1 kW set was produced and the wavelength range broadened to include 300 and 350 m. To address the problem of achieving selectivity in a spectrum crowded with activity, Telefunken developed with some success a valve receiver fitted with reaction (*Rückkopplungempfänger*), although one

Figure 8.5   German quenched spark field unit, 1911
Source: Eichorn 'Wireless Telegraphy', 1906

could question whether this did not add to the difficulty in unskilled hands, since it was directly coupled to the aerial.

The British, after initial trials using a Marconi commercial spark transmitter, began to overcome the deficiencies in the operation and use of wireless equipment as new sets became available for the forward troops. Three types of wireless set were produced for the army: the BF (British Field) set, the Wilson set and the loop set; all of which incorporated spark transmitters. The first two were fairly substantial installations used mainly at headquarters and corps locations, and only the loop set could be considered as 'portable'. The BF sets, popularly known as 'boy scout' wireless sets, were 1.5 kW Marconi spark sets mounted in lorries or limbered wagons and used mainly at brigade headquarters. The BF and Wilson sets had communication ranges of a few kilometres, depending on the height of the aerial, and required a set of heavy accumulators needing a 'carrying crew of six, in addition to an operating crew of three' [4]. Accumulators, and particularly the charging of them, presented a major problem in mobile operations, alleviated later by the availability of petrol engine dynamo sets. The loop set had a 20 W spark transmitter, working on wavelengths between 65 and 80 m, and was designed for use by unskilled operators; it was more portable and less conspicuous than the BF and Wilson sets. Its transmitting aerial was a tubular loop of 1 m^2, and the receiving aerial consisted of two insulated ground wires some 10 m long. It was first used at the Battle of Loos, where its limited communication range of about 2 km was no disadvantage. The larger spark sets used one of three wavelengths, namely 350, 450 and 550 m, which were allocated separately to each corps in the Army, a procedure which minimised interference between them as well as avoiding affecting artillery reconnaissance transmissions from the Royal Flying Corps equipment which, by arrangement, transmitted on wavelengths up to 300 m. The CW valve sets used later by the artillery had a large number (about 30) of possible wavelengths, which varied from 600 to 2000 m, although in practice this range was limited to 700–1400 m where a short aerial had to be erected.

At the beginning the performance and reliability of the spark sets was not particularly good, and they could not be tuned. The position improved with the appearance of the first CW tuned sets to incorporate thermionic valves in 1917. They were much more selective, though subject to jamming by spark sets situated anywhere within a radius of 500 m. The very low, short aerials which were needed (often just an insulated cable lying on the ground) enabled them to be installed in forward positions under enemy observation. The earliest of these, the W/T trench set, was almost exclusively allocated for artillery purposes and formed the vital link between the forward observation posts and the battery positions. In its Mk III form the trench set consisted of a single-valve oscillator having tuned anode and grid circuits, with the anode supply controlled by the Morse key. The receiver contained a grid detector with reaction control and a single-valve amplifier with a transformer telephone connection. These units, together with a selector box, heterodyne wavemeter, low aerial and power supplies, formed the bulky but efficient portable field set shown in Figure 8.6, which was used by both the Army and Air Force as a mobile forward communications system. It had a range of up to 3500 m when the low, single-wire aerial of about 30 m in length was used, raised only about 1 m above the ground. One of the earliest locations for these sets was at Passchendaele, where the terrain made cable-laying extremely difficult. One set there maintained communication with Ypres, some 11 km away, throughout the battle and, as noted by Captain Schonland, 'despite very heavy shelling gave prompt information of targets for the artillery for many days until blown up by a shell. A new one was installed, a little way back, and the same communications kept up' [18]. Wireless communication was used effectively from the end of 1916 to maintain contact between the headquarters of battalions, brigades and divisions, to supplement the telephone and telegraph in case of failure through lines broken by shelling, or where it proved difficult to lay buried lines. This did not go unnoticed by the German Army, which at that time had made little use of wireless for line communication in that sector. In a captured report on the Somme operations by General von Arnim, he asked for portable trench wireless stations 'like the British', and within a few months of this battle the new Telefunken *Funkensendern* trench sets made their appearance on this front [19].

It was not until CW transmitters became available in 1917 that wireless

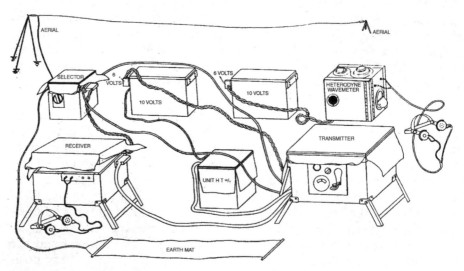

Figure 8.6   The wireless telegraphy trench set, Mk III, 1918
Source: Public Record Office

communication for the forward troops began to make any useful contribution, but they failed to become available in large numbers before the Armistice in August 1918. A typical distribution of all the various types of wireless equipment over the front for the Allies is seen in Figure 8.7, which shows the First Army front just before the final offensive of August 1918 with large numbers of buzzer telegraphs engaged in the most forward positions. These were linked to a 50 W mobile spark set, which was in turn in contact with a 120 W spark set, and so on back through higher levels of control, to a link established with a 1.5 kW light motorised set within range of base headquarters.

8.2.4 Two military engagements

Wireless was also applied in other theatres of the war, often with considerable success. Two specific areas of engagement – Mesopotamia and the drive towards Baghdad in the Middle East, and the war in East Africa – are described below. During the early phases of the operations in Mesopotamia, communications with the control headquarters in India were by naval wireless from a guardship anchored at Fao, where the troops had landed. Messages from the landing forces were sent by army wireless pack sets to the guardship and then relayed to Karachi. A signal company accompanying the brigade laid cables to the divisional and brigade headquarters inland, and operated mainly with vibrators and sounders, supplemented with heliographs. The pack set was a Marconi 0.5 kW spark set incorporating a synchronous rotary spark gap in the transmitter and a carborundum crystal detector in the receiver. It was a fairly substantial affair, requiring a 6 hp Douglas motorcycle engine to drive its dynamo, and installed with two jointed 10 m wireless masts and a 30 m aerial strung between them. It operated on three fixed wavelengths of 500, 600 and 700 m, the middle one being the international shipping wavelength, useful for maritime cooperation in the essentially amphibious nature of the initial operation. The set had a range of about 80 km and was considered highly successful, although it was occasionally plagued with atmospherics caused by tropical storms. From the beginning of the campaign wireless communication was accepted with confidence by the military authorities, unlike the mixed reception it received on the Western Front. The main operation centres in Mesopotamia were linked by telegraph lines manned by operators drawn from the Indian Government telegraph service. They were given a uniform and a courtesy military rank, although they remained civilian employees. They brought with them standard Indian pattern telegraph equipment, including central battery Duberne sounders which broke the circuit when the key was pressed, thus enabling a linesman equipped with a sounder and a key to tap in anywhere to establish communication, a practise established from the days of the American Civil War (see Chapter 4).

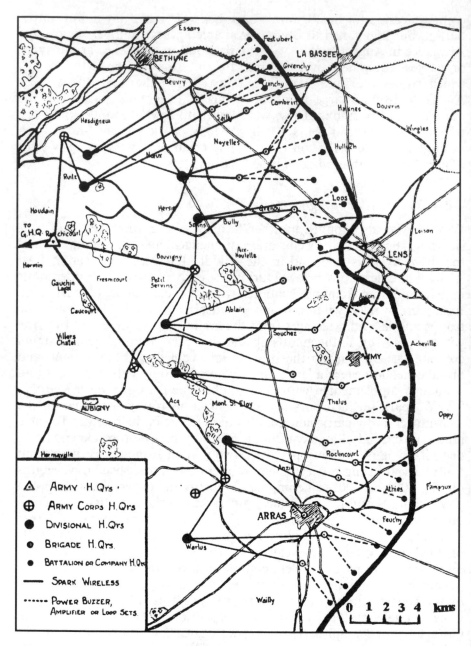

Figure 8.7   Wireless equipment distributed over the Allied front in August 1918
Source: Schonland, Wireless World, 1919

By March 1915 reinforcements reaching the army in its advance towards Baghdad included a corps signal company and some new 1.5 kW spark sets, each installed in a covered wagon drawn by four horses and carrying a 6 hp Douglas engine, dynamo and aerial mast equipment. Again, three fixed wavelengths were used, centred on 1000 m, allowing a communication range of between 250 and 500 km to be achieved. After the fall of Baghdad a separate telegraph route was built from Basra, a distance of 650 km, using heavy galvanised wires carried on standard Indian Government issue iron poles surmounted by porcelain insulators. These were used for the dual purpose of carrying a telephone trunk circuit, operating in quadruplex, and a Wheatstone telegraph recorder which allowed the network to accept paid telegram traffic. A large 30 kW Marconi synchronous rotary spark set was installed at Basra to communicate with shipping and also to provide a link to similar stations which had been established in Bombay and Karachi. In the newly captured Baghdad a Poulsen arc station was installed, and towards the end of the war a similar Poulsen station was erected in Karachi. The campaign in Mesopotamia was significant in the use made of wireless, which was generally more effective there than in any other theatre of the war; the terrain made landlines difficult to install, so wireless became the only practical method of communication. Operations were aided by the absence of enemy transmissions and, more importantly, enemy reception of the transmitted messages, so that, at least initially, the ability to operate 'in clear' raised no problems about the security of the information transmitted.

A second area where military wireless communication achieved a large measure of success was in East Africa, again through necessity, as telegraph lines laid through the bush were frequently ineffective and subject to damage and sabotage. Most of the telegraph lines that were laid followed the railways and were difficult to protect. The German telegraphists apparently regarded the lines as a source of matériel for their own purposes, and often removed large sections of the line to replenish their stores! Signal lines were also liable to be tapped, rendering cipher operation essential. Nevertheless, by December 1915 as much as 2500 km of new cable had been laid in the area, and communication secured across the country from Dar-es-Salaam in the east to Lake Tanganyika in the west. Outside this route the few available wireless sets proved essential for communication between the general headquarters of various Allied forces located in neighbouring states, as well as for the lines of communication within East Africa. Once the port of Dar-es-Salaam had been captured by the Royal Navy and its German wireless station put out of action, the British were able to erect their own military transmitting station to act as a communications centre for subsequent operations. In general the equipment used by the Allied troops was similar to that used in the Middle East, with the valuable addition of some Telefunken CW sets contributed by the South African wireless section, which provided a superior service when used in the African bush terrain.

8.3 The inter-war years

The belated appearance of CW sets on the Western Front during the closing years of the war heralded the demise of the spark set for military communications and the development of a new range of powerful valve transmitters which were to occupy the energies of designers for the next two decades. The key to these new sets lay in the construction of a range of thermionic valves capable of delivering to the aerial a power of several hundred watts. Initially a series of experimental valves were produced for medium-power sets required by the services, and given numbers 31 to 35 when used by the Royal Navy (a full list of these experimental valves and their characteristics is found in Reference 7). These valve types were later used in a new series of Army sets, given an identification letter as A, B or C, according to range. Of these only the A set, a portable ground set of 120 W output with a range of 10 km, and the C set, carried in a vehicle, with an output power of 5 W and a range of 32 km, were produced. A series of medium-range sets for armoured fighting vehicles were also produced and designated MA, MB and MC. These were designed for telephony transmission, but with a disappointing range of less than 10 km. Experience with these and other experimental transmitters prompted a re-evaluation of requirements for army communication equipment, and in 1929 an important review was carried out leading to the specification of a number of new set types. Originally six types, numbered 1 to 6, were considered: No. 1 set for infantry and artillery brigades, No. 2 set for communication between division and brigade, No. 3 set for corps communication, No. 4 set for GHQ and the armies, No. 5 set for the lines of communication, and No. 6 set for worldwide communications (Army Chain). Nos 1, 2, and 3 sets covered a relatively narrow frequency band, and the total range of the three sets was from 1.36 to 6.66 MHz. In 1929 it was not thought that the use of reflected waves would be feasible for forward tactical communications, and the sets were designed for ground wave operation only – a decision which was to have a deleterious effect on subsequent military operations. These proposals formed the starting point for the army wireless requirements in 1939–45, which were to a large extent met, and the range was added to as the conflict proceeded.

8.4 War on the ground, 1939–45

The outstanding characteristic of the Second World War on land was its mobility. With a few exceptions the kind of trench warfare experienced in the 1914–18 conflict never materialised, and the communications techniques that developed were mainly those of wireless telegraphy and telephony, supported by line communications for major signals traffic in the controlled areas.

At the beginning, however, in the static phase of hostilities in Europe before the collapse of France, communication systems were still based on buried lines with wireless used as an emergency reserve. The lines carried more telephone than telegraph circuits between various headquarters, with teleprinter operation superimposed on the telephone lines. Techniques for burying cables had improved, with mechanical trench excavators provided by the Royal Engineers, and a grid cabling system used which had been found so effective in the closing stages of the First World War. This pre-planned scheme was soon rendered invalid as the war entered a mobile phase with the fall of France and the subsequent evacuation from Dunkirk and other locations.

Apart from its use during the retreat from France, discussed below, the value of wireless first became clear in offensive operations intended to recapture Narvik from the Germans in May 1940. This required reliable communications links to be established between the United Kingdom, the Royal Navy and units of the army carrying out field operations. A newly released and experimental 250 W transmitter (a prototype No. 12 set) was used to provide communications across the North Sea, while smaller mobile sets, the army sets Nos 2 and 9, were used by the fighting formations. For the purpose of this operation these operated with radiotelephony,[4] using ground-wave communication over a limited range of less than 70 km. After the initial recapture of Narvik it became apparent that the only submarine cable in the area, linking Narvik to Bodo, was being used by Norwegian enemy agents (quislings) to communicate with the German forces in the south. The cable was accordingly located by the navy on the seabed, raised, and cut by removing a section to prevent repair. Thereafter the evacuation of this bridgehead was carried out in reasonable secrecy. During the defensive activities in France and Belgium associated with the retreat to the Channel ports, wireless became the essential link between scattered units, the Navy and the War Office in London. One pre-war cable was kept in operation throughout the final stages of the evacuation. This was the submarine cable from Dover, terminating in La Panne, adjoining Dunkirk harbour, which was augmented

---

[4] In 1910 the term 'radio' began to be used in the United States and was soon adopted elsewhere in place of the term 'wireless'. For some time, however, and during the period discused here, 'radio' was generally used to indicate communication by telephony, with 'wireless' retained where telegraphy was employed, although both terms are used indiscriminately in the literature.

by further terminal equipment transferred from elsewhere and played a vital role in the operation.

8.4.1 Line working

Line transmission within the United Kingdom provided the backbone of communications between military sites, and especially the RAF bomber and fighter stations with their anti-aircraft defence installations, being established across the country. Strenuous efforts were made by the Post Office to maintain connections during bombing raids. The severity of the problem may be seen in the recorded damage in the central part of London where on one occasion a single high-explosive bomb caused the severance of 5200 working circuits. Airfield damage to line communications, although affecting a large number of cables, could usually be repaired within 24 hours, despite the extreme difficulties encountered at the site of the bombing. (A detailed account of this little-known but vital work is contained in the Post Office archives [19].) In anticipation of the problems expected to be caused by bombing, invasion or sabotage, an extensive network of wireless telegraphy stations had been established by the Post Office before 1939. This consisted of 32 emergency stations located at the headquarters of 12 regions, into which the country had been divided for defence purposes, together with stations at several large towns. A number of frequencies in the marine mobile frequency band had been allocated for this purpose, and some of the stations could also operate in the high-frequency bands. The power of the stations did not exceed 500 W and was generally about 200 W. An important innovation pioneered by the Post Office was the operation of point-to-point wireless telegraphy links using single sideband radio telephone equipment. With the cooperation of the American Telephone & Telegraph Company, a technique was evolved whereby several Morse and teleprinter channels could be carried on each of the single sideband radio telephone channels. This method of enhancing the number of message channels supported on a single telephone channel was later taken over by the Army and RAF and widely used in overseas locations [20].

The need for multiple working of transmission lines for telegraph and teleprinter communication within the United Kingdom had been fully appreciated by the Allied forces before 1939, when a large number of communication channels were seen as imperative for successful defensive operations. To facilitate this provision a Defence Teleprinter Network (DTN) was set up throughout the United Kingdom to meet the requirements of the three services. This involved the creation of a number of telegraph installations, some of them quite large, in different parts of the country. These were manned switchboard centres, containing both teleprinters and reperforators for automatically transmitting data and messages to the next link in the exchange network. The provision of new lines and centres for the DTN grew during the conflict, and by the end of 1944 some 7000 telegraph channels were being provided [21]. In addition the individual services arranged their own separate mobile telegraph and teleprinter systems based on the same frequency-division multiplex (FDM) switching methods used by the Post Office. Here a number of telephone or teleprinter messages were amplitude-modulated, each to a different carrier frequency, and the modulated signals added together to form a composite message signal. They were subsequently separated at the receiver by bandpass filtering and demodulated in the usual way.

This method was adopted by the military in a number of versions, generally taking the form of a modulated carrier system known as Apparatus Carrier Telephone or ACT(1+4), which became used throughout the campaigns in North Africa, Italy and Western Europe. This provided one audio channel and four carrier channels over a single pair of heavy copper wires. A signal (usually speech) from each of the four derived carrier channels modulated one of four carrier frequencies, 6.0, 9.2, 12.5 and 16.0 kHz, and propagated this in one transmission direction. In the opposite direction the modulated signals are first group-modulated at a new carrier frequency before going down the line, producing a new set of carrier frequencies of; 19, 23.5, 25.8 and 29 kHz. This allows four speech signals to be carried in both directions along the line without mutual interference, with separation in the two directions effected by channel filter sets. For lengthy lines the attenuation for these high frequencies (high, that is, for line transmission) was alleviated by inserting repeater amplifiers every 150 km, each having a gain of about 30 dB. A similar FDM system was used for line telegraphy and, because of the lower bandwidth needed, six duplex working telegraph instruments could be supported in the same frequency band as that employed normally for telephony. The ACT(1+4) system was used frequently with the audio frequency channel carrying multiplexed telegraph signals in this way, while the four speech channels remained available without mutual interference, being carried on the same wire.

For shorter lines laid for a specific operation, and often in difficult terrain, the earlier line telegraph equipment proved effective. The Fuller-phone, for example, developed in 1916, was much improved and used extensively in the jungles of South-East Asia, albeit with frequent attention to the interrupter which still retained its mechanical form [4]. Laying field cables for these single lines across rivers, minefields and thick jungle was sometimes achieved by unorthodox means such as shooting the line across an obstacle by means of mortar bombs or rockets, a procedure that was effective up to 100 m. Experiments in which cable was laid across thick jungle by paying it out from an aircraft were reasonably successful [22]. The scale of the task of laying single and multiplexed lines for the Allied forces during a single operation can be seen in the requirements needed for the Italian campaign, in which over 82 000 km of wire was supported on 29 000 poles, exclusive of additional buried lines and extensive repairs to the civilian post office network carried out during the liberation of Italy.

Line working for the American forces came under the jurisdiction of the US Signal Corps. They had initially abandoned Morse code in 1938 for line transmission, and advocated instead extensive use of teleprinters together with R/T for wireless communication. This view was not to survive the entry of the United States into the war. Much use was made of teleprinters for non-mobile land operations, particularly during the European mainland operations towards the end of the war, and radio speech communication when VHF FM sets became available in 1944. But advanced ground tactical communications and aircraft operations made extensive use of W/T, and the Signal Corps developed line and wireless communication equipment accordingly. A field set Type TG-5 incorporating a buzzer and manual keying was in wide use, and high-speed keying and recording equipment was used for automatically transmitting and receiving Morse code signals over wireless networks. A variety of telegraph terminal and relay equipment combined a number of teletype, telegraph and, in some cases, speech channels using a carrier transmission system, described earlier [23]. The Signal Corps devised a robust four-channel cable, known as 'spiral four' (WC-548), which began to be manufactured and used in vast quantities by the Allied forces, either buried or simply laid on the ground and capable of conveying teleprinter and telegraph signals long distances without interference from signals on adjacent line sources [24]. At a later stage in the war the Signal Corps began to use FM VHF radio sets for tactical communication in place of wire lines. A broadband FM set, the Type AN/TRC1, was used in conjunction with speech-carrier equipment CF1 and CF2, to provide multi-channel teleprinter communication during the Normandy landings, and became an important communication system during the subsequent European campaign [25].

8.4.2 The African campaigns

In the Western Desert, another major location for land operations, wireless was the primary means of continual operational communication, supporting a quasi-permanent line route along the North African coast which was only intermittently available. Among the difficulties encountered was the fundamental design of the wireless equipment as solely a ground-wave communication device. This originated in the first two decades of the century, when not much was known about the function of the ionosphere in reflecting waves in the higher frequency bands. In the Western Desert the distances over which messages had to be exchanged were often too great for ground communication with the low power available, and although aerials suitable for reflected waves were improvised, the sets did not have a wide enough frequency range for the optimum frequency for sky-wave transmission to be selected. The position improved later in the conflict when new sets were supplied to the army units. A further problem which affected tank communication was that power for the wireless equipment was drawn from the tank battery. The equipment was frequently unusable, since recharging relied on running the engines and was nullified by lengthy periods without movement. Until separate batteries were installed specifically for the wireless equipment spare batteries needed to be available for this purpose.

By the time of the El Alamein offensive in 1943 the position was much changed. The equipment was better designed for the purpose and more plentiful; supply and maintenance had improved and staff were more experienced. A small robust and reliable wireless set fitted to tanks consisted of a short-wave transmitter with a nominal range of about 24 km, a small ultra-short-wave receiving set for local communications between groups of tanks, and an intercommunication amplifier for the crew. The headquarters signals office was made completely mobile with communications equipment, switchboard and duplicate equipment mounted in converted commercial vehicles. The equipment used for this latter purpose was mainly the Marconi SWB8 or SWB11 transmitter. Over 380 such wartime conversions were prepared at Marconi in Chelmsford for the Army during the war [26]. This procedure was to be adopted as standard practice in all the European theatres of war, culminating in outstanding designs for the 'Golden Arrow' and the Army No. 10 multichannel sets, to be described later. In North Africa the coastal cables were improved from Alexandria to Mersa Matruh to contain 14 pair buried cables and a network of field cables installed by cable ploughs. This cable network was supported by a parallel wireless network, the combination providing extremely reliable communication for the rest of the war in North Africa. As the forces advanced through Tunisia, a permanent line route of cables kept up with the rate of advance. On arrival at Tripoli the end of the submarine telegraph cable to Malta was found hidden but undamaged and put into use immediately, thus completing arrangements for telegraphic communications in the Mediterranean area.

8.4.3 Communication systems

The progress of the war on all fronts had shown the need for high-speed and mobile communications capable of dealing with the massive quantity of operational signals now required to maintain the initiative for any major action. Two new wireless developments were borne out of this requirement, both highly mobile and capable of transmitting a large quantity of information over long distances. The first was the 'Golden Arrow' communications system first used in Tripoli for direct communication with Cairo during the North African campaign. It consisted of a Marconi SWB8 high-power mobile transmitting station providing 3 kW to the aerial, together with a number of receivers, and was used for long-distance communication direct from the tactical head-quarters at the front. The five vehicles and two trailers that transported the system were, when travelling together, dubbed the 'Golden Arrow' after the London-to-Paris boat train, well known to service personnel in the war years. The largest vehicle, containing the receiving equipment and some of the 22 operating staff, was semi-articulated with a long van body. A military-type Wheatstone system was carried, with several keyboard perforators for preparing the transmitter tape, communications receivers and a receiving undulator (siphon recorder). The undulator worked at operating speeds of up to 150 wpm, although for recording the output tape was drawn across in front of a typewriter and transcribed by eye at speeds up to about 30 to 35 wpm. This method had several advantages over complete automatic working, not least in minimising recovery time in the event of a faulty transmission. The transmitter was housed in a separate vehicle, connected to the receiving vehicle by telephone. The frequency was adjusted by means of interchangeable tank circuit coils and aerial switching. The two transmitting aerials used were suspended between 20 metre collapsible masts, one for night operation (low frequency) and the other for daylight working (high frequency). Power was provided by two diesel generator units mounted in trailers, giving a 27 kVA three-phase supply at 230–400 V [27]. A number of these mobile units were constructed for operations in various locations in North Africa, and during the landings in Sicily a 'Golden Arrow' station was run ashore on landing craft late in the evening and was handling messages to and from the War Office in London by 09:00 the next day [28].

The second development was a highly secure transmission system which became available during the last phases of the war, using techniques derived from radar transmission. As the war progressed on all fronts, more emphasis was put on telephony, teleprinter and wireless transmission for all short-distance links. The American forces contributed a four-channel single-sideband radio teleprinter system used for communication between headquarters and the War Department, together with a single-channel VHF link serving the same purpose, the AN/TRC1, described earlier, which provided a major wire line replacement system for the later operations in Europe. Towards the end of 1944 a few No. 10 army sets, a beam wireless system of advanced design, made their appearance in the Italian campaign. The No. 10 set had a distinct advantage over the AN/TRC1 in the rapidity with which it could be brought into use, and in addition it provided almost completely secure operation for the first time in military history without the use of signal encoding. These sets were manufactured in quantity and used effectively during the final operations in Europe, mainly to bridge gaps in landline communication across water (including the English Channel) and in difficult terrain. The system brought together a number of advanced UHF communications techniques in its design. It operated on centimetriec waves at a frequency of 4500 to 5000 MHz (6–7 cm wavelength), and the method of generation used gave it an exceptionally wide bandwidth. It provided eight speech channels, the signals being impressed on the carrier by a system of pulse width modulation [29]. The carrier itself was generated by means of a split-anode magnetron, the equipment using waveguide techniques in both transmitter and receiver, the latter conveniently making use of a 45 MHz intermediate frequency amplifier culled from pre-war television experience. The complete equipment was carried in a four-wheeled trailer, with generator, test equipment and operating positions; the sending and receiving aerials were dishes mounted on the roof (Figure 8.8). The sets were used in pairs over a narrow line-of-sight transmission path of between 30 and 80 km. A transportable tower mounted on the roof of the trailer was sometimes used to gain additional range and if possible one of the sets was situated at a high elevation. This was possible in Austria, for example, where a set was located at Stuhleck, an 1800 m mountain, giving satisfactory reception at Kahlenberg, some 86 km away [30]. Much use was made of the relay capability of the sets to follow the line of advance during the final drive into Germany, with as many as six relay stations in use at one time. This ensured that Field Marshall Montgomery's tactical headquarters

Figure 8.8   The No. 10 army set
Source: The Royal Corps of Signals Museum, Blandford, Dorset

was never out of touch with the whole of his 21st Army Group or with the War Office in London.

8.4.4 Across the Channel

The invasion of Northern Europe in 1944 demanded a wide range of different types of communication equipment, many of which had been developed since the war began. In addition to the standard equipment developed for tank, infantry and long-range communications there were a number of sets produced for special purposes. One of these was the ubiquitous Type 76 set, originally produced for commando formations, and used extensively during the early stages of the invasion over the Normandy beaches and by the parachute troops. The set was small, light and robust enough to withstand being dropped by parachute, but powerful enough to enable communication to be maintained at distances of up to 400 km. By this time communications forward of brigade were conducted almost entirely by wireless telephony, but further back telegraphy was used extensively and, with skilled operators, the traffic could be handled more quickly and reliably by Morse code [31] One area where telegraphy was used exclusively was in the operation of submarine cables. Earlier, following the defeat of the Axis forces in the Mediterranean, the cables linking North Africa with Malta and Gibraltar and the existing cable to the United Kingdom were restored to their original effectiveness. Several of the French and Italian cables were also retrieved from the seabed and adapted for services to Tunisia and Sicily, and the cable ships were kept busy at this work throughout the Mediterranean to lay the new cables that were needed.

New cables and terminals were established in conjunction with D-Day, in particular two new ones direct from Southbourne to Longues and from Swanage to Querqueville in Normandy [32]. The Post Office had four cable ships in operation: *Monarch, Alert, Ariel* and *Iris*. The *Alert* and *Iris* laid the first submarine cable to the Normandy beachhead two days after the initial landing. These were hazardous operations, since the stationary cable ships formed sitting targets. *Alert* was sunk with all hands lost while repairing an Anglo-Belgian cable in February 1945 and *Monarch* was lost a few months later.

Other cables were reinstalled between England and the Continental coast as the invasion front widened. These are indicated in Figure 8.9, which also shows the beam wireless links, as dotted lines, established by using the No. 10 army set described earlier. By the end of August 1944 there were a total of 27 speech and 39 telegraph channels to the Normandy coast, with further cables installed later to the French, Belgian and Dutch coasts, some 664 km in total length before D-Day and a further 990 km after the invasion [33].

The traffic carried by these lines was enormous, not only because of

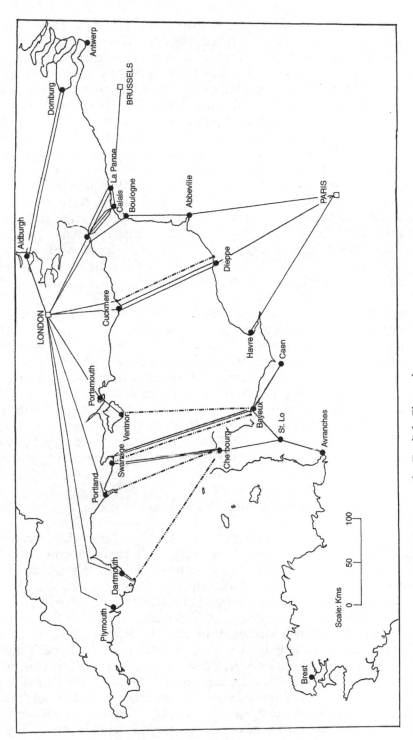

Figure 8.9 Invasion communications across the English Channel
Source: Nalder 'History of the Royal Corps of Signals', 1958, reproduced by permission of the Royal Signals Institute

the enormous requirements of administration, allocation of resources and operations control, but also to maintain morale with invading forces and public at home. Elaborate arrangements were made to accommodate press telegrams, and private telegrams between the troops and their relatives in the United Kingdom. One of the line communication problems that arose as the armies moved into France was the acute shortage of high-speed circuits capable of supporting teleprinters: the Baudot system, prevalent in France, was not at that time compatible with British equipment. While the speech plus duplex line-sharing systems, used throughout the war, were applied to this problem a valuable addition to this equipment was a development of a Voice Frequency Telegraph System, considered briefly earlier. This replaced the d.c. telegraph signal with a tone frequency system using trains of interrupted continuous impulses within the audio frequency band, making it possible to fit six separate duplex telegraph circuits into the same band width as one speech (voice) signal. With the frequency spacing between channels reduced to 60 Hz twelve teleprinter messages could alternatively be sent along the same pair of wires [34].

During the actual landings, communications depended almost entirely on wireless, and special wireless ships were adapted for this purpose. The first was the merchantman S.S. *Bulolo*, which was refitted in 1942. The transmitting and receiving aerials were located in different parts of the ship and separated by the maximum possible extent, and by a careful selection of frequencies it was found possible to work simultaneously 15 transmitters and a large number of receivers without mutual interference. By including signals personnel from all three services on board ship, military commanders were able to achieve complete operational control in advance of the setting up of headquarters in the landing area.

8.5 Army wireless in the Second World War

In 1939 the radio industry in Britain was fairly well placed to meet the wireless requirements for the three services, but was ill-equipped to respond to the design problems encountered. Fortunately the services had their own well-developed establishments which had been carrying out research into wireless for service purposes for the previous decade. These were the Admiralty Signals Establishment, the Signals Research & Development Experimental Establishment, the Radio Department of the Royal Aeronautical Establishment, the Air Defence Experimental Establishment and the Bawdsey Research Station, although this last was concerned solely with the development of the Radar Home Chain, and was the first of several research stations concerned with the new techniques of radiolocation (as it was then called). All these establishments were responsible for research, development and actual design of new equipment, and were able to assist British industry in gearing their factories into profitable wartime production with considerable success [41]. Later in the war some production of British service equipment was carried out in the United States under the 'Lend-Lease' arrangements, and use was made of a number of American designs originally produced for commercial or United States services use.

The diverse nature of operations in the Second World War demanded a number of different types of wireless communications equipment. Many of these were designed for mobility and carried by armoured and non-armoured transport, tanks and infantry (man-packs). They can be designated medium installations where the transmitted aerial power was less than 1 kW, and short range where the transmitted power was less than 25 W. The man-pack had a very short range and an aerial power of less than 2 W. Others, specifically for long-range communication, had aerial powers in the kilowatt region and were fixed installations with large aerial arrays. In addition to these some special wireless sets became available, often in large numbers, to armoured fighting and other vehicles for communication with aircraft and for multi-channel radio link purposes. At least one receiver, the Type R308, was developed specifically for intercepting German Army tank transmissions in the range 20–145 MHz, which used VHF FM (not then in use by the British Army).

Throughout the war most of the infantry and tank sets operated in the HF band below 9 MHz, despite the congestion of the numerous frequency channels operating at that time. It was only gradually realised that, while the attenuation of transmissions in the 30–150 MHz VHF band would be high under, for example, jungle terrain conditions, this was more than offset by lower noise levels, particularly at night. Towards the end of the war there was a movement towards the use of the higher frequencies, already used for aircraft communications, and to apply FM in telephony transmissions for the army.

Most of the army operations in the field in 1939–45 required the individual units of an army group, for example a tank squadron, to tune their transmitters to the same operational frequency at a time when the supply situation made the bulk use of quartz crystals impracticable. The process, referred to as 'netting', was carried out by tuning the receiver at the substations to obtain the maximum signal from the control station. The frequency of the transmitter oscillator at each substation was then adjusted to give zero beat with the incoming carrier signal, which completed the netting process. Towards the end of the war a more plentiful supply of crystals became available, and crystal control was widely used to simplify the netting procedure. This did, however, mean that very large numbers of crystals needed to be held to cover alternative frequencies: a special pack containing over 900 crystals was required by a single army division [36].

8.5.1 Allied wireless equipment

Various Allied army transmitting and receiving sets have been mentioned in connection with specific campaigns, and an extended list is given in Table 8.1. In this section the design and performance of the most common of these sets are discussed. First, it will be helpful to list the numerical designations of the specifications that were in use. In the British Army they followed the initial designations introduced after the 1929 reorganisation, whereby the type of complete communication set or transmitter was indicated by the last digit:

1   Short-range brigade and artillery sets
2   Short-range division sets
3   Medium-range corps and mobile sets
4   Intercommunication (GHQ/corps) sets
5   Long-range (base/GHQ) sets
6   Long-range Army Chain sets
7   Special types
8   Infantry (man-pack) sets
9   Armoured fighting vehicle sets
0   Multi-channel radio link sets

Letters were sometimes added to indicate variations on the basic type and an initial number added to designate a replacement type: Type 18, for example, was the first replacement for Type 8. The entire nomenclature was revised in 1947 since by that time many sets were being employed outside their original tactical roles, and a new coding was extended to include letters and figures to indicate power output, operational frequency range and version [22]. The US Army used a Signal Corps designation SCR for wireless/radio sets and a two-letter designation, TA, TC, etc. for line equipment. Towards the end of the conflict the SCR nomenclature was replaced by a new system reflecting the need to standardise on equipment types among the armed services. This was the Army–Navy, or AN, designation.

Whereas in the First World War the wireless sets (mainly spark transmitters and crystal detectors) were designed as well as manufactured by industrial concerns, principally the Marconi Company, during the inter-war years the military organisations became increasingly involved in specification and design. By the late 1930s specifications were being proposed by the military users and put into effect by the military design organisations in cooperation with the manufacturers, of which there were now several actively engaged in producing military equipment. The US armed forces, although as concerned with specification as the British, contracted out to their commercial organisations the task of designing and producing the equipment, which often appeared with a service nomenclature to include the name of the manufacturer. To assist in the

Table 8.1 Allied army wireless sets, 1939–45

Type	Frequency (MHz)	Range (km)	Power (W)	Purpose
1	4.2–6.66	13	0.5	General-purpose set
2	1.875–5.0	30	10	Division communications
3	1.3–3.4	160	400	Mobile set
5	2.4–20.0, 0.2–0.6	3000	2,000	Point-to-point communication (Army Chain)
6	3.0–25.0	worldwide	1,500	Army Chain
9	1.875–5.0	30	10	Tank communication
10	4.55–4.76	80	0.4	Line of communication
11	4.2–7.5	32	4.5	Mobile set
12	1.2–17.5	96	25	Line of communication
12HP	1.2–17.5	1500	350	Command set
15	4.0–22.2	worldwide	20,000	Man-pack set
18	6.0–9.0	16	0.25	Tank communication
19	2.0–8.0, 19.0–31.0,	24	12	Brigade communication
21	4.2–7.5	3.2	0.8	
22	2.0–8.0	32	1.5	General pupose
23	1.12–13.55	1600	250	Mobile set
33	1.2–17.5	640	250	Line of communication
36	10.0–60.0	40	25	Anti-aircraft communication
42	1.6–12.8	40	10	Man-pack set
46	3.4–9.1	16	1.5	Man-pack set
48	6.0–9.0	16	0.25	Man-pack set
53	1.2–17.5	800	250	Mobile set
63	3.0–20.0	1600	1000	Line of communication
68	1.75–2.9, 3.0–5.2	16	0.25	Battalion headquarters
76	2.0–12.0	480	9	Long range CW communication
R107	1.2–17.5	640	—	Receiver used with sets 12, 23, 33 and 53
R109	2.0–12.0	500	—	Receiver used with set Type 76
R206	0.055–30.0	—	—	Intercept receiver
R308	20.0–145	—	—	VHF receiver
SCR274-N	0.2–9.0, 20	240	50	Command set
SCR293	20–27	5	50	Tank and mobile communication
SCR299	20–48	500	400	Mobile set
SCR300	40–48	10	0.5	Man-pack set
SCR399	2–18	400	400	Tank communication
SCR499	20–26	400	400	Parachute Corps
SCR510	20–30	5	50	FM tank communication
SCR597	2–20	560	300	FM mobile set
FT4331	3.0–20.0	1000	10000	Line of communication
ET4336	2.0–20.0	480	250	General-purpose set
BC447	2.0–18.0	3000	300	Army Chain set
BC460	2.0–18.0	160	200	Command set
BC610E	2.0–18.0	500	300	Mobile set
SWB8E	3.0–22.5	8000	3500	Point-to-point communication
SWB11E	3.0–22.5	worldwide	10'000	Point-to-point communication
AN/TRC	70–100	160	36	Mobile set for radio relay

production and distribution of all these products a number of inter-services committees were established and were extremely successful in meeting the service requirements during the Second World War.

The most powerful transmission sets available to the British Army at the commencement of the Second World War were the long-range Type 5 and 6 transmitters equipped with high-speed Morse sender and keyboard perforator, allowing keying speeds up to 600 wpm to be achieved. In 1951 they were replaced by an even more powerful Type 15, a single-sideband transmitter especially adapted to worldwide multi-channel teleprinter code transmission [37]. All these long-range wireless transmitters operated on frequencies between 2 and 25 MHz, using sky waves, and generally incorporated directional aerial arrays. They formed part of a limited series of 'Army Chain' stations, located at strategic posts around the world and linking Aldershot to garrisons at Hong Kong, Cairo, Jubbulpore (Jabalpur) and Singapore. Two other transmitters used by all three services were the Marconi SWB8E and SWB11E transmitters having similar performance to the Type 5 and 6 but rather greater power output, applied to a specially designed and directional rhombic aerial array. The smaller of these, the SWB8E, was employed in the Golden Arrow mobile station discussed earlier. For medium-range mobile installations, typically those required for corps command vehicles, a range of at least 150 km for CW Morse operation was required, a range later demanded for modulated speech transmission also. Medium frequencies between 1.36 and 20 MHz were used for this purpose, but in many of the earlier receivers the tuning range was narrower than this.

For point-to-point communication, set No. 3 was available at the outset of the war (it was designed in 1929), and although it produced a fairly high output power its range was not found acceptable for later operations, and neither was its frequency coverage acceptable for the large number of communication channels later to be demanded by operational requirements. Two developments which did much to remedy these deficiencies were the No. 33 set and the No. 53 set which replaced it. These covered a much wider frequency range, provided a higher power output and were reliable but bulky in comparison with the American SCR299, 399 and 499, which became the standard medium range group in later periods of the conflict when 'Lend-Lease' supplies of US equipment began to be available in quantity. The SCR299, manufactured by the Hallicrafter Company, was a 400 W transmitter having the extensive range, for a mobile set, of several hundred kilometres using short-wave transmission. It was transported in a vehicle towing a trailer containing its power supply generator. The SCR499 achieved some notoriety owing to its ability to withstand parachute dropping in difficult terrain [23]. High-speed telegraphy in the form of teleprinter and teletypewriter code was by then essential to handle the telegraph traffic requirements of the armies, and a valuable contribution was made through a modification of the SCR399 set, so that it could be employed as a relay station in the 26 MHz band, for this allowed it to work with the United States teletypewriter,[5] then in wide use with the Allied armies [4].

The requirements of combined operations with airborne formations led to the production of a transportable medium-range transmitter, the Type 76, mentioned earlier, which could be dropped by parachute or transported in a light vehicle. This operated in the 2–12 MHz band and was designed for CW Morse transmissions. It was one of the first British military sets to make use of a crystal oscillator to provide up to six fixed transmission frequencies which, by appropriate anode tuning in the power amplifier, could be doubled to give twelve spot frequencies from the fundamental and second harmonic of the crystal oscillator. This type of 'spot frequency set-up' was valuable for mobile use when fully trained operators were not available. It was normally used as a battery operated transmitter from a 12 V supply, its HT supply provided by a rotary converter, but provision was made for mains operation in a suitable fixed station location. The receiver normally associated with this transmitter was the Type R109a, an eight valve superhet covering the same frequency range of 2–12 MHz and with a vibrator supply unit [38].

Armoured units required a reliable short-range set which could operate in a rugged environment. Several types were produced as the war progressed, but initially Types 9 and 11 provided the backbone of short- and medium-range wireless communication for tanks in the first years of the war. They operated in the HF band between 1.875 and 7.5 MHz, with transmission confined to the ground wave only. A master-oscillator, power-amplifier (MO-PA) configuration was used with the MO sufficiently stable to avoid the use of quartz crystals, except as a means of calibration for the control sets during the 'netting' procedure. This operation was sometimes difficult to carry out, but was much simplified in the Type 11 combined transmitter and receiver. This set incorporated the novel feature (at the time of development) that the MO frequency and the receiver tuned frequency were mechanically kept the same so that it automatically receives on the frequency to which the transmitter is set. Types 9 and 11 were later replaced by Types 19 and 22, both having a frequency band coverage of between 2 and 8 MHz; smaller and lighter, they became the standard medium-power tank wireless set after 1942, and were manufactured in the United States, although designed in Britain. The Type 19 provided three different sets in one assembly: an HF transmitter for working over distances of up to 24 km, a small VHF set for local communications and an intercommunications amplifier for internal communications between members of a tank crew.

As the combined Allied forces expanded, so did the need for many

---

[5] The British teleprinter and the United States Teletypewriter were incompatible, which caused some problems in combined operations.

more communication channels. The British, with their general policy of maintaining field communications by ground waves at frequencies below 30 MHz, were in some difficulty in finding sufficient numbers of channels and attempted to avoid interference between them with a reduced separation of only 4 or 5 kHz. This was not sufficient for the large numbers of operational channels now required. The US forces used frequency modulation and generally higher frequencies, in the band 20–48 MHz, with greater use of telephony for communication with and between armoured vehicles. While the US Army used vehicle-mounted SCR299 and SCR399 transmitters for long range AM communication, as discussed earlier, for short-range transmission reliance was placed on frequency modulation using two new series of compact sets developed in 1942, the SCR508 and SCR608. Frequency modulation was used extremely effectively by the US Army for its armoured vehicles, beginning in 1941 with the improvisation of an FM set, the SCR293, based on a civilian police radio designed by Fred Link. Many of these 'Link radios' were pressed into service before the Armored Corps received the more robust transmitters of the '500' and '600' series. These included the short-range SCR510 tank set which featured ten preset crystal FM channels (out of a possible eighty) in the frequency range 20–30 MHz. The use of these push-button crystal-controlled sets marked the beginning of easy communication, long advocated by Colonel O'Connell of the US Signal Corps with his declaration of, 'We're all through with radio; hereafter we want communication.' He meant instantaneous R/T communication, which was to come with widespread use of such FM systems towards the end of the war [39]. The superiority of FM over AM in short-range military communications was demonstrated in a major tank engagement towards the end of the war in Europe. A number of SCR510 transmitters were fitted to aircraft and used as airborne jamming transmitters during operations in the Battle of the Bulge in the Ardennes during December 1944, when the German AM tank communication on the frequency band 27–33 MHz was severely dislocated by FM transmissions on similar frequencies, with no serious deterioration of FM reception over the same frequency band.

A number of sets were produced in the United States and in Britain for special purposes, such as for multi-channel radio links, anti-aircraft artillery communication, ground-to-air operation and a large number of short-range portable sets, 'man-packs', were produced for the infantry. The most successful of the man-pack sets were Types 46 and 68, both of which were waterproofed and robust. The sets were crystal-controlled for both sending and receiving which made them ideal for rapid adjustment in an active field situation. Towards the end of the war the Allied forces standardised with two compatible man-pack sets, the American SCR300 and a British equivalent Type 31. Both employed frequency modulation and operated in the frequency band 40–48 MHz. The SCR300, the original 'walkie-talkie', as it was known to the US military, weighed less than 17 kg. It was operated almost exclusively as a R/T transreceiver using frequency modulation in the frequency band 40–48 MHz and with a transmission range limited to about 10 km. As with the bigger SCR293 and the SCR500 series, it owed its origin to work on civilian police communication in the United States, and very large numbers were produced during the Second World War for the Allied forces. An even smaller pack set, used primarily for parachute troops, was the SCR536, which covered a similar frequency band but weighed just 2.25 kg.

8.5.2 German Army wireless

A feature of the German Army equipment, as with similar wireless equipment used in their navy and air force, was its uniform date of design and method of manufacture. Unlike their radar equipment which developed continually during the later stages of the conflict, their wireless communication apparatus showed little change during the war, having been carefully designed in the 1936–40 period to satisfy the needs of a mobile army, not expected to participate in a long war.

Almost all German Army wireless sets were made from a lightweight magnesium alloy. The rigid construction consisted of a box-like assembly of thick alloy castings with subunits similarly designed and bolted onto the main framework. While this method of construction simplified maintenance and led to a marked stability in operation, it also gave the design little flexibility, with the result that very little change in design occurred throughout the war. Other features of note were the wide use of high-quality ceramic insulation and dust-iron cores for the RF inductances, both a result of the exceptional progress made by the electronics industry in Germany in the pre-war period [40]. A considerable attempt to standardise on the valves used in a given piece of equipment facilitated the supply of spares and provision of maintenance. German Army sets used two main systems of identification: a simple numbering system, like the British, preceded by Fu (with an original meaning of 'spark', but now used to mean 'radio'); and a more informative system in which a leading number gives the transmitting power in watts followed by some letters indicating its function and type, for example 10W.S.c for a 10 W sender type c used for vehicle communication.

Finding sufficient frequency allocation for different sections of the army was as much a problem for the Germans as it was for the Allies. The choice was made on the basis of functional needs. For long-range working with both fixed and mobile stations, low and medium frequencies up to about 7 MHz were used. Artillery units used 3–7.5 MHz for communication and 25–27 MHz for gun ranging. Inter-tank communications were carried out between 27 and 33 MHz, while equipment for armoured vehicles used 20–25 MHz, both employing amplitude modulation for

R/T, although some use was made of CW Morse transmission when MCW was employed. The German tank wireless system was developed by Telefunken as early as 1937. With this equipment the problem of effective inter-tank communication had been solved for the German Army before the outbreak of war, and was a vital factor in its success in many different campaigns prior to 1943. The standard Telefunken tank set, Type 10W.S.c, consisted of a 10 W ultra-short-wavelength transmitter, and a compact superhet receiver, Type Fu2, both constructed in a sturdy die-cast aluminium frame [41]. The transmitter used three identical pentodes operating as master oscillator, frequency doubling power amplifier and a final power amplifier. A further pentode, used as a triode, served as a modulator, either for speech or as a MCW oscillator. The transmission range was limited to about 6 km, adequate for inter-tank communication but limited for airborne communications, although the reception range was considerably greater. A superhet receiver was installed. Infantry pack sets used a different set of frequencies depending on their purpose: when in support of armoured units 20–25 MHz; for the front line, higher frequencies of around 100–150 MHz; and for ground-to-air cooperation, frequencies in the band 42–48 MHz. The Germans made little use of FM, despite their extensive interest in the technology for civilian broadcasting, except in their fixed communication chain of decimetre-wave relay stations. These worked on a fairly long wavelength of 50–60 cm and provided only two channels, one for telephony and one for two-tone telegraphy. They also eschewed crystal control, apart from the use of crystals for calibration, relying on the excellent mechanical features of their tuning mechanisms, ensured a remarkably stable performance for all their transmitting and receiving equipment. This was particularly apparent in the receivers, generally of superhet design, in which an elaborate turret system was used for changing wavebands [42]. In some of their transmitters a special form of crystal calibration (*Leuchtquartz*) was incorporated in the equipment. This consisted of a small bar of quartz supported between two electrodes in a glass envelope containing neon. This was connected via a blocking capacitor across the transmitter power amplifier, which, when tuned to the correct quartz frequency, caused the neon to glow, and could then be observed by the operator. A list of the most important German Army wireless equipment in use at the time is given in Table 8.2. Not included are a very large number of intercept receivers used by the army's intercept, or 'Y' service, which covered all frequency ranges from 10 to 305 MHz.

Nearly all the transmitting apparatus was manufactured by Telefunken (AEG) to fairly standard but pre-war designs. An exception was a lightweight UHF pack-set operating in the range 90–110 MHz, Type Feld.Fu.b, which was used in fairly large numbers from 1941 onwards. The receiver in this set had a simple two-valve regenerative circuit, with poor selectivity compared with its British and American counterparts,

Table 8.2 German Army wireless sets, 1936–45

Type	Frequency (MHz)	Range (km)	Power (W)	Purpose
Transmitters				
Fu5	27.2–33.3	6	10	Commander's tank, assault guns, used with Fu2 receiver
Fu7	42.1–47.8	50	20	Commander's tank, used with Fu3 receiver
Fu8	1.12–3.0	50	30	Commander's tank, self-propelled guns, used with Fu4 receiver
Fu11	0.2–1.2	80	100	Ground Transmitter used with Torn E.b receiver
Fu12	1.12–3.0	80	80	Reconnaissance vehicle (Lynx), used with Fu4 receiver
Fu16	23–25	40	10	Self propelled guns
Fu19	3.0–7.5	65	15	Mobile transmitter
5W.S.c	0.95–3.15	58	7	General MF Gd. transmitter used with Fu2 receiver
10W.S.c	22–33	6.5	6.5	Armoured vehicle, tanks, used with Fu2 receiver
20W.S.c	27–33	4.8	20	Tank formations, generally used with Fu4 receiver
30W.S.a	1.1–3.0	80	30	Armoured cars, used with Fu4 receiver
80W.S.a	1.1–3.0	200	80	Tank division, used with Fu4 receiver
100W.S.a	0.2–1.2	322	100	Ground administration transmitter
1000W.S.b	1.1–6.7	1127	1000	Army and corps staff transmitter
1500W.S.b	1.1–6.7	1170	1500	Army and corps staff transmitter
Receivers and transceivers				
Fu1	0.1–6.97	—	—	Pack receiver
Fu2	27–33	—	—	Tank receiver
Fu3	42–48	—	—	Tank receiver
Fu4	0.84–3.0	—	—	Armoured vehicle receiver
Fu15	23–25	—	—	Assault gun receiver
Torn.E.b.	0.1–6.97	—	—	Command receiver
Torn.Fu.b1	3–5 (T) 3–6.7 (R)	40	0.65	Mobile set
Torn.Fu.a2	33.8–38 (T/R)	14.5	1.0	Mobile set (battalion)
Torn.Fu.g	2.5–3.5 (T/R)	24	1.0	Mobile set
SE469A	3–5 (T/R)	97	15	Artillery fire control
Feld.Fu.b	90–110 (T/R)	6.5	0.15	Infantry pack set

Type 31 and SCR 300 respectively. Later in the war receiving sets for fixed and mobile locations in the UHF band were often commercial designs, typically a Rhode u. Schwarz communications receiver manufactured for civilian use.

8.6 British Army training and recruitment

The geographical range of the Second World War was immense, as was the reliance placed on technology by the combatants, and on both accounts all three services needed a constant supply of trained communication specialists, particularly telegraphists. This was met, initially and in part, by recruiting appropriately qualified and experienced civilians, and here the cooperation of the Radio Society of Great Britain was most welcome, leading to the formation of the Royal Naval Wireless Auxiliary Reserve and the RAF Civilian Wireless Reserve, already discussed in Chapter 7. But the number of skilled and knowledgeable technicians recruited in this way was never enough, and recourse was made chiefly to the civilian and the services training schools, although the Post Office was able to provide large numbers of trained Morse telegraphists for the army [43].

The technical press contributed their particular help. With the outbreak of war the need to recruit and train large numbers of telegraphists for the armed forces was reflected in the pages of that section of the technical press which catered for an enthusiastic band of technologists and radio amateurs. Many included details of the design and construction of thermionic valve practice oscillators which formed the subject of numerous articles. Chief among such publications was *Wireless World*, which from 1940 to 1945 carried numerous articles and news reports covering the techniques of learning the Morse code and the design of training apparatus. This was considered at two levels: the initial process of learning the Morse alphabet and the attainment of high-speed transmission. Numerous articles proposed methods of achieving these objectives, accompanied by a constant stream of letters from readers, some extolling preferred methods of handling the Morse key and others commenting on the advantages/disadvantages of the methods taught by the British services against those preferred by their American allies (who generally rested the forearm on the table and achieved transmission by wrist action alone). Such minutiae as to whether dots and dashes should be conveyed aurally by the sounds 'dit' and 'dah' or by 'de' and 'dah' during the learning process were exhaustively discussed. These divergent views were summarised in all their aspects in an Editorial to the August 1940 issue, which concluded with the comment that

When the learner passes on the real thing, and has to work his key in the cramped quarters of an aeroplane or in the reeling cabin on a small patrol boat in a seaway, he may have to forget some of the things he has learned. But, if he has been trained to avoid slovenly methods, he will adapt himself to circumstances, and form an efficient link in one of our most vital communications [44].

The gramophone companies advertised for sale in the pages of the technical journals their Morse practice records, some having a background of interfering signals to simulate realistic conditions. The BBC in this period, transmitted a series of news bulletins in Morse code, directed to enemy-occupied countries, in English, French and German, at a slow speed of 12 wpm, on frequencies of 49.39 and 261.1 m. Such transmissions were also popular with students learning the code in the British Isles [45]. The sales of a *Wireless World* booklet, *Learning Morse*, exceeded 250 000 in 1942 alone.

However, it was the service training schools that instructed the greatest number of telegraphists and technicians in all branches of the armed forces. For the Army the training was principally carried out at the Army School of Signals by personnel of the Royal Corps of Signals, who provided technical training in telegraphy, wireless, line equipment and construction. In a separate branch of the school training was given for officers and NCOs of the Royal Artillery, Royal Engineers, Infantry Regiments and the Royal Electrical & Mechanical Engineers. The throughput for the school was high, with some 8000 students, ranging in rank from Brigadier to Lance-Corporal in 1944 alone, when some 224 courses of 37 different types were undertaken [46].

References

1   HEZLET, A.: 'The electron and sea power' (Peter Davies, London, 1975)
2   AUSTIN, B.A.: 'Wireless in the Boer War', IEE Conference '100 years of radio', 5th–7th September 1995, London
3   COCHRANE, Major J.E.: 'Wireless telegraphy for military purposes', *in* 'Yearbook of wireless telegraphy and telephony' (Wireless Press, London, 1913), pp. 378–85
4   NALDER, R.F.H.: 'History of the Royal Corps of Signals' (Royal Signals Institution, London, 1958)
5   CUSINS, A.G.T.: 'Development of army wireless during the War', *Journal of the Institution of Electrical Engineers*, 1922, **59**, p. 763
6   BAKER, W.J.: 'A history of the Marconi Company' (Methuen, London, 1970)
7   'First annual report of signals schools for 1917' (Admiralty, 1918) Doc. ADM/186/753, Public Record Office, Kew, Surrey
8   KEEN, R.: 'Wireless direction finding' (Iliffe, London, 1941), p. 39
9   MEISSNER, A.: 'Die deutsche Elektronindustrie in den Kriegsjahrer' (Telefunken, Berlin, 1919)
10 ROUND, Cpt. H.J.: 'Direction and position finding (discussion)', *Journal of the Institution of Electrical Engineers*, 1920, **58**, pp. 247–8
11 GRIFFITH, P.E.: 'Battle tactics of the Western Front' (Yale University Press, New Haven, Conn., 1994), p. 171
12 FULLER, A.C.: 'The Fullerphone: its application to military and civil telegraphy', *The Electrician*, 1919, **82**, pp. 536–8
13 PRIESTLEY, R.E.: 'The signal service in the European War of 1914–1918' (Chatham, 1921), p. 112

14 'The Telefunken system of wireless telegraphy', *The Electrician*, 1911, **68**, pp. 213–16

15 EICHORN, G.: 'Wireless telegraphy' (C. Griffin & Co., London, 1906)

16 TRENKLE, F.: 'Die deutschen Funknachrichtenanlagen bis 1945' (Hüthig Buch Verlag, Heidelberg, 1946)

17 SCHMIDT, G.: 'Die deutsche Elektrotechnik in den Kriegsjahren', *Elektrotechnische Zeitschrift*, 1919, **40**, Heft 11, pp. 113–14; Heft 26, pp. 309–11

18 SCHONLAND, B.F.J.: 'W/T. R.E. — an account of the work and development of field wireless sets with the armies in France', *Wireless World*, July 1919, **7**, p. 174

19 HARBOTTLE, H.R.: 'The provision of line communication for the fighting services during the War' (1946), PO Archives, POST 56/121

20 ANGWIN, A.S.: 'Telecoms in war', *Engineering*, 1947, **159**, pp. 298, 305

21 NANCARROW, F.R. 'The telecommunications network for defence', *Post Office Electrical Engineers' Journal*, 1946, **38**, pp. 121–5

22 GRAVELY, Col. T.B.: 'Army signal communications, the Second World War 1939–1945' (Royal Corps of Signals, London, 1950)

23 TERRETT, D.: 'The Signal Corps: the emergency', Official Publication AS 742/11(8) (Government Printing Office, Washington, DC, 1956)

24 COLTON, J. 'The spiral-four system', *Electrical Engineering*, 1942, **64**, (5), p. 175

25 MARKS, W.S.: 'Radio relay communication systems in the U.S. Army', *Proceedings of the Institute of Radio Engineers*, 1945, **33**, pp. 502–22

26 GODWIN, G.: 'Marconi 1939–1945: a war record' (Chatto & Windus, London, 1946)

27 'Army's highest powered mobile station', *Wireless World*, April 1944, **50**, pp. 106–7

28 'Army mobile wireless station', *Engineering*, 1944, **157**, p. 227

29 'Multi-channel pulse modulation — Army Wireless Set No. 10', *Wireless World*, 1946, **52**, pp. 187–92, 282–5 (see also *Ibid.*, **51**, pp. 383–4)

30 SAUNDERS, A.B.: 'Wireless Set No. 10 in Austria 1945–46', *Royal Signals Quarterly Journal* (new series), April 1947, **1**, (2), pp. 67–76

31 'Army signalling equipment', *Engineering*, 13th April 1945, **159**, p. 298

32 RHODES, J.: 'P.O. cable communications for the invasion armies', *Post Office Electrical Engineers' Journal*, 1946, **38**, p. 136

33 DUNSHEATH, P.: 'Presidential address to the IEE', *Journal of the Institution of Electrical Engineers*, 4th October 1945

34 CRUICKSHANK, W.: 'Voice frequency telegraphs', *Journal of the Institution of Electrical Engineers*, July 1929, **67**, (385), pp. 813–31

35 'Official history of radio and radar' (1950), Doc. AVIA 46/271, Public Record Office, Kew, Surrey

36 HICKMAN, J.B.: 'Military radio communications', *Journal of the Institution of Electrical Engineers*, 1947, **94**, (3A), pp. 60–73

37 MEULSTEE, L.: 'Wireless for the warrior' (G. Arnold, Broadstone, Hampshire, 1995)

38 'Army Set — Type 76', *Wireless World*, 1945, **51**, pp. 137–9

39 THOMPSON, G.R.: 'The Signal Corps: the test' (Government Printing Office, Washington, DC, 1957), p.71

40 BAUER, A.O.: 'Some aspects of unconventional engineering during Inter-

bellum' (Foundation Centre for German Communication and Related Technology 1920–1945, Diemen, Netherlands, 1999)

41 'German tank wireless' (1942), Doc. AVIA 23/808, Public Record Office, Kew, Surrey

42 FARRAR, W.: 'German army wireless equipment', *Royal Signals Quarterly Journal* (new series), April 1947, **1**, pp. 62–6

43 NALDER, R.F.H.: 'The history of British army signals in World War II', *Journal of the Royal Corps of Signals*, 1953, p. 243.

44 'Editorial: Morse operating technique', *Wireless World*, August 1940, **46**, p. 347

45 'News in Morse', *Wireless World*, 1942, **48**, p. 122

46 'Army signal training and equipment', *Engineering*, 20th April 1945, **159**, p. 305

Chapter 9

Military telegraphy at sea

9.1 Wireless experiments at sea

The first direct involvement by the British Navy in the use of wireless telegraphy came with experiments made by Captain Henry Bradwardine Jackson (later to become First Sea Lord, Sir Henry Jackson), then in command of HMS *Defiance* at Devonport in 1895. HMS *Defiance* was then the location of the Navy's torpedo school, and with its requirement to teach the rudiments of remote electrical control of underwater mines, the torpedo school became the natural venue for the early natural tests of wireless telegraphy. Jackson had become aware of the work of Jagadis Bose in Calcutta and Oliver Lodge in England from their published papers, and set up his own equipment, modelled on their work, on board the *Defiance*. This consisted of an untuned spark transmitter worked from a 12 V battery, and a receiver incorporating a home-made coherer. Later, at a War Office meeting, Jackson met Marconi, who was also engaged in naval wireless transmissions at this time. Both were at about the same stage of development in 1896, but whereas Marconi had his sights firmly fixed on long-distance transmission, Jackson had navy communications in mind and continued to improve his equipment until he was successful in establishing a link between the *Defiance* and his Admiral's house at Plymouth, a range of about 6 km [1]. The experiments were continued and recorded by several other naval officers, using Jackson's equipment, after he left the *Defiance* for another appointment a year later.

Marconi carried out a series of shipboard demonstrations in the English Channel a few years later, in 1899, with the cooperation of the French Government, which lent him a naval gunboat, the *Ibis*, for this purpose. These tests were followed by the Admiralty with interest, encouraged perhaps by the early results of Jackson's work. In the same year the Wireless Telegraph & Signal Company, recently established by

Marconi, was invited to install its equipment on three warships of the Fleet, *Juno, Alexandria* and the cruiser *Europe*, for the 1899 naval man-oeuvres in the English Channel, with Marconi on board *Juno* and in charge of the wireless trials. Communication was established using the spark transmitters on board over a distance in excess of 100 km, a record at the time. Captain Jackson, who had been appointed by the Admiralty to liaise with Marconi for the exercise, reported that the trials were very satisfactory and recommended that wireless should now be installed on each of Her Majesty's warships and that an exercise programme be arranged [2]. Initially 32 Marconi sets were ordered, together with 11 sets constructed by the Admiralty to Captain Jackson's specification. They began to be put into service in 1900, and inten-sive training courses were established at Devonport and Portsmouth to train the wireless telegraphy operators needed [3]. By the end of the year the Royal Navy had more than 50 sets in regular use in ships and on shore stations. This marked the beginning of a wide inter-est in the use of wireless by navies throughout the world, with the market shared by Marconi, Telefunken, Ducretet-Rochefort and Popov-Ducretet.

In Britain the development of wireless telegraphy had so far been carried out by the Navy's Torpedo Branch. In 1906 a new department, the Wireless Telegraphy Branch, was established within the Navy, which enabled men to specialise and become operators or technicians in the new technology for a slight increase in pay. In the same year range trials were carried out between HMS *Vernon* (the shore naval training estab-lishment in Portsmouth) and HMS *Furious*, on passage to the Mediter-ranean. Ranges of 1000 km were obtained regularly by day and up to 2,000 km by night. A wavelength of 750 m was found to be the most satisfactory. As wireless communication began to be used operationally on board ship, the re-radiation of signals induced in the ship's rigging was observed to affect the efficiency of reception, which initiated a pro-gramme of refitting and re-routing such structures. The effect of local thunderstorms on the wireless equipment was also causing concern, and in 1903 a command went out from the Admiralty that 'during a severe thunderstorm the Wireless Telegraphy gaff should either be lowered or else the aerial should be disconnected from the receiver and connected to the hull of the ship' [4].

The standard pattern of equipment in use by the Navy at that time was a spark transmitter incorporating a 1.5 kW alternator and induc-tion coil, and became known as the Service Mark 1 installation. It was a naval modification of a Marconi set (Type T18), then being manu-factured in quantity at Chelmsford and described in Chapter 6. An alternator, rather than a d.c. generator as used in the early experiments, was chosen because the modulation of the damped wavelength at an audible frequency, related to the rotational speed of the alternator, made the received signals much easier to discern, using headphones, from interference on neighbouring wavelengths. The adaptation of Marconi designs to meet service requirements was largely due to the influence of Jackson, who had been lobbying for the standardisation of wireless equipment in the Navy for some time. Cruisers were equipped with the Type 2 standard ship set, a 1.5 kW spark set with a range of 160 km; destroyers were fitted with the Type 4, a 1 kW spark set with a range of 80 km. Initially submarines were fitted with similar equipment but, with only a small aerial carried, the transmission range was limited to 48 km.

Similar installations of wireless telegraphy equipment were being carried out by most of the world's other naval forces. They relied mainly on equipment designed and installed by the Marconi Company through its various overseas companies, although substantial orders were placed with the Telefunken company by the German, American, Swedish and Japanese fleets to supply their new design of quenched spark transmission. The approximate numbers of warships fitted with spark transmitters in 1906 for the major naval powers are given in Table 9.1 [5].

Table 9.1 Ships fitted with wireless telegraphy apparatus in 1906

Britain	United States	France	Japan	Germany	Italy
220	114	80	66	61	53

Source: The Archives of the National Maritime Museum Library

With so many spark transmitting stations now in operation, the Navy was becoming increasingly aware of the need for some form of standardisation between the multiplicity of frequencies in use in the fleet and shore stations. These different frequencies were referred to as 'tunes', and specified in wavelengths (often expressed in feet). The first to be designated were Tune A and Tune B, of 100 and 270 m respectively, which had been allocated by Marconi during his work on the use of a separate tuned coupled circuit to transfer the spark signal to the aerial. By establishing tight coupling between the two circuits, two separate wavelengths were produced simultaneously – Tune A and Tune B. First considered as beneficial in providing two possible reception opportunities, this was later discouraged and a looser coupling was specified to limit the transfer of energy to the aerial circuit to a single transmitting frequency. The idea of installing two separate receivers on board ship to receive two separate 'tunes' was, however, retained in the earliest naval equipment [6]. Other tunes adopted by the navy included:

Tune D	212 m (1415 kHz)
Tune R	788 m (380 kHz)
Tune S	1000 m (300 kHz)
Tune T	1273 m (235 kHz)
Tune U	1515 m (198 kHz)
Tune V	1727 m (174 kHz)
Tune W	1970 m (150 kHz)

with certain wavelengths established by International Conferences for the Safety of Life at Sea:

Tune P	300 m (1000 kHz)
Tune Q	600 m (500 kHz)

This rationalisation of naval frequencies was carried out early in the First World War, and by 1915 the British Naval Wireless Instruction listed 28 specific different frequencies which could be used, ranging from 109 m (2743 kHz) to 4260 m (70 kHz). The shorter waves were used for auxiliary wireless sets and by destroyers, submarines and small ships, while the long waves were employed by shore stations and capital ships. By about 1930 the Admiralty ceased to use Tunes and referred instead to the transmitting wavelength in metres and, increasingly, the frequency in kilohertz.

Despite the careful attention paid to allocating frequencies, the large number of shipboard spark transmitting stations now operating was beginning to cause considerable mutual interference, particularly in the crowded home waters around Britain and in the Mediterranean. While the change to quenched spark generation offered some relief, other methods of transmission were sought. The Royal Navy and the US Navy were beginning to explore the potential of arc transmission and the use of more selective receivers. In 1907 an experimental arc transmitter was installed in HMS *Vernon* and in HMS *Furious*, and at the same time a crystal detector was brought in to replace the 'Maggie' magnetic detector. This was followed in 1916 by a standard Poulsen arc set fitted to certain ships in the British and American fleets. The first to benefit by the increased range and reduced transmitting bandwidth was the submarine service in the Royal Navy, which had previously secured only a limited range with the spark transmitters then available. This was the Type 14 arc set, a 3 kW transmitter, which increased the submarine range to about 500 km and considerably increased their effectiveness at sea [7]. The use of these arc sets constituted the first major seaborne experiments in the use of continuous waves for naval communication. In the early 1900s arc sets were replaced by thermionic valve systems, which were much more reliable and transmitted a narrower bandwidth signal, resulting in less mutual interference, as described later in this chapter.

However, it was the Russian Navy that was the first to use wireless in an actual wartime engagement. The navy began to take an interest in wireless telegraphy in 1895, at about the same time as Jackson was carrying out his experiments on the *Defiance*. A Russian physicist, Aleksandr Popov, initiated a similar series of experiments with the cooperation of the Russian Navy on board the cruisers *Russia* and *Afrika* in Kronstadt harbour. The results obtained with regard to distances of transmission and wavelength used were remarkably similar to Jackson's, the only operational difference being the method of reception used. Whereas Popov made use of a telephone receiver to give audible signals, Jackson used a sensitive galvanometer to provide visual signals. In January 1900 the Russian Navy had an opportunity to experience the value of the new communications technique when the battleship *Grand-Admiral Aprexin* ran aground on the island of Suursaari in the Gulf of Finland. Popov was able to set up a wireless link between the island and the nearest mainland town, Kotka, some 50 km away. The link remained in operation for several months after the battleship was refloated, and was used to direct the rescue of a group of fishermen stranded on an ice-floe in the Baltic. Although the value of the Suursaari installation was sufficient to persuade the Russian Naval Ministry to install wireless apparatus on its ships, this was not carried out until just before the outbreak of war with Japan in 1904.

The Russo-Japanese war of 1904–5 was the first in which naval wireless communication was used extensively. Both sides were equipped with the latest wireless telegraphy installations on their respective warships. The Russian fleet was equipped with German Slaby-Arco transmitters and receivers, and one auxiliary vessel, the *Ural*, carried extra wireless equipment to permit both short-range communication for inter-ship signalling and long-range equipment for contact with shore stations. There were at that time a large number of shore W/T stations in Europe and the Far East, all operating on long wavelengths and capable of establishing two-way contact with ships in their vicinity. Major stations which the Russian fleet expected to contact on its voyage to Japan included a Telefunken coastal station having a 1000 km range situated in Manchuria, similar stations set up by the Marconi Company at St Petersburg (then the capital of Russia) and at Vladivostok in the Far East. In October 1904 the Baltic Fleet assembled at Libava (or Libau, now Liepāja) in the Baltic Sea. It consisted of 59 ships, under the control of Admiral Rozhestvensky, fitted out to commence the 30000 km voyage to travel round South Africa, across the Indian Ocean, and past Indo-China. It was expected to reach the East China Sea in May 1905 (Figure 9.1). The fleet was able to maintain continuous contact with headquarters in St Petersburg through a mixed communications link, consisting of wireless to one of the shore stations and relaying the transmissions via telegraph landline to the capital. Japan, on the other hand, had relied on the

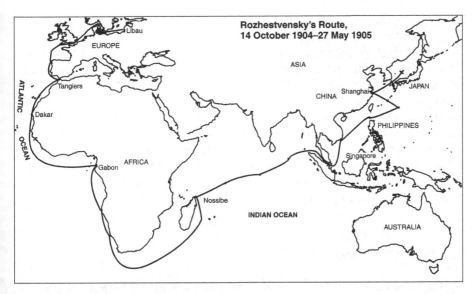

Figure 9.1   Admiral Rozhestvensky's voyage to the Far East
Source: de Arcangelis, 'Electronic Warfare', 1985

Marconi Company to equip its fleet, not directly but through a Japanese copy of the Marconi installation made by Professor Shunkichi Kimura, somewhat simplified to single transmission frequency and a plain aerial [8, 9]. This had a slightly smaller range than the Russian equipment, but Japan's need was not so immediate as the only naval engagement took place in Japanese waters, near the Tsushima Strait in the Sea of Japan, where the Russians were eventually forced to surrender their fleet and concede defeat [10].

Rozhestvensky's aim had been to proceed to Vladivostock, drop anchor and refit his fleet after the three-month journey before commencing an attack on the Japanese fleet. In the event the detection of his fleet by the Japanese, making use of the Russian wireless transmissions and visual sightings, made this impossible. Wireless communication was of value to both sides, particularly the Japanese, as it made it possible to patrol key sea passages with single ships which were able to maintain contact with the flagship by wireless, providing accurate information on the approach of the Russian fleet. The Russians were able to maintain contact with their naval headquarters in St Petersburg during the voyage, and could assess the location of the Japanese ships with some accuracy, but in the end it was the superior tactical use made of intelligence, rather than the relative performance of the two opposing sets of equipment, that resulted in the first naval wartime victory in which wireless communication was used to the full [11]. This view was supported by Telefunken, which commented on the more skilful use made by the

Japanese of their sets, an opinion which considerably enhanced Marconi's reputation in naval circles [12].

Although slight, the use of wireless by the Royal Navy during the Boer War of 1899–1902 was significant. It came about almost by accident when the Navy in South Africa inherited five Marconi sets, originally brought out by the Army but discarded as being of little value to the land campaign (see Chapter 8). When installed in the ships of a squadron blockading Lourenço Marques (now Maputo) they proved effective, despite continuing problems with interference. Communication was established between ships separated by up to 100 km and by using the shipboard equipment to relay messages across the bay for the Admiralty at any time of the night or day, a successful blockade was enforced.

9.2 War at sea, 1914–18

By the time of the outbreak of war between the major powers in Europe in 1914, seagoing wireless apparatus had reached new levels of effectiveness. Although the transmitters still relied on spark generation, they were now accurately tunable, and their associated receivers benefited from rather better techniques of detection than the coherer. The Marconi Company had installed its magnetic detector on board ships (see Chapter 6), and Telefunken made use of an electrolytic detector [13]. Both sides were quick to install crystal receivers when they became available in the early war years (they were first used experimentally by the Navy in 1904), and selectivity was enhanced. Separation of the desired signal from interfering signals was a much desired improvement, since under wartime conditions the jamming of Admiralty signals by enemy transmissions on nearby frequencies had become the norm. To overcome this, not only were improvements to receivers made, but duplicate transmissions on different wavelengths began to be made from several different stations at the same time. This technique had already been used by the Admiralty, an order, 'W/T Messages from Admiral of the Fleet', having been issued in December 1912 giving details of the conditions under which duplicate transmissions should be carried out and the Admiralty stations involved, 'the order not to go into any signal books but should be regarded as strictly secret for emergencies only' [14]. Despite the improvements in wireless equipment and procedures used aboard naval vessels, reservations about the use of wireless telegraphy in active service were frequently expressed by the Admiralty. In a minute transmitted to the fleet in 1915 headed, 'W/T Excessive Use', their Lordships drew attention to

the excessive and unnecessary use of W/T throughout the Fleet at home and abroad, which not only hampers our own operations, but also materially assists the enemy. Cable, landline, visual signals and messengers are to be considered as the normal means of communication, W/T being reserved for conveying urgent orders or information which it is undesirable should await until usual communication can be established [15].

A large number of shore stations were available to support the naval activities for both sides, some of which have been referred to earlier [16], and the British were beginning to build at home and in France a chain of direction-finding stations with a directional accuracy of 1.5°. These stations were put to good effect in charting the passage of the Zeppelins in their raids on England (see Chapter 8). They were also used to locate enemy submarines when they surfaced to begin transmission to their home base. The German submarines were fitted with powerful spark transmitters working on a wavelength of 400 m, and when they were detected particulars of their position and course could be sent to warn shipping in the area. To keep in touch with their fleet the Germans had erected a series of powerful shore stations just before the war. The high-power station at Nauen was at the time the most powerful telegraph transmitter in the world, containing Telefunken 200 kW high-frequency alternators and large aerials (see Chapter 6). On the eve of the declaration of war this station was responsible for alerting all ships of the German merchant fleet at sea to seek shelter in German or neutral ports, an action which must have been responsible for reducing the effect of the British naval blockade on Germany at the beginning of the war. The Allies were able to make use of numbers of shore stations in home waters but there was a lack of suitable shore stations for naval use in the South Atlantic, the Indian Ocean and across to Australia. Considerable effort was expended by the Marconi Company to erect a chain of stations in these areas. By June 1915 stations had been installed at Ascension, the Falkland Islands, Bathurst, Ceylon, Durban, Demerara, the Seychelles, Singapore, Aden, Hong Kong, Mauritius and Port Nolloth. These were of similar design, incorporating rotary spark transmitters and an aerial consisting of a curtain of wires suspended from 100 m masts to give them sufficient range to reach the next station in the chain.

Almost two years of sporadic naval activity for both sides followed the declaration of war. The importance of wireless communication to the combatants is reflected in the number of naval raids made by both sides against coastal shore wireless and telegraph stations. One of these in December 1914 was notable in indirectly causing the destruction of the German armoured cruisers *Scharnhorst* and *Gneisau*, and two light cruisers. Vice Admiral Graf von Spee, having used his squadron to good effect at the Battle of Coronel in November when he defeated and sunk an inferior British force off the coast of Chile, was turning for his home port in Germany, having seriously depleted his ammunition supply, when he decided to attack and destroy the undefended Falkland Islands wireless station at Port Stanley. This delayed his departure sufficiently to allow a much superior British naval force under Admiral Sir Doveton Sturdee to reach his squadron and inflict severe damage, from which the German High Seas Fleet never completely recovered [17].

Although Germany retained the use of its high-power telegraph transmitter at Nauen throughout the war, its contact with the United States was limited to diplomatic communication after its American transmitter at Sayville was taken over by the United States Navy in 1915. The Nauen station still proved invaluable for communicating with the German naval and merchant fleets throughout the world. The overseas stations – principally Kamina in Togoland, Windhoek in South-West Africa and Sayville in America – were intended to form part of a German imperial wireless chain, but this plan was disrupted in August 1914 when the 100 kW station at Kamina was destroyed by the German Navy to avoid capture and Windhoek fell to the South African forces on 12th May 1915. The other overseas stations, which were to form a network of telegraphic communication to link with their colonies and protectorates, were forced out of action one by one. On 9th August 1914 the station at Dar-es-Salaam was destroyed by bombardment. Three days later the station at Yap in the Caroline Islands was similarly destroyed. Towards the end of the month the Australian Navy captured the stations at Samoa, Nauru in the Marshall Islands, and Herberbshihe on New Pommera Island. Further actions followed as the war proceeded in Africa and the Far East. The station at Douala in Cameroon was seized in September; and finally the Japanese occupied the Kiaohsien station in November.

With both sides in the naval encounters monitoring each others' wireless transmissions, the need for secure codes was paramount and resulted in a growth of techniques in cryptanalysis and the practice of recording all enemy transmissions from both warships and shore stations. It was the detection and decoding of an increasing amount of transmissions from the German High Seas Fleet concerning ship movements that eventually led to a major naval encounter between the British and German navies at the Battle of Jutland. For a number of years both fleets had avoided direct conflict. Towards the end of May 1916 this was to change as both fleets left their respective harbours, heading out into the North Sea. The German fleet had a problem in emerging from the narrow confines of Wilhelmshaven without revealing its intentions. Vice Admiral Reinhard Scheer, commanding the German fleet, decided on a simple wireless telegraphy deception that proved extremely effective. As the German flagship, the *Friedrich der Grosse* left harbour, its call-sign was transferred to a shore station which continued to transmit the usual fleet signalling messages on the same frequency as the flagship. Subsequent direction-finding operations by the British seemed to confirm lack of movement of the German fleet, which by this deception bought itself several hours of undetected movement. This was the first of a number of wireless deception activities which contributed to the final indecisive nature of the battle. The British were more skilled in the art of deciphering coded messages than the Germans were, but Admiral John Rushworth Jellicoe was reluctant to take full advantage of this intelligence, knowing the opportunities that the enemy had for wireless deception, and tended to rate visual communication above wireless signalling [8]. After Jutland both sides reconsidered the possibilities of decisive fleet action by large numbers of capital ships acting in unison. The effectiveness of wireless signalling and intelligence contributed to stalemate in this type of operation, in which neither side could manouevre its forces into a position for a surprise attack. The focus of naval operations shifted to aggressive submarine activity and convoy protection, a development which reached its apogee in the naval engagements of the Second World War.

9.2.1 Shipboard wireless equipment

A range of Marconi spark transmitters had been installed in different classes of ship for the British Navy since 1906. The smaller ships of the line used the 1.5 kW Type 1 set already discussed. A Type 2 standard ship set, a 1.5 kW set with a range of 160 km, was fitted in cruisers, often with an auxiliary Type 9 set, a 1.25 kW set with a range of 90 km. The smaller Type 10 1 kW spark set fitted in submarines had a surface range of 48 km. This was the first reliable wireless set to be fitted to a submarine, and its use lent considerable value to the submarine offensive. The transmitter/receiver circuit for this set is shown in Figure 9.2. It was powered by a motor-alternator driven from the ship's power supply, and the whole equipment was installed in a 'silent' cabinet. The transmitter was designed to transmit any wavelength from 114 to 247 m, and a crystal receiver was fitted. To amplify the weak signals received, the set incorporated a 'Brown's relay', which was in use for a number of army and navy receivers before the thermionic valve amplifier came into use [18]. The principle of this ingenious device is shown in Figure 9.3. A steel rod replaced the disc in a telephone receiver, and the vibration of this rod caused by the received signal compressed the granules of carbon in the microphone M, included in a second battery circuit. The larger fluctuations of current produced by the second circuit were transferred to a pair of high-resistance telephones, producing correspondingly louder signals. A number of separate receivers incorporating Brown's relays were fitted, particularly in the larger ships, to enable reception on several wavelengths simultaneously [19]. Most of these used the new crystal detector developed by the Marconi Company.

During the last two years of the war the wireless equipment used by the navy did not alter very much. A range of Marconi arc transmitters were fitted, having powers of between 3 and 25 kW and designated naval

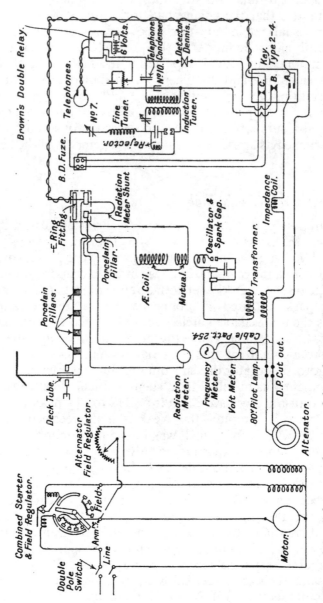

Figure 9.2   Schematic diagram of the submarine Type 10 set
Source: Public Record Office

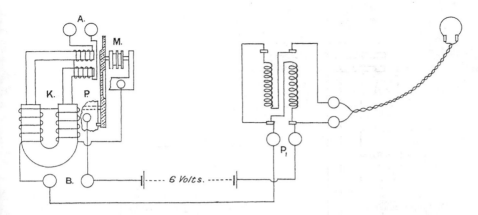

Figure 9.3   Brown's sensitive relay
Source: Public Record Office

Types 14 to 18, and a high-power Poulsen arc set (Type 20) operating at 100 kW was introduced for shore use. Heterodyne valve receivers began to be fitted throughout the fleet, and a low-power valve transmitter came into use in 1917 for fire control and aircraft communication. This was the Type 31 single-valve transmitter designed by the Admiralty Signal School for active service (Figure 9.4). The single tuned aerial circuit could be adjusted over a number of fixed frequencies between 882 and 1000 kHz, with power obtained from a 0.25 kW dual motor-generator set. In operation only two transmission frequencies were used, known as 'divisional' and 'subdivisional' waves. The aerial circuit was tuned to a frequency approximately midway between the two frequencies and was not altered when there was a change from one frequency to the other. The problem of keying high-power valve transmitters was solved by using a keying relay that performed many functions, a procedure that was followed by most of the navy transmitters thereafter. With the Type 31 transmitter the relay would sequentially make or break the positive HT supply, connect the aerial to either the receiver or the transmitter, and make or break the grid leak resistance to achieve the actual transmitter keying. A list of most of the Royal Navy transmitting and receiving sets used between 1914 and 1918 is given in Table 9.2.

Wireless equipment in German wartime vessels was similar in specification and quantity to that used by the British fleet. The larger capital ships had three main locations for equipment, (and in most cases the receiving and transmitting equipment was kept in separate rooms with some sort of noise lagging to screen the noise of the spark transmitter from the receiving room [20]. Telefunken was the main supplier of quenched spark transmitters operating over a range of 130 to 300 m, while Lorenz were responsible for the arc sets installed in the larger ships. The German U-boats carried a 1 kW quenched spark set transmitter and

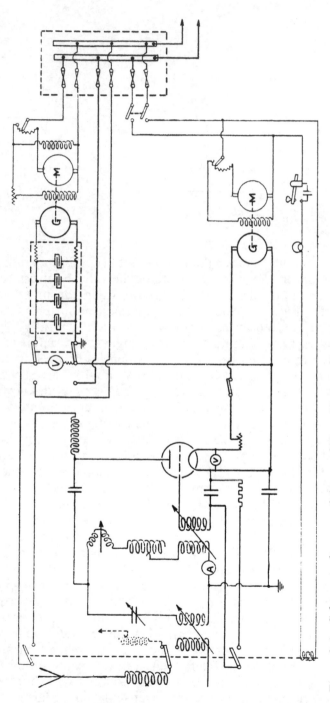

Figure 9.4   Schematic diagram of the naval Type 31 transmitter
Source: The Naval Museum, HMS Collingwood, Fareham

Table 9.2 Royal Navy transmitting and receiving sets, 1914–18

Type	Frequency (kHz)	Range (km)	Power (kW)	Where fitted
1	150–1000	800	1.5	Standard ship spark set
2	150–1000	160	1.5	Cruiser spark set
3	2380	8	1.0	Auxiliary spark set
4	1408	80	1.0	Destroyer spark set
5	1000	8	0.5	Landing party set
7	82–17	1600	100	Shore station set
8	70–82	2900	150	Malta shore station set
9	1910–2380	48	1.25	Auxiliary spark set
10	1215–2632	50	1.0	Submarine set
14	187.5–125	500	3.0	Arc set (submarine)
15	187.5–107	500	3.0	Destroyer arc set
16	187.5–107	500	3.0	Auxiliary arc set
18	100–66.67	1000	25	Arc set (capital ships)
20	82–173	4000	100	Arc set (shore station)
22	60–70	3000	150	Synchronous spark set
30	142.8–461.5	64	1.0	CW valve set, landing party
31	882–1000	10	0.25	Standard ship valve set
A	70–2752	—	—	Magnetic detector set
C	70–2752	—	—	Crystal set
D	70–2380	—	—	Crystal set for Type 3 transmitter
F	70–2752	—	—	Crystal set for Type 10 transmitter

a Perikon crystal receiver, preceded by a two-valve high-frequency valve amplifier. The transmitter had a range of about 160 km at a wavelength range of 300 and 820 m. For their capital ships Telefunken had manufactured, just before the war, a quenched spark gap transmitter having a radiated power of 15 kW, which was generally installed together with an auxiliary spark set of 1.5 kW. Transmitters employing thermionic valves were not available before the end of hostilities, although Telefunken had constructed a 1 kW experimental valve set in 1912 for use in submarines [21]. Extensive use was made of valves in receivers for amplification, detection and heterodyne work. The most common German receiver in use by the military in the First World War was the E81, manufactured by Telefunken in Berlin. This used crystal reception with a choice of five different crystals, selected by a rotary holder, to ensure continuous reception. A separate two-valve high-frequency amplifier preceded the detector, and the set could be connected to a heterodyne apparatus for wireless telegraph reception. A later Telefunken set, the E183, was a two-valve receiver covering the range 250–12 500 m by switched inductances. This receiver incorporated a mechanical 'ticker' rather than a heterodyne oscillator for W/T reception. A similar set manufactured by Lorenz, the

MAT21, contained a crystal detector, a valve amplifier and two forms of ticker, one of which was a rotary device; this was found to be more reliable than the interruptor ticker, which required very careful adjustment (a disadvantage which had dogged the performance of the British Fullerphone, discussed in the previous chapter).

9.2.2 The Naval wireless telegraph network

The difficulties of establishing an Imperial Wireless Network encountered by the Marconi Company before 1914 have been discussed earlier. The Royal Navy had always been unenthusiastic about this quasi-commercial activity, though stopping short of active opposition, and after the government had dropped the scheme it noted that 'the whole chain was part of a policy of preparation for war which, unfortunately, had not advanced far enough when the present war broke out to produce anything like the comprehensive scheme intended, the merit of which would have lain in its worldwide character' [22]. In fact the Admiralty had been actively considering its own worldwide naval network, aided by the Marconi Company and the Post Office, and planned to erect high-power wireless telegraphy stations in various locations around the globe with the limited objective of achieving full control over its far-flung ships and overseas bases without using governmental or non-naval transmitting stations. The proposed naval scheme for 1915 included two high-power stations at Ascension and the Falkland Islands, and eleven medium-power stations at other overseas locations. These were to be supplemented by four existing naval stations at Horsea, Cleethorpes, Gibraltar and Malta. A list of stations with some details of their capabilities is given in Table 9.3. They were all Marconi-designed installations, and had a Marconi engineer in charge, who was given the wartime rank of Sub-Lieutenant RNVR. The entire chain became operational in 1916 and provided a reliable link between the Admiralty in Whitehall and its vessels abroad. It made the Navy completely independent of existing commercial networks, and was of inestimable value during the war at sea in 1915–18.

After the Armistice in 1918 the cost of maintaining the stations in peacetime was considered too high, and many of them were closed or released from naval control. The only stations remaining in operation for the Navy were the four medium-power stations at Aden, Matara, Singapore and Stonecutters [23]. In subsequent years the development of the Marconi beam stations for the governmental Imperial network, and the availability of the low frequency GPO station GBR at Rugby, enabled the Admiralty to maintain contact with its overseas bases and ships by sharing these services with other organisations, although the Rugby station was to be reserved for the Admiralty during the entire 1939–45 conflict.

Table 9.3 Admiralty W/T chain of transmitting stations, 1915–16

Location	Type	Frequency (kHz)	Range (km)	Power (kW)	Method
Ascension	22	60–70	3000	150	Synchronous spark
Falkland Islands	23	60–70	3000	150	Synchronous spark
Aden	21	150–500	1600	300	Synchronous spark
Mauritius	21	150–500	1600	100	Poulson arc
Durban	21	150–500	1600	30	Synchronous spark
Port Nolloth	21	150–500	1600	30	Synchronous spark
Seychelles	21	150–500	1600	30	Synchronous spark
Ceylon	21	150–500	1600	30	Synchronous spark
St Johns	21	150–500	1600	30	Synchronous spark
Singapore	21	150–500	1600	30	Synchronous spark
Hong Kong	21	150–500	1600	30	Synchronous spark
Gambia	21	150–500	1600	30	Synchronous spark
Guiana	21	150–500	1600	30	Synchronous spark
Jamaica	1	150–1000	1300	14	Spark
Bermuda 1	1	150–1000	1300	14	Spark
Bermuda 2	20	82–173	3000	25	Poulson arc
Horsea	20	82–173	4000	100	Poulson arc
Cleethorpes	7	82–173	1600	100	Spark
Gibraltar	7	82–173	1600	100	Spark
Malta	8	70–82	2900	150	Spark

9.2.3 Cable operations

Many of the operations targeting submarine cables carried out by both sides were simply destructive, but missions were mounted to raise and resplice cables, effectively turning an enemy cable into a friendly one. Typical of these was the operation carried out early in the 1914–18 war when the cable from Borkum, off the German coast, to the Azores was lifted by a British cable ship, cut and spliced to a new length of cable, and diverted to Porthcurno. It was part of an early aggressive action by the British Navy in August 1914, on the second day of the conflict, in which the cable ship *Telconia* raised five German submarine cables from the seabed off Emden and linked them to Falmouth, thus denying overseas telegraph communication to the German Government except through neutral countries. This was to have serious repercussions towards the end of the war, when in 1917 the German foreign minister, Artur Zimmerman, sent a telegram in cipher to Count von Bernstoff at the German Embassy in Washington for relaying to Von Eckhardt, his minister in Mexico. This effectively promised a German military alliance with Mexico in the event of the United States entering the war against Germany, and so encouraged Mexico to attempt to recover its 'lost' territories in

Texas, New Mexico and Arizona. In the absence of a direct German cable route the message had to be sent via Sweden, and inevitably, over British cable routes. It was decoded, and the information contained in the 'Zimmerman telegram' was released to the US State Department and became a major factor in bringing about the entry of the United States into the war in 1917 [24]. After 1917 the two existing cables from Germany to America were cut and diverted to France and Britain, and the re-routing was made permanent after the war by the Treaty of Versailles [25].

9.3 The shore stations

The first coastal stations for communication with ships in British waters were installed and operated by Marconi, together with a small number operated by Lloyd's of London, also with Marconi equipment. They were spark stations with an aerial power of about 0.25 kW, transmitting on wavelengths of 150–180 m and 450–540 m. By 1901 Marconi stations were in operation at Withernsea, Caister, North Foreland, Lizard, Niton, Holyhead, Rosslare, Crookhaven and Malin Head. These and other marine shore stations were licensed by the Postmaster General to the Marconi Company or to Lloyd's Shipping Company, both providing a service to mariners. However, in 1912 the licences were withdrawn and all stations around the coast of Britain were purchased by HM Government for operation by the Post Office, apart from a few of the smaller stations operated by the railway companies in conjunction with their ferry services. A year later the stations were upgraded, replacing the spark sets with synchronous rotary spark transmitters, or in some cases with Poulsen arc transmitters. New stations were established, and all provided a 24-hour distress watch with mandatory 'silent periods' of 15 to 18 and 45 to 48 minutes past each hour. This was laid down by the 1912 International Radiotelegraph Conference, which demanded that all operators of transmitting stations on shore or at sea should listen at these times for possibly weak distress calls. Several of the stations were equipped with direction-finding apparatus, but this was used mainly to assist ordinary reception in the face of interference or jamming, rather than to provide a direction-finding service to ships.

All these stations would become available for Admiralty use in times of war. Shore stations already in operation by the Admiralty included four high-power stations: at Horsea Island, Cleethorpes and abroad at Gibraltar and Malta. Their design derived from cooperation between the Marconi Company and the Admiralty Wireless School at HMS *Vernon*, and each employed a 100 kW rotary spark set for communication with overseas naval stations at Singapore, Ceylon and Halifax (Nova Scotia) on a frequency of 86.3 kHz (Gibraltar) or 90.2 kHz (Malta). They were supported by a low-power station at Ipswich for communication in home waters on 88.2 kHz. The function of these stations was point-to-point communication with one another rather than direct communication with Admiralty ships. The latter function was carried out by shore stations abroad which were in constant contact with the naval squadrons in their local waters. They thus acted as terminal points for wireless traffic coming from the powerful point-to-point Admiralty stations, which they would relay to their local ships [6]. In 1915 the Admiralty established sixteen new auxiliary stations around the coast, from Fair Isle and St Kilda in the north to Newhaven and the Scilly Islands in the south, and around the coast of Ireland. The purpose of these stations was to communicate with a large fleet of yachts, trawlers and drifters assembled by the Navy to form an auxiliary patrol and minesweeping service as the main anti-submarine measure for home waters. Few of the trawlers and drifters had wireless at that time, and HMS *Vernon* designed a special set which did not need a trained operator and which could transmit one of three standard signals when a handle on the built-in generator was turned. This foreshadowed the 'lifeboat' sets which came into prominence in the Second World War (see later). After the First World War a number of the Admiralty's coastal stations with enhanced ship-to-shore capability were taken over by the Post Office (Figure 9.5). At about this time valve receivers were installed in the stations to replace crystal sets, and the synchronous spark transmitters were gradually replaced by valve transmitters designed for tonic train transmissions (modified continuous wave), a change which did much to improve reception for vessels not equipped with a heterodyne receiver. A full direction-finding service was by then operated from Niton on the Isle of Wight [26].

In the 1930s experiments were carried out using short-wave transmissions to communicate with ships equipped to receive them. A temporary service was established at Devizes, after which the short-wave transmitter was transferred to Portishead to become a successful prototype for such installations. The advantage of ship-to-shore transmissions on the shorter wavelengths was the long distance that could be reached using a comparatively small transmitting power, enabling suitable, affordable equipment to be carried by merchant ships. Unlike transmissions from the long-wave transmitter, GBR at Rugby, the short-wave transmissions enabled ships in distant waters to acknowledge receipt of a message sent from Portishead, a facility denied to them by their inability to transmit far enough with the long-wave transmitters they possessed which, out of necessity, could only operate with short aerials. Two transmitters were established at Portishead, each consisting of a single silicon valve oscillator (a Mullard valve, having a maximum dissipation of 3.5 kW) coupled directly to the aerial, with transmissions on wavelengths of 36.54 m (8210 kHz) and 26.59 m (11 280 kHz). With suitable choice of these two wavelengths, a working range of about 3200 km could be obtained over a

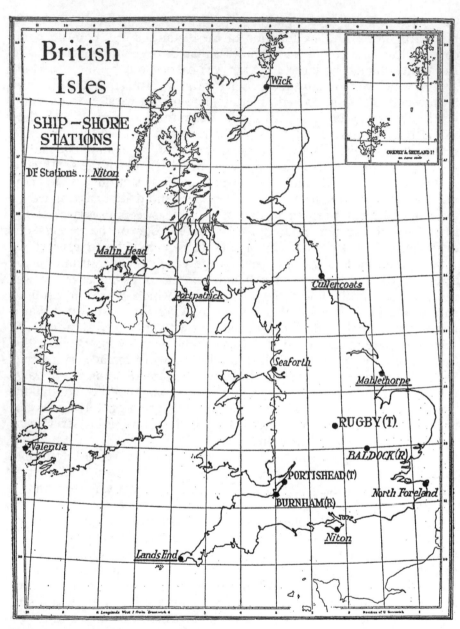

Figure 9.5   Post Office coastal stations, 1919
Source: Post Office Electrical Engineering Journal 19, 1926

24-hour period. The power supply to the oscillators was only partially smoothed, and the inherent 300 Hz ripple provided an ICW transmission which gave the telegraph signal a characteristic note recognised by the ship's operator [27]. One immediate consequence of using this new equipment at Portishead was that it became the only transmitting station able to maintain daily contact with the exploration ship HMS *Discovery* during its work in the Antarctic during 1930.

A major task for the coastal stations in wartime is the reporting of distress communications from ships at sea and from aircraft in difficulties over the ocean. Five categories of Morse coded distress calls were recognised, recorded and passed on to the naval and RAF authorities. These were all distinguished by transmission of the international distress call SOS, but in four of the categories an additional code was transmitted:

SSSS   = Attack by submarine
AAAA   = Attack by aircraft
RRRR   = Attack by surface warship raider
QQQQ   = Attack by disguised merchant ship raider

The SOS call itself was for British ships in difficulties, not known to be caused by enemy action, and also for foreign ships in distress. The number of calls recorded around Britain's shores during the Second World War was impressive: over three thousand from ships and over one thousand from aircraft in the period 1939–44, one-third of the calls from ships in distress related to attacks by enemy submarines [28]. Many lives were saved as a result of this service, aided by the bearings taken from the Post Office direction-finding stations situated around the coast on the emergency transmitters operated from dinghies and lifeboats. This information was passed to naval headquarters, where the position of the vessel or lifeboat could be determined and rescue arranged.

9.4 The inter-war years

Wartime experience and the development of valves and valve transmitters after 1918 led to a rapid deployment of powerful transmitting sets in naval ships during the two decades of peace which followed. Spark and arc sets were considered obsolete, and numerous valve transmitters were designed to carry out a variety of functions for both ship and shore stations. These continued to carry a serial numbering system devised in the early 1900s[1]; a list of the transmitters available in the 1930s is given in Table 9.4. A number of these were withdrawn or modified towards the end of this period, and those marked with an asterisk represent the

---

[1] Not to be confused with the Army system, in which the numbers indicated a specific operational function (see Chapter 8).

Table 9.4 Royal navy transmitting and receiving sets, 1930

Type	Frequency (kHz)	Range (km)	Power (kW)	Where fitted
Transmitters				
24	60–150	3200	30	Shore station
30	142–461	64	0.01	Landing ships (army)
33	115–330	160–320	0.5	Ship's 3rd office
34	64–330	160–400	0.8	Ship's 2nd office
35	75–500	960–1280	8.0	Ship's main office
36*	67–500	1600	2.0	Ship's main office
37*	100–1365	960	0.75	Ship's 2nd office
	5700–26 000		0.15	
38*	86–666	960	3.0	Flotilla leader
39	375–500	1600	2.0	Destroyers
	7500–16 700			
43*	1875–2308	80	0.4	Ship's 3rd office
47*	107–666	1280	2.0	Patrol submarine
	7500–16 700			
48	60–16 500	1600	2.0	Capital ships
81	750–1500	80	1.0	Cruiser fire control
83	2300–5000	16	1.0	Destroyers
Receivers				
B11	1500–23 000	—	—	Capital ships
B12	150–1500	—	—	Capital ships
B13	15–2000	—	—	Standby receiver

* In use at the outbreak of war in 1939.

wireless transmitters remaining in common use by the fleet at the out-break of war in 1939.

The high-powered sets available before 1923 were low-frequency transmitters operating at frequencies below about 3000 kHz. They were suitable for long-range working but had several disadvantages, not least because of the excessive driving power required. In 1924 and 1925 experiments were carried out on Horsea Island with the new short-wave transmissions sent from Portsmouth to determine reception conditions and assist in the process of establishing high-frequency installations for the Royal Navy. Signals transmitted under a variety of conditions and at various times were received on board several naval ships, then cruising to Mediterranean and Far East locations. The tests were considered to be highly successful, and resulted in a number of short-wave naval transmitters being installed in ship and shore stations by 1926, some of which are included in Table 9.4. The availability of a short-wave system made it possible for ships in any part of the world to communicate directly with one another and with the Admiralty at certain times of the day, supplementing the wireless relay network system already put in place by the Admiralty. Valve receivers had become common on board ship by this time, with the valve displacing the crystal rectifier, and the increasing use of CW transmitters in the Navy led to the need for heterodyne reception to render the Morse signal audible on the head telephones of the receiving set. Low-power arc oscillators had been used for heterodyne purposes, but their action was irregular and the availability of the valve oscillator for this purpose was widely welcomed [29]. The receivers generally used aboard ship before the outbreak of war in 1939 were tuned radio frequency (TRF) sets, typically a three-valve receiver with plug-in coils in which a multi-band HF amplifier was followed by a valve detector and a two-valve LF amplifier. A separate heterodyne valve oscillator was included for Morse reception and provision was made for changing the HF amplification to cover both the short- and long-wavelength bands [30]. The superhet receiver, with its increased sensitivity and selectivity, did not make its appearance in the Navy until just before the Second World War. When it did the Admiralty took advantage of civilian development in communication receivers to install these on board ship, with suitable modifications for the new role they had to play.

Up to about 1934 the transmitting and receiving offices on board the Navy's capital ships were located together, as were the aerials, often erected around a single mast. This led to enormous difficulties, effectively preventing reception on all medium and high frequencies when the transmitter was in operation, and necessitated electrical protection at the receiver input terminal, usually by a key-operated short circuit. On larger ships it became possible to mitigate these effects by grouping all the transmitting aerials around one mast and the receiving aerials around another, and this became standard practice by the mid-1930s. This was not entirely straightforward, since for some time additional transmitting equipment had been located in several widely separated onboard locations as a precaution against damage in action. The main W/T office was located under armour protection in the centre of the ship, with a second and third auxiliary office to the fore and aft, as shown in Figure 9.6. A fourth office housing the direction-finding equipment was generally mounted on the superstructure.

9.5 War at sea, 1939–45

The rapid expansion of the Royal Navy at the commencement of the war brought with it two major problems: fitting the new ships, particularly the converted trawler ships and ocean liners, with the appropriate communications equipment, and enlisting and training the men who would operate them. Fortunately for the larger capital ships, those that had the

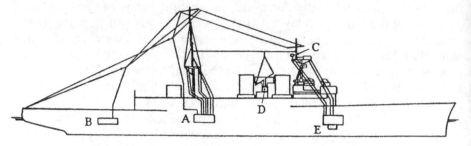

A. Main wireless-telegraph office, central receiving room, and coding office.
B. Second wireless-telegraph office.
C. Alternative direction-finder (frame coil).
D. Direction-finder office.
E. Auxiliary-sets office.

Figure 9.6   Position of W/T offices in capital ships
Source: Shearing, Journal of the Institution of Electrical Engineers 74, 1934

greatest requirement for wireless equipment, an arrangement known as the Centralized Wireless Scheme had been prepared in 1937 and was rapidly being implemented in the fleet. This laid down the conditions under which shipboard wireless equipment was to be run, and led to considerable simplification in the type and layout of equipment expected to be installed in a warship, thus facilitating the training of operators and their transfer from ship to ship. The system provided a common a.c. power supply (400 V, 50 A, 3-phase) and central control circuit for all W/T offices and their equipment. It both standardized the range of equipment which could be used on board ship and permitted the installation of civilian designs, such as communication receivers, which would normally operate from the public 240 V a.c. supply in Britain. Initial requirements for the Navy were met in the early stages of war by considerably increasing the production of just three types of transmitter: Types 49 and 60 for destroyers and larger ships, and the Marconi Type TW12 and its associated receiver for the new trawlers and motor torpedo boats.

Very soon, however, the role of the navy in the new conditions of warfare in 1940 demanded the rapid development of new equipment especially designed for the convoy system, for protecting convoys was now a prime task of the Navy in securing Britain's lifeline of food and matériel from its overseas suppliers, particularly the United States. The problem was to obtain secure and reliable communication between the naval escort ships and the merchant ships, and the latter carried a diversity of wireless equipment having a very varied capability. A satisfactory solution would later be found in the development of high-power VHF R/T equipment, but in 1940 and throughout the war this was not an option, and an acceptable solution was sought in the design of low-power MF transmitters and receivers. The commercial Marconi Type TW12 and TV5 transmitters, together with the American Hallicrafter HT11 transceivers, solved the immediate problem. Commercial receivers were widely used for shipboard naval reception. These were mainly the Eddystone Type 358 receiver, known in the Navy as the B34, and the Marconi CR100 receiver, the B28 (Table 9.5 and Figure 9.7). Both were modern superhet receivers covering a wide frequency range of 40 kHz to 30 MHz, and with the very desirable attribute of being able to be set quickly and precisely to a new reception frequency, as was required in active fleet operations. This ability became vital for convoy operation where the use of crystal controlled transmitters and crystal monitored receivers became necessary for rapid communication without delay, even after wireless silence had been maintained for days, which demanded a retuning process in much of the older equipment [31].

The ubiquity of wireless and radio transmission in naval warfare now demanded a large number of transmission channels and transmitters with several alternative modulation methods to be carried. Whereas in 1939 a battleship would probably carry nine transmitters and receivers on board, by 1945 some 16 transmitters and 23 receivers would be needed to communicate with merchant ships in convoy operations, aircraft and landing craft, in addition to the normal means of communication with other ships in the fleet, shore stations and the Admiralty in London. The installations, in various shipboard locations, had to be

Figure 9.7   The Marconi CR100 receiver (Navy Type B28)
Source: The Naval Museum, HMS Collingwood, Fareham

robust, simple to adjust – particularly in frequency, reliable under all climatic conditions, and resistant to vibration and shocks of gunfire. Also, cables and aerials needed to be well insulated against the severely wet conditions and sea spray experienced by ocean-going vessels [6]. This

Table 9.5 Allied navy transmitting and receiving sets, 1939–45

Type	Frequency (kHz)	Power (kW)	Where fitted
Transmitters			
36	67–500, 1600	2.0	Ship's main office
37	100–1365, 960,	0.75	Ship's 2nd office
	5700–26 000	0.15	
38	86–666, 960	3.0	Flotilla leader
39	375–500, 1600,	2.0	Destroyers
	7500–16 700		
43	1875–2308	0.4	Ship's 3rd office
47	7500–16 700,	2.0	Patrol submarine
	107–666		
48	60–16 500	2.0	Capital ships
49	3000–20 000,	0.75	Destroyers
	100–500		
52	900–1350	0.015	Transportable
55	1500–19 000	0.015–1.0	Submarines
57	3000–20 000	2.0–3.0	Capital ships
59	100–500	2.0	Capital ships and aircraft carriers
60	100–17 000	0.15	Cruisers and destroyers
60EQR	100–17 000	0.25	Emergency use on all classes of ship
65	1200–2000	0.06	Landing craft
86	100 000–156 000	0.08	VHF emergency set
87	100 000–156 000	0.05	VHF ship/ship, ship/air communication
89Q	1500–20 000	0.35	Cruisers and destroyers
601	1500–24 000	0.05	General purpose
602	1500–24 000,	0.05	Emergency transmitter
	200–500		
603	1500–24 000	0.45–0.65	Destroyers and above
604	200–500	0.4–0.5	General purpose
605	1500–24 000,	0.45–0.65	Cruisers and above
	200–500		
5F	14 000–23 000	0.35	Add-on VHF unit for transmitters 48/49
SCR274N	200–9000	0.05	Command set
SCR300	40 000–48 000	0.005	Amphibious operations
SCR597	2000–20 000	0.30	FM mobile set
TCK-7	2000–18 100	0.40	Submarine, destroyers and cruisers

Receivers and transceivers

B19	14–13 500	—	TRF receiver for destroyers
B28	40–30 000	—	Marconi CR100 receiver
B29	15–550	—	TRF submarine receiver
B34	40–30 000	—	Eddystone 358 receiver
B40	650–30 000	—	Naval design to replace B28
B41	700–14 700	—	Superhet receiver for submarines
B50	550–23 000	—	Superhet receiver for destroyers
TV5	500–3 000	0.2	Transceiver for trawlers
TW12	375–3000	0.05	Transceiver for trawlers
394	150–3158	—	TRF receiver associated with Type
HT11	40–30 000	0.05	Hallicrafter transceiver
TBS-8	60 000–80 000	0.05	Convoy communication
TCS-12	1500–12 000	0.075	Portable and mobile services

resulted in an exceptional mechanical design for naval equipment, a heavier structure than would be experienced for example in the more mobile army equipment or the lightweight assemblies used in aircraft.[2] This also accounts for the comparatively few alternative models being used by the Navy where changes in design could often be accommodated by modifications to the installed equipment. Some of the basic units used in the Second World War are discussed below, commencing with the Royal Navy's equipment.

9.5.1 *Allied wireless equipment*

For most of the war period wireless communication was carried out at transmitting frequencies between 100 and 20 000 kHz, and receiver frequencies covering a somewhat wider range. Towards the end of the war, and into the post-war refurbishment of naval equipment, very high frequencies of the order of 100 MHz were used, and proved ideal for short-distance operation, where modulated speech transmission replaced Morse code in convoy operations. At the commencement of hostilities the transmitters used with the centralized wireless system fitted to capital ships were naval Types 57, 59 and 60, plus the reliable but older Type 49. The Type 57 was a high-power HF system consisting of a master oscillator and power amplifier covering the frequency range 3.0–20 MHz in five sections with partial crystal control. Two pentode silica envelope valves were connected in a push–pull circuit as a power amplifier, inductively coupled to the aerial circuit to provide 2–3 kW of aerial power for W/T transmission. This reduced to 0.5–1 kW when R/T modulation was

---

[2] For example the Type 59 transmitter, used for capital ships, aircraft carriers and cruisers, weighed in at 2.5 tonnes, with an additional 0.25 tonnes for spares.

applied. The Type 59 provided similar facilities but in the MF band from 100–500 kHz and had a full-power output of 2 kW CW or 0.5 kW for MCW with a range of about 1000 km on CW. Types 57 and 59 both used relay keying to shift the output away from the aerial to an artificial load impedance. This maintained a constant load on the supply system and at the same time ensured a uniform keying wave-shape. The Type 60 was a low-power transmitter for use as an auxiliary transmitter or for smaller ships. This was also a two-stage MO–PA transmitter, having an output power of 10–15 W for CW and 4–5 W for MCW or R/T, and a frequency range of 100–17 200 kHz. The Type 49 had been in use in the fleet since 1934, and could be operated either as a two-stage MO–PA transmitter covering the frequency range 100–500 kHz or as a single-stage self-excited oscillator for use in the HF band of 3–20 MHz, and in both cases limited to CW or ICW Morse transmission.

A considerable improvement in shipboard communication equipment took place in 1943 with the release of the Type 600 series of transmitters. These were HF/MF transmitters capable of operation using CW, MCW and R/T. The largest of them, the Type 605, had a power of 450–650 W and was fitted to cruisers and larger ships. An HF version covering the range 1.5–24.4 MHz, the Type 603, was fitted to destroyers. Two low-power HF/MF transmitters, Types 601 and 602, both with an output power of 50 W, were smaller, the latter type often being carried as an emergency transmitter since it could be operated from batteries if the ship's main power supply failed. Submarine operation which had its own problems of limited space, compounded by difficulties in fitting suitable aerial arrangements for HF transmission, was served at the beginning of the conflict with the Type 55 transmitter, a medium-power HF transmitter covering the two frequency ranges 1.5–3.0 MHz and 3–19 MHz. The power output varied from as little as 15 W on the lower-frequency band to several hundred watts on the higher band. Transmission was limited to Morse using CW or ICW. A compact medium-power transmitter of RCA design, the Type 89Q, was later fitted, which had a power output of 350 W in the range 1.5–20.0 MHz, capable of CW or R/T operation. The receiver installation for submarines was generally a TRF set, the Type B29, fitted with a direction-finding loop aerial; it was later replaced by the Type B41, a superhet receiver with improved frequency stability.

A series of Admiralty-designed receivers appeared from 1937. There was a TRF four-valve receiver with reaction, the Type B19, covering low to medium frequencies in the range 40 to 13 500 Hz and containing one stage of RF amplification, a detector and a two-stage audio frequency amplifier. This was replaced later by a superhet receiver covering a similar frequency range, the Type B50. The frequency range for this receiver was covered by a set of six coils which were switched into position by a turret mechanism, the first to be used in a naval receiver. Two stages of radio frequency amplification preceded the superhet receiver, and both were employed when high immunity from interference was desired. These Admiralty-designed receivers were not produced in sufficient numbers during the war years, and use was made of two British commercial receivers, the Eddystone 358 receiver and the Marconi CR100 receiver, known to the Admiralty as Types B34 and B28 respectively; the latter was used in very large numbers for both ship and shore installations. They were replaced in 1946 by a naval design, the Type B40, which covered a similar frequency range and had similar facilities but was of more rugged construction [32].

An important piece of equipment carried on naval ships, used to accurately synchronise transmission frequencies for the 'netting' procedure, was the wavemeter. Wavemeter Types G61 and G62 were included in most ship's installations to cover ranges of 1–24 MHz (G61) and 15–2500 kHz (G62). Both wavemeters carried quartz crystal oscillators having a long-term stability of about 5 parts in 10^6, achieved by placing the crystal in a thermostatically controlled oven. The transmission frequency was measured by comparing it with the frequency signal provided by the wavemeter or one of its harmonics by a 'dead-space' heterodyne method.

Long distance ship-to-ship communication on VHF, referred to earlier, was not achieved fully until the war had been in progress for two years. A stop-gap transmitter, the Type 5F, was designed in 1940 and fitted as a high-frequency attachment to several of the older transmitters, such as the Type 49, from which it obtained its power supply. This combination proved adequate for the lower frequencies of the VHF band up to about 23 MHz, radiating CW or ICW signals. Later, specially designed naval VHF sets became available, based on the design of aircraft sets which had been using these frequency bands for some time. The Type 87 transmitter was a low-power set, originally designed for RAF ground stations and adopted for naval use in ship-to-ship communication in the frequency band 100–156 MHz. It was used for both MCW and R/T, providing an output power of 40–50 W. VHF sets were also required for emergency use and these were generally transceivers, often limited to R/T operation only. The standard emergency VHF set fitted to most ships and submarines after 1940 was the Type 86M, an American design, originally fitted to aircraft. It was controlled by a set of four plug-in crystals covering the range 100–150 MHz, and operated on R/T only.

The US Navy went through a series of equipment reviews following the First World War. The result was a new set of shipboard naval transmitters and receivers, initially low-power, low-frequency sets, but extending to HF sets from 1925. The network of high-power arc and HF alternator transmitters set up by the Navy Department to interconnect with its major bases in 1918 had also been replaced by powerful valve transmitters, capable of communicating with its fleet throughout the world. By 1930 the equipment installed throughout the US Navy included a series of sets covering a wide frequency range and it was with this equipment that the fleet entered the war after Pearl Harbor. Several of the Navy sets were originally designed for Signal Corps use, and were described earlier [33]. Indeed, the policy of the Marine Corps in 1935 was 'to use radios developed for the Army and Navy, and not secure special units for its own use' [34]. As discovered by the British Navy in their first few years of war, new sets were necessary for new operations during the 1939–1945 conflict, particularly for convoy escort duty and to support landings in the Pacific. These included the short-range FM sets of the SCR500 and SCR600 series described in the previous chapter, and the pack-sets SCR 300 used for amphibious operations. The main navy sets were those in the TCK, TBS and TCS series. The TCK sets, typically the TCK-7 set covering a frequency range of 2–18 MHz, were medium- and high-frequency sets designed for use in submarines, destroyers and cruisers and having a nominal output power of about 400 W for CW and somewhat lower for telephony operation. The TBS transceiver was produced by RCA later in the war (in 1944) and covered the VHF band from 60–80 MHz in four crystal-controlled frequencies common to both transmitter and receiver. This made it valuable for short-range convoy operations where its main use was as an R/T transmitter and as a MCW telegraphy transmitter operating at the low power of 50 W. The TCS series were also complete transmitting and receiving sets designed for use in portable and mobile services such as motor boats, tanks and other vehicles, where severe vibration and shock was often experienced. Typical of these was the low-power HF transmitter, widely used at sea, the Type TCS-12; a 74 W combined W/T and R/T unit covering the frequency band 1.5–12 MHz, and installed with a superhet receiver covering the same frequency band. Some United States sets were also installed in British naval ships during the war, including the low-power Hallicrafter transceiver and the Type 89 medium-power HF transmitter, which could be crystal controlled at spot frequencies between 1.5–20 MHz. A command or short-range transmitting set, again originally developed by the Signal Corps, was the Type 274N (N for navy). This was a manually tuned set, having an exceptionally long range (240 km) for its weight (34 kg), and was used in Allied navy ships and aircraft in large numbers.[3] Later in the war a new and combined nomenclature for military sets came into use for army and naval forces. This was the 'AN' designation, taking the form AN/xyz-nn, where AN, 'Army/Navy', shows that the equipment is military, xyz specifies the installation, equipment type and purpose, and nn is a code indicating the model [35].

---

[3] Western Electric manufactured 100000 by the end of 1944, and this was only one of a number of US producers of the 274N.

9.5.2 German wireless equipment

The design of German shipboard equipment was well under way in 1934 by two manufacturers, Telefunken and Lorenz, taking advantage of the considerable progress made in Germany in ceramics, iron dust cores and modular construction in the 1930s [36]. By 1939 a comprehensive range of communication equipment was available for the German Navy (Kriegsmarine), covering all classes of ship, initially for the MF/HF bands but later for VHF also, using equipment developed by Lorenz and a Danish manufacturer, Hagenuk.

As with other wireless equipment prepared for the military in Germany in the years before the outbreak of war, production was built up very rapidly so that the Kriegsmarine was well equipped by 1939, but with equipment which did not change much in design as the war progressed. The early transmitters from Telefunken (the principal supplier) carried the designation 'Spez' followed by a model number. A major piece of equipment at that time was the short-wave transmitter type Spez 884S, which with its variants formed the most widely used general purpose transmitter for capital ships and coastal stations. The Spez 884S was a 1.5 kW transmitter covering the MF frequency range 1.5–7.5 MHz, in 21 steps, with a capability of operation with CW and MCW. It had extremely good frequency stability, and also found use with motorised army columns on the Russian front. For ships with short action radius, such as torpedo boats, minesweepers and auxiliary ships, Telefunken produced the Spez 958 and 959 series, a 200 W transmitter covering the LF and HF bands, 100–600 kHz and 1.7–7.5 MHz with facilities for R/T as well as CW and MCW. Ships with a greater range, the cruisers and battleships such as the *Graf Spee* and *Scharnhorst*, carried 800 W transmitters of the Spez 400 series. Several transmitters were generally carried, situated in various locations in the larger ships, and between them covering a range of frequencies in the LF, MF and HF bands [37]. For cramped situations and for emergency purposes Telefunken developed a small 200 W long-wave transmitter, the Spez 2113 and, for the HF band, the Spez 406, limited to CW and MCW transmission. The latter was used in U-boats, together with a type E437 receiver fitted with loop aerial and direction-finding stages. In 1936 Lorenz brought out a new series of small transmitters which were installed in destroyers, light cruisers and in land stations operated by the Kriegsmarine. These were the *Ehrenmalsender* series known as the Lo 200 and Lo 800 series, representing the final stage in transmitter development before the outbreak of war. They are included in Table 9.6, which gives details of some of the Kriegsmarine communication equipment fitted up to 1945.

From 1939 a numbering system was used by the Kriegsmarine in which the first one or two letters indicated the manufacturer (e.g. T for Telefunken, Lo for Lorenz and Ha for Hagenuk) followed by numbers

Table 9.6 German Navy transmitting and receiving sets, 1935–45

Type	Frequency (kHz)*	Power (kW)	Where fitted
Transmitters			
Spez 367S	MF	1.5	Coastal stations and mobile use
Spez 402/3/4	LF/MF/HF	0.8	Cruisers and battleships
Spez 406S	3750–15 00	0.2	Submarines
Spez 427S	LF	0.15	Submarines
Spez 884S	MF	1.5	Capital ships and land stations
Spez 9589S	LF/MF	0.2	Torpedo boats and minesweepers
Spez 2003S	LF	1.5	Cruisers and battleships
Spez 2113	LF	0.2	Submarines and emergency use
Lo40K39	3000–16 600		
	5000–16 600	0.04	Submarines and small ships
Lo200L36	LF	0.2	Destroyers and light cruisers
Lo200K36	MF	0.2	Destroyers and light cruisers
Lo20FKL36	HF	0.2	Destroyers and light cruisers
Lo800L36	LF	0.8	Capital ships and land stations
Lo800K36	MF	0.8	Capital ships and land stations
Lo800FK36	HF	0.8	Capital ships and land stations
T200L39	LF	0.2	Auxiliary cruisers and submarines
T200K39	MF	0.2	Auxiliary cruisers and submarines
T200FK39	HF	0.2	Auxiliary cruisers and submarines
T800L39	LF	0.8	Capital ships
T800K39	MF	0.8	Capital ships
T800FK39	HF	0.8	Capital ships
Lo5000L41	LF	5.0	Land stations
Lo5000K41	MF	5.0	Land stations
Lo5000FK41	HF	5.0	Land stations
Receivers and transceivers			
E436	LF	—	Submarine receiver with D/F loop
E437	1500–25 000	—	Receiver for submarines
Lo6L39	75–1500	—	Early TRF receiver
Lo6K39	1500–25 000	—	Early TRF receiver
T8L39	75–1520	—	Superhet receiver
T9K39	1500–25 000	—	Superhet receiver
T7KL39	166–1428, 3300–20 000	—	'All-wave' superhet receiver
Ha5K39	3000–6000	0.005	Small ships, VHF transceiver
Ha15K42	3000–6000	0.015	Torpedo boats, VHF transceiver
Lo1UK35	41 500–45 750	0.007	Ship-to-ship VHF transceiver
Lo10UK35	37 800–45 500	0.01	Ship-to-ship VHF transceiver
SERK230	3300–5000	0.005	Emergency transceiver
Spez 636F	300–600, 3000–6000	0.07	Small boat transceiver

* LF, 100–600 kHz; MF, 1.5–3.5 MHz; HF, 3.0–23.8 MHz

indicating the transmission power in watts, some letters indicating the frequency band (e.g. L for long wave, K for short wave, UK for very high frequency) and finally the year of release for Kriegsmarine use. Variants of a basic design were produced for a small number of different powers, for example 200, 800 and 1500 W, and similarly for alternative frequency ranges. Thus a single design, such as the Lo200L36, could appear in at least seven different versions, depending on frequency band coverage and power.

A new series of main transmitters for the Kriegsmarine became available in 1939. These were the Telefunken T200FK39, a 200 WHF transmitter covering a frequency range of 3–23 MHz, with similar versions for the LF and MF bands, fitted to auxiliary cruisers. Similar 800 W transmitters in the series T800L39 were fitted to the larger ships. The Kriegsmarine had deeply distrusted the early superhet receivers, due to their inability to remain in tune during action, and preferred the well-established six-valve TRF receiver, the EO3750 manufactured by Lorenz, with its three tuned HF stages, reaction detector and audio amplifier [37]. With the appearance of new designs by Telefunken in 1939, including the nine-valve T9K39 superhet receiver (for receivers, the number following the manufacturer's initial denoted the number of valves used), the Kriegsmarine changed its view and thereafter the more selective superhet receiver was fitted to most navy ships. The T9K39 and its companion receiver the T8L39, covered the MF/HF and LF bands respectively, but their release was closely followed by an 'all-wave' set, the T7KL39, which covered the ranges 166–1428 kHz and 3.3–20 MHz, the unit having an optical projection for the tuning control dial which made it extremely precise in operation. Supply ships and U-boats were limited to 200 and 150 W, and both carried the smaller transceiver SZ3875 designed by Lorenz which operated on the VHF band [38]. Lorenz had designed a number of small transceivers which were used in small boats and carried as emergency equipment in the larger boats and capital ships, an example being the 0.4 W crystal controlled SERK230 set covering the MF band 2.3–5.0 MHz, fitted with a TRF receiver. The most widely used of these small transceivers was, however, the Telefunken Spez 636F, which was designed in 1931 and had a long history of use in small boats. It was rated at 40 W for R/T and 70 W for W/T, and was available in two forms, one for the HF band 3–6 MHz and another for the LF band 300–600 kHz. The compact set was later modified to become the Luftwaffe transmitter, Type FuG5, produced in large quantities for use in aircraft and in motorised transport.

For communication with the U-boat fleet, the Spez 406S, developed by Telefunken in 1936, became the major system until 1944. This three-stage 200 W transmitter was manufactured by Telefunken, and transmitted on a frequency range 3.75–15.00 MHz and was capable of working with CW and MCW for W/T. It was usually fitted with the Spez 2113 HF transmitter, a 150 W transmitter which extended the range to cover the LW band 300–600 kHz. The associated receivers were the Type E437 (1.5–25 MHz) and E436, the latter covering the LW band and connected to a direction-finding loop. Towards the end of the war the U-boats were fitted with a new four-stage transmitter, the T200FK39, covering a wider frequency range 3.0–24.0 MHz and used for long-distance communication during the lengthy excursions to Japan noted below, and also a new receiver, the T8K44 (1.5–25 MHz). A problem with long-wave transmission from submarines was in providing the lengthy aerial needed. In 1943 trials were carried out with a 20 m trailing aerial, floating on the surface, and connected to the submarine by a 250 m long coaxial cable, a system which enabled communications to be maintained with the submerged vessel over a distance of 50 to 100 km.

During the war the Kriegsmarine attempted to set up W/T connections with the Japanese naval authorities in Tokyo in order to assist Far East shipping communications and to support a possible German submarine operation in the Indian Ocean. This proved difficult to arrange, but eventually in late 1943 German 1.2 kW transmitters were established at Tokyo, Singapore and Jakarta, with a submarine control station at Penang, all manned by German personnel. In Germany the powerful Norddeich 20 kW transmitter formed part of this communication W/T network, providing an exchange of W/T traffic between Berlin and Tokyo, as well as a naval service to ships on their journeys between the two countries using a short-wave transmission on 24 and 36 m [39].

9.6. Cable ships and cables

As in the First World War, the submarine cables linking Britain with its overseas allies and dependencies was vital in coordinating its policy, strategy and supplies. Invaluable work was done by the small number of cable ships whose task it was to maintain and improve submarine cable communication as the war progressed. The total length of submarine cable maintained by Cable & Wireless alone was over 250 000 km, and although a number of the cables became cut or damaged, this lifeline was secured by ensuring that one or more alternative routes for all major destinations was available throughout the entire period of the war. This was necessary for security purposes. The strategic importance of cable communications, which could not be tapped,[4] contrasted with wireless circuits which could, and accounted for the high priority given to cable ship activities throughout the war.

---

[4] Experiments were carried out by the Admiralty to determine whether the current carried by a submarine cable could be induced in a 'pick-up' cable laid alongside it for a considerable length. No useful result was obtained, and there is no evidence that the Germans were any more successful.

The Germans had little need for such extensive cable communication since their lines of communication were internal to Europe. Only two German cables were in operation at the start of the conflict, one from Emden to the Azores and the other from Emden to Lisbon, and both were promptly cut by the afternoon of 3rd September 1939 [40]. The ultimate destination of the European end of the Emden–Azores cable, which now lay at the disposal of the Allies, became the subject of considerable controversy in 1942. While the British authorities wished to divert it to the south coast of Britain to play a part in the worldwide network of submarine cables now being operated for wartime service by Cable & Wireless, the Americans wished to obtain sole use of this newly acquired cable and use it after the war for private United States commercial purposes. A compromise was reached which allocated to the Americans a submarine communications channel through the cable, but this facility would be withdrawn when necessary in the interest of the war effort [41].

The wartime role of the Allied cable ships was threefold: recovering and repairing damaged cables, locating and severing enemy cables, and locating and diverting enemy cables for Allied use. All these activities were difficult, time consuming and dangerous since the ships were slow and incapable of manoeuvring to avoid enemy action. The technical procedure for locating the region of a break in a cable could often be carried out on shore before the ship set out on its recovery mission (see Chapter 5). This enabled the depth of the cable at that point to be determined from the charts, from which the length of new cable that needed to be taken on board could be determined. The process of grappling the cable off the seabed and splicing a new section is a lengthy one, and some form of defence provided by accompanying naval vessels was often possible. But this was not always completely effective and several cable ships were lost between 1939 and 1945 – two during the Normandy landings. At the outbreak of war Cable & Wireless, the main company concerned with these operations, owned seven cable ships and the Post Office a further four. Several other cable ships acquired by the Allies as the war progressed were added to this small force, which was augmented by the Admiralty's own cable ships and those of its American allies. Repairs to submarine cables, whether damaged by accident or by enemy action, took the cable ships all over the world. One such ship, CS *Cambria*, spent six months in repair work in British and French waters, then travelled to mid-Atlantic, to Halifax, via the Cape Verde Islands to the West Indies, along the Brazilian coast and from there to West Africa, returning later to Recife in Brazil. At each of these locations a number of repair operations on damaged cables were undertaken.

A new task for the cable companies and the cable ships emerged during the early stages of the war; to defeat the German magnetic mines, which were beginning to be laid by enemy aircraft in large numbers in the approaches to harbours and ports. The task was to provide lengthy loops of cable, sometimes over 100 km in length, to be used for minesweeping outside British harbours at home and overseas [42]. The process was known as the LL sweep, and involved towing a floating loop of cables between a pair of minesweepers. A heavy electric current was pulsed through the cables, situated well astern of the minesweepers, and the circuit was completed by the sea-path between them. This proved effective in firing the new German magnetic mines, and played a valuable part in keeping busy wartime harbours free from damage.

The cable fleet was especially active in the English Channel during the invasion of Normandy, as discussed in the previous chapter. Before D-Day the fleet laid at different points round the coast eight major cable links, amounting to almost 700 km of cable. The cable ships *Alert* and *Iris* laid the first submarine cable to the Normandy beachhead just two days after the initial landing, and continued to lay and repair cables between Britain, France and Belgium for the remainder of the war, while sustaining heavy casualties due to the static nature of their work.

9.7 British naval wireless training

The need for specialist training of wireless personnel was understood as early as 1899, when Captain Jackson recommended the establishment of naval schools to train operators, and later mechanics, at HMS *Defiance*, *Vernon* and *Vulcan* [43]. At the outbreak of war in 1914 the Admiralty needed a great number of skilled W/T operators for its ships and shore establishments, partly to extend the W/T watch period to 24 hours on almost five hundred ships then fitted with wireless, and also for the many auxiliary vessels and the training stations now being established. Already the bigger ships required a larger number of operators to man the wireless rooms, of which there were now more than one, in different locations, on each vessel, and to meet the shift requirements for this exacting task, which could not be carried out by other ratings. Experienced volunteers called upon to join the Royal Naval Wireless Auxiliary Reserve (RNWAR) came from civilian positions in the Post Office and the Merchant Navy, and from among the amateur wireless fraternity. The Wireless Club of London (the forerunner of the Radio Society of Great Britain) and the Institution of Electrical Engineers offered to help [44]. The RNWAR applicants were expected to have reached the age of twenty-one and not be older than thirty, and to already hold a Certificate for Wireless Operators from a recognised company or the Post Office. Training was to be given to selected applicants in Naval W/T procedure on a short course lasting just two or three weeks, with the rank of chief petty officer telegraphist offered upon its successful conclusion [45]. This was an attractive offer, and quickly attracted experienced volunteers to fill the navy posts. The Admiralty also secured the services of all available Marconi operators from the training schools and from the mercantile marine. This, of course, led in turn to a shortage of marine operators, a difficulty which was solved only as the output of the training schools, both Marconi and private, increased. The Marconi Company, through its training courses at Bush House, King's College and the Birkbeck Institute in London, was able to keep up the flow of young recruits, and in 1915 alone produced over five hundred graduates ready to receive further operational training from the Admiralty and the merchant service.

These efforts were in the nature of emergency responses to the outbreak of war. The Admiralty had already started its own permanent schools for training W/T operators at Scapa Flow, the Royal Naval Depot at the Crystal Palace in London, and a much extended Signal School at HMS *Vernon* [46]. At these various locations the general procedure was to give ten weeks of training in signals followed by sixteen weeks of training for ordinary telegraphists. By the end of 1917, some 3600 naval officers and ratings had undergone this basic course of W/T instruction. The Morse code taught to the recruits in 1914–18 had been extended by the Admiralty to include a set of signs representing special flags and pennants so continuing an old naval tradition in a new form. Each of these new signals (which were actually pairs of identifying letters) was preceded by a linking sign, AA. Thus a battle cruiser flag (BC) would be transmitted as one word, AABC, although some signs such as a pilot flag were sent as one letter sign, PJ, with no intervening space. Together with a number of procedural signs, already in use in marine Morse communication, and a set of manoeuvring signals, the naval wireless operator candidates at that time would be expected to reach a higher state of proficiency than existed in several of the other services [47]. The requirements for merchant shipping became particularly acute after 1915, partly because serving wireless operators were being recruited by the Royal Navy, but more specifically because of changing requirements for continuous watch and the equipping of certain merchant ships which had hitherto not been fitted with wireless, namely vessels of between 1600 and 3000 tonnes. W/T communication was seen as essential for the convoy system, introduced early in the war, and although Marconi and other training schools increased their output of trained operators it was some time before the deficiencies could be met. The new active war status of the civilian wireless operators who transferred to the merchant marine and Admiralty shore stations at home and abroad from Marconi and Lloyd's installations was recognised by special 'on war service' insignia, and the same recognition was given to those recruited from Marconi establishments for work in Post Office stations. In March 1915 these transferred employees were issued with full RNVR uniforms, together

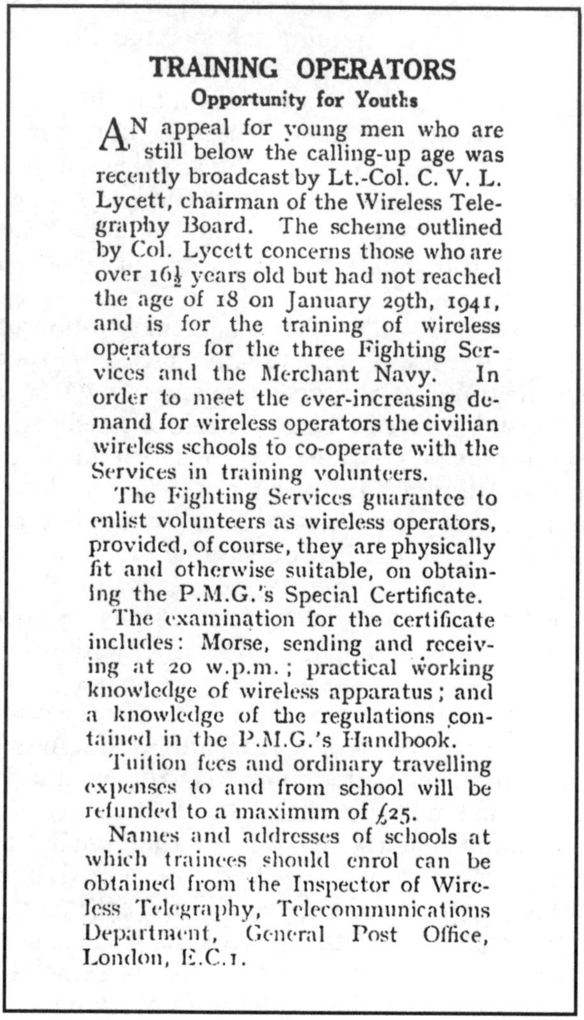

TRAINING OPERATORS
Opportunity for Youths

AN appeal for young men who are still below the calling-up age was recently broadcast by Lt.-Col. C. V. L. Lycett, chairman of the Wireless Telegraphy Board. The scheme outlined by Col. Lycett concerns those who are over 16½ years old but had not reached the age of 18 on January 29th, 1941, and is for the training of wireless operators for the three Fighting Services and the Merchant Navy. In order to meet the ever-increasing demand for wireless operators the civilian wireless schools to co-operate with the Services in training volunteers.

The Fighting Services guarantee to enlist volunteers as wireless operators, provided, of course, they are physically fit and otherwise suitable, on obtaining the P.M.G.'s Special Certificate.

The examination for the certificate includes: Morse, sending and receiving at 20 w.p.m.; practical working knowledge of wireless apparatus; and a knowledge of the regulations contained in the P.M.G.'s Handbook.

Tuition fees and ordinary travelling expenses to and from school will be refunded to a maximum of £25.

Names and addresses of schools at which trainees should enrol can be obtained from the Inspector of Wireless Telegraphy, Telecommunications Department, General Post Office, London, E.C.1.

Figure 9.8   Notice appealing for young volunteers for telegraph operator training
            with the Post Office
Source: *Wireless World, July 1941*

with certain allowances and pension rights in recognition of their war service in regions of enemy activity, a concession much appreciated by the operators at a time when the stigma of being a 'non-combatant' was a serious matter [48].

The need for a large influx of wireless operators into the Royal Navy and merchant fleet was no less urgent in 1939, and similar incentives were offered to secure the numbers required. Under wartime conditions, the Admiralty recommended, all ocean-going vessels of UK registration should have at least two wireless operators on board, and this requirement alone demanded an additional 2500 operators. The Post Office was concerned with the output of private wireless schools (of which there were 26 when the war started), together with their own training schools in various parts of the country. Some five thousand operators qualified from these schools in 1940, and efforts were made to increase the number of trained shipboard operators by limiting the requirement to a single class of certificate only [28]. Over ten thousand of these single-class certificates were issued during the conflict. To meet the demand for operators early in the war, former shipboard operators were appealed to to return to the sea and, as in 1914–18, radio amateurs were asked to help. The amateurs were already trained operators and desired by all three services and the merchant navy, but their numbers were limited. Nor were the young men, still below the calling-up age, neglected. A special appeal was made on behalf of the Post Office by the Wireless Telegraphy Board for men between the ages of sixteen and a half and eighteen to volunteer for a course of instruction leading to an award of the PMG Special Certificate (Figure 9.8). The Navy's own training establishments were augmented, and in 1939 began a close cooperation between training branches in Shotley, Portsmouth, Devonport and Nore to improve the training of signallers and signal officers [49].

References

1 'Wireless telegraphy in the Navy', *The Army and Navy Illustrated*, (589), 2nd March 1901
2 JOLLY, W.P.: 'Marconi' (Constable, London, 1972)
3 POCOCK, R.F.: 'The origins of maritime radio' (HMSO, London, 1972)
4 'Wireless telegraphy 1903–1907' (Admiralty, 1903), Doc. ADM144/20, Public Record Office, Kew, Surrey
5 'Wireless telegraphy: British and foreign miscellaneous information, ship and shore stations', (Admiralty Intelligence Department, 1906), National Maritime Museum Library, Accession Number PBB2845
6 SHEARING, G., and DORLING, J.W.S.: 'Naval wireless telegraph communications', *Journal of the Institution of Electrical Engineers*, 1930, **68**, p. 259
7 'Communications in the Atlantic' (1914), Doc. ADM1/8403/430, Public Record Office, Kew, Surrey
8 DEVEREUX, A.: 'Messenger gods of battle' (Brassey's (UK), London, 1991), p. 57
9 WAKAI, N.: 'Dawn in radio technology in Japan', *in* Proceedings of the IEE Symposium '100 years of Radio', Conf. Publ. No. 411 (IEE, London, 1995)

10   ARCANGELIS, M. de: 'Electronic warfare from the Battle of Tsushima to the Falklands' (Poole, Dorset, 1985)

11   NOVIKOV and PRIBOI: 'Tsushima', trans. Eden and Cedar Paul (Allen & Unwin, London, 1936)

12   NESPER, E.: 'Die Drahtlose Telegraphie und ihr Einfluss auf der Wissenschaftverkehr unter besonderer Berüchsichtegung des Systemes Telefunken' (Berlin, 1905), pp. 88–9

13   PHILLIPS, V.J.: 'Early radio wave detectors' (Peter Peregrinus, London, 1978)

14   'W/T messages for Admiral of the Fleet' (1912), Doc. ADM1/8379/143, Public Record Office, Kew, Surrey

15   'W/T training 1914—1918' (Admiralty, 1915), p. 652, Doc. ADM 131/121, Public Record Office, Kew, Surrey

16   See 'The yearbook of wireless telegraphy and telephony' (Wireless Press, London) for the years 1914–18

17   MARDER, A.J.: 'From Dreadnaught to Scapa Flow', vol. 2 (Oxford University Press, Oxford, 1970)

18   'Wireless telegraphy handbook for Type 10 (submarine) sets' (1915), Doc. ADM 186/810, Public Record Office, Kew, Surrey

19   'Admiralty Signal Division history', vol. 1 (Royal Naval College, Greenwich, 1926), Doc. ADM116/3403, Public Record Office, Kew, Surrey

20   'Reports on interned German vessels', part 4 (1919), Doc. ADM186/793, Public Record Office, Kew, Surrey

21   MEISSNER, A.: 'Die deutsche Elektroindustrie in den Kriegsjahren' (Telefunken Publication, Berlin, 1919)

22   Letter from Admiralty to Post Office, 13 October 1914, Doc. ADM 116/1409, Public Record Office, Kew, Surrey

23   'Admiralty Signals Division history', vol. 2' (1919), Doc. ADM116/3404, Public Record Office, Kew, Surrey

24   GRAMLING, O.: 'Associated Press: the story of news' (Associated Press, New York, 1940)

25   COGGESHALL, I.S.: 'An annotated history of submarine cables' (University of East Anglia, Norwich, 1993), p. 70

26   'E.H.S.': 'The Post Office Wireless Service', *Post Office Electrical Engineers' Journal*, 1925, **19**, pp. 58–66

27   GILL, A.J., and MacDONALD, A.G.: 'Portishead short-wave transmitter', *Post Office Electrical Engineers' Journal*, 1930, **23**

28   'Post Office reports, Wireless Telegraphy Section 1939–1944' (1944), Doc. 56/142, British Telecommunications Archives

29   GOSLING, R.S.: 'Development of thermionic valves for naval use', *Journal of the Institution of Electrical Engineers*, 1920, **58**, p. 670

30   SWAN, F.B.: 'Valve amplifiers for shipboard use', *in* 'Yearbook of wireless telegraphy and telephony (Wireless Press, London, 1920), pp. 924–35

31   ANDERSON, W.P., and GRAINGER, E.J.: 'Long, medium and high frequency communication to and from HM ships', *Journal of the Institution of Electrical Engineers*, 1947, **94**, (3A), pp. 46–58

32   'The user's guide to wireless equipment' (Admiralty Signals Division, 1958), Doc. ADM234/49, Public Record Office, Kew, Surrey

33 TERRETT, D.: 'The Signal Corps: the emergency', Official Publication AS 742/11(8) (Government Printing Office, Washington, DC, 1956)
34 WOODS, D.L.: 'A history of tactical communication techniques' (Arno Press, New York, 1974), p. 231
35 KEELEY, J.: 'American military equipment designations', *Radio Bygones*, August/September 1999, (60), pp. 23–5
36 BAUER, A.O.: 'Some aspects of unconventional engineering during inter-bellum' (Foundation Centre for German Communication and Related Technology 1920–1945, Diemen, Netherlands, 1999)
37 TRENKLE, F. 'Die deutschen Funknachrichtenanlagen bis 1945' (Hüthig Buch Verlag, Koblenz, 1946)
38 BAUER, A.O.: 'Receiver and transmitter development in Germany 1920–1945', *in* Proceedings of the IEE Symposium '100 years of Radio', Conf. Publ. No. 411 (IEE, London, 1997), pp. 76–82
39 'Cooperation with Japan', *in* German Naval Report NID 24/T (1945), p. 62, Doc. ADM223/51, Public Record Office, Kew, Surrey
40 GRAVES, C.: 'The thin red line' (Standard Art Book Company, London, 1946), p. 9
41 'Use of Imperial and ex-enemy world-wide cable networks 1940–1944' (1944) Doc. ADM116/5137, Public Record Office, Kew, Surrey
42 'Cable laying and charter of cable ships 1939–1945' (1945), Doc. ADM116/5433, Public Record Office, Kew, Surrey
43 JACKSON, H.B.: Letter to Admiralty, 10th August 1899, Doc. ADM116/523, Public Record Office, Kew, Surrey
44 APPLEYARD, R.: 'The history of the Institution of Electrical Engineers (1871–1931)' (IEE, London, 1939), p. 213
45 'Establishment of reserve W/T operators' (1914), Doc. ADM8395/335<#>523, Public Record Office, Kew, Surrey
46 JELLICOE, J.R.: 'Grand Fleet 1914–1916' (Cassell, London, 1919), p. 189
47 'RNWAR wireless signalling instructions' (Admiralty, London, 1935), Admiralty Doc. PBC1023
48 BLOND, A.J.L.: 'Wireless telegraphy and the Royal Navy 1895–1920', Ph.D. thesis, University of Lancaster, 1993 (BLOND/A/PHD/HY93)
49 'W/T signal branches amalgamation' (1939), Doc. ADM116/1<#>210<#>523, Public Record Office, Kew, Surrey

Chapter 10

Military telegraphy in the air

Airborne telegraphy for the military was borne out of the necessity for immediate reporting of events seen from the air, vital when travelling over enemy territory before radar and other navigational techniques became available to the flyer. Initially there were many problems to overcome: poor range of transmission, high noise level in the cockpit, suspicion of the often fragile wireless equipment by the pilot, and the weight added to the payload of the aircraft. Extra weight was less of a problem when lighter-than-air craft were used. In the early days of flying, installing wireless equipment on board airships and balloons proved a valuable way to carry out air-to-ground transmission experiments, and the Royal Flying Corps (RFC), formed out of the Royal Engineers in 1907, owed much to the airship for this work at a time when dirigibles formed part of the Military Wing.

10.1 The dirigible

The first wireless company of the Royal Engineers was formed at Farnborough in 1907, primarily to investigate the possibilities of airborne equipment for military use, and for this purpose was attached to the RE Balloon School. In one of the first reports of the company, Lieutenant C.I. Aston, who carried out some of these early experiments, noted that 'In May 1908, a free run was made in the Army Balloon, *Pegasus*, in which a receiver of wireless had been installed. When the balloon was over Petersfield very good signals were received at Aldershot Wireless Station some 20 miles [32 km] distant.' [1] The first army airship, *Beta*, was fitted with wireless in 1911. Messages were transmitted from the airship to the Royal Aeronautical Establishment (RAE), Farnborough, over a distance of 48 km, and good ground reception was achieved. To obtain acceptable ground-to-air transmission it became necessary to stop the airship's engines, not only because of the high noise level, but also because of the intense interference the ignition produced in the receiver. Later, screening of the wireless equipment and bonding of engine parts remedied this defect. Several more experimental airships were fitted with wireless in 1912 and 1913, the ships given further letters of the Greek alphabet, *Gamma, Delta, Eta*, and so on. *Delta*, carrying a 0.5 kW Marconi spark transmitter, succeeded in establishing clear communication with the North Foreland telegraph station when over 200 km from it, a range considered acceptable for a practical service. These experiments foreshadowed the production of several active-service airships for the army, commencing with the dirigibles *Astra-Torres* and *Parseval*, which patrolled the English Channel during the crossing of the British Expeditionary Force in 1914, reporting ship movements by wireless telegraphy [1].

Airships at that time were fitted with Marconi spark transmitters, in a specially produced lightweight version of the 0.25 or 0.5 kW spark system, with a carborundum detector for reception. This was before valve amplifiers became available, and reception of weak signals on board the airships was an uncertain operation. An early success in wartime operation was the use of such equipment aboard an airship in the African campaign, which led to the sinking of the German cruiser *Königsberg* in the Rufiji river in German East Africa [2]. The use of spark transmitters was considered dangerous aboard the airships of the time, which were kept aloft by huge envelopes filled with highly inflammable gas and, after an initial exploratory use, not a great deal of use was made of airborne transmitting apparatus until 1917, when it became possible to equip them with valve transmitters. Instead receivers only were carried aboard the airships, and to compensate for the lack of air-to-ground communication the pre-1917 dirigibles usually carried a complement of carrier pigeons! The first airship to carry a Marconi-designed valve transmitter, in 1917, was the No. 9 Rigid Airship from Howden in Yorkshire. This was successful but, apart from the spasmodic use of tethered observation balloons fitted with wireless telegraphy equipment, airborne wireless development for the Allies was by that time concentrated entirely on producing equipment for aircraft, and no further development for dirigible use was made in the last two years of the war.

The German High Command had no such inhibitions in the use of spark transmitters in dirigibles, and wireless equipment, including spark transmitters, was carried aboard their gas-filled Zeppelins from the out set. The equipment at that time used low and medium frequencies, and the need for large aerials presented no problem for the huge Zeppelins, some of which were several hundred metres long. The balloon itself was not always used as a support for the transmitting aerial, and instead a trailing aerial was used, some 35 m long. Neither did weight present a difficulty. Transmission equipment weighing 300 kg or more, often a converted navy transmitter, was installed in a special soundproof cabin which added to the payload. The Telefunken quenched spark transmitter used, which had already proved effective in a naval context, was able to transmit on three possible wavelengths, 150, 200 or 250 m, later expanded to include 300 and 350 m. These wavelengths were selected because they were lower than those used by the German army for its trench transmitters, in order to avoid the problem of interference from the ground stations when operating close to the front. The bigger aircraft and Zeppelins carried a 1 kW transmitting station and longer aerials (60–100 m); Meissner, one of Telefunken's chief designers, remarked in 1917 that, 'for the traffic from England to the German coast 1 kW energy is sufficient' [3]. The wireless navigation method used by the Zeppelins was to transmit signals at pre-arranged intervals which were picked up by two separate directional stations at opposite ends of a long base-line; the position of the airship was thus triangulated and transmitted immediately to the airship by W/T. This procedure was not always effective, and Zeppelins were frequently 'lost' over the North Sea. It also carried the grave disadvantage that the 'fixing' signals could be detected by Allied ground stations, as described in Chapter 8. Towards the end of the war the Germans experimented with a revolving directional wireless telegraph transmitter on the ground from which bearings could be taken by a receiver fitted to the airship, but this did not become available in service before the Armistice [4].

10.2 War in the air, 1914–18

The Royal Flying Corps (RFC) entered the war in 1914 with very limited knowledge of the operational role of wireless in wartime [5]. The Corps had been formed in 1907 as a branch of the Royal Engineers, with the Royal Naval Air Service (RNAS) following in 1912. (They were later to amalgamate to form the Royal Air Force in 1918.) The first airborne spark transmitters and ground-based receivers sent to France at the beginning of the conflict enabled some tests in air-to-ground communication to be carried out under actual war conditions. These were moderately successful in daily reconnaissance flights during the Battle of Mons, and later during the Battle of the Marne, but these operations underlined the very limited capacity of the combat aircraft to support the weight of the wireless apparatus then in use. The Marconi Company had been working on this problem for several years, and early in 1915 the first improved receivers and airborne spark transmitters were shipped from their works for service with the RFC in France.

The Royal Navy was also interested in airborne communication, and experiments were conducted by the newly formed Royal Naval Air Service in 1912, using a Short S41 hydroplane. The equipment carried was a

30 W spark transmitter and a small crystal receiver, and was used in the naval manoeuvres of that year. Good communication was established over a range of about 10 km. On one occasion the aircraft's engine failed and it was obliged to land in the sea, but a Morse message to HMS *Hermes* enabled the aircraft to be located and the crew rescued – the first air–sea rescue to be effected through wireless transmission. By 1913, twenty-six seaplanes had been fitted with French Rouzet transmitters for wireless telegraphy trials, operating on wavelengths of 91–400 m. The Rouzet receiver accompanying the transmitter was not considered satisfactory by the Navy, and the transmitters were fitted with crystal sets purchased from the department store Gamages Ltd of High Holborn! The Rouzet transmitters were later installed in mobile vehicles to maintain continuous communication between monitors in the field and the French artillery using captive balloons for spotting work [6, 7]. The spark wireless telegraphy system devised by Lucien Rouzet for the French forces was 'lighter in weight for its aerial power output than anything previously seen', and was widely used for aircraft work at the beginning of the war, particularly by the RNAS. It was a simple, sturdy device consisting of a helical tapped tuned coil shunted by a synchronous rotary spark gap, having several gaps in series, and powered by a 0.25 kW alternator driven from the aircraft engine. In the circuit (Figure 10.1), the Morse key was included in the primary circuit of the step-up transformer from the alternator. The repetitive spark train frequency was fixed at about 800 Hz, easily heard above interference at the receiver. Although powerful (Rouzet transmitters with output powers up to 750 W were produced), the set was not regarded very favourably by the RFC because there was the risk that its exposed spark gap would start a fire on the aircraft in the presence of leaking petroleum [8]. The RFC made greater use of the British Sterling spark transmitter with its spark gap contained in a neat, gas-tight box (see later), until the Marconi Company was able to develop a valve transmitter and receiver later in the war. An early operational use of the RNAS version of the Sterling spark transmitter, the Type 52, was in aircraft carried on the *Arc Royal* in 1915. These were employed in aerial observation work at the Gallipoli front with considerable success [9].

The RFC, however, used wireless mainly for communication between spotting aircraft and ground batteries, which was carried out extremely effectively:

The aircraft were attached to corps command and took up artillery officers daily from each division to fly over the German batteries. Information was relayed to the ground and the direction of artillery fire directed in this way from the air for up to an hour, despite anti-aircraft shelling.

A wireless telegraphy code was devised during the Battle of the Marne in 1915 by Lieutenants James and Lewis to provide the information

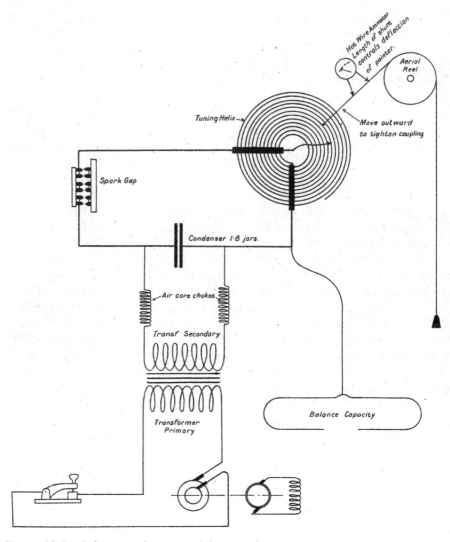

Hot Wire Ammeter
Length of shunt
controls deflection
of pointer.

Aeriol Reel

Tuning Helix

Move outward to tighten coupling

Spark Gap

Condenser 1·8 jars.

Air core chokes.

Transf Secondary

Balance Capacity

Transformer Primary

Figure 10.1   Schematic diagram of the French Rouzet transmitter
Source: Public Record Office

required by the gunners without transmitting unnecessary details [10]. A clock code was used in which the target (usually German artillery) was taken as the centre of a clock and imaginary lines were circumscribed around it at distances of 10, 25, 50, 100, 200, 300, 400 and 500 m to define range. Direction was indicated by a clock notation with twelve o'clock always taken as true north from the target, and other bearings designated accordingly. An observer noted where the rounds were falling with reference to the imaginary circle, with distance from the centre identified by a letter of the alphabet to give the range and the clock notation for the direction. A direct hit was signalled as 'OK', and there were other coded signals. Messages from the battery or any other ground station were signalled to the observer in the aircraft by means of white strips which were laid on the ground to form letters of an agreed code (for some time wireless reception in aircraft was not operationally acceptable and only air-to-ground transmission was available). The value of aircraft for locating the position of the exploding shell was quickly appreciated, not least because of the ammunition saved by excessive ranging shots, and increasing demands were made for more and more aircraft to be fitted with equipment for this purpose.

10.2.1 British airborne equipment

By the end of 1915 the weight of the airborne equipment for the RFC had been reduced to about 9 kg and installation became possible in many types of aircraft. 'When wireless machines were increased in number artillery observation came into its own', reported Brigadier-General Stokes, who commanded the 27th Division Artillery in February 1915. He stressed 'the enormous advantage of wireless', and cited instances of a sixfold increase in target accuracy [1]. This was using the Sterling spark transmitter, shown in Figure 10.2, manufactured by the Sterling Wireless

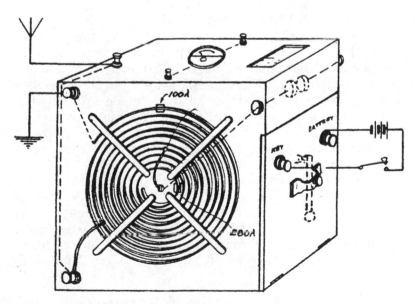

Figure 10.2   The Sterling spark transmitter
Source: Honan, Wireless World, 75, 1920

Company of London, then at the forefront of civilian wireless development [11]. The Sterling set, the most widely used airborne spark transmitter produced in Britain at the time, was a compact device contained in a steel box, measuring 200×200×130 mm, excluding the power supply which consisted of three 2 V accumulators. The set was a short-range instrument, expected to be used with a non-amplifying crystal receiver, and the combination was effective under fair conditions over a range of 32 km. Its short range was an advantage since it was used mainly for artillery observation and did not interfere with other airborne sets used in the same way over an extended front [12].[1] The spiral tuning inductor seen in the diagram was tapped to couple it to a trailing aerial wire and, by short-circuiting some part of the helix, transmission on a wavelength range between 100 and 280 m was obtained. The battery was keyed to the primary of a step-up transformer connected through a vibrator, with the secondary coupled to the spark circuit. With careful adjustment of the spark gap and vibrator contacts, reception could be favourable: Vincent-Smith enthusiastically noted that when 'adjustment of the spark gap and primary make and break [was] very carefully attended to . . . the rasping guttural splutter gave place to a pleasing flute-like note' [5]. A hot-wire ammeter indicated aerial current. To obtain a maximum indication the two taps on the helical coil would be adjusted before the machine became airborne – a procedure which led to only an approximation to the desired wavelength. This was not as disadvantageous as it might seem, since the selectivity of the receivers used at the time was poor – but this did become a problem as more aircraft transmitters came into use. The Sterling set in its various versions was also used for army signals, and continued to be used for a short time after the cessation of hostilities in civil aviation, and as a favoured experimental device in the radio amateur fraternity. The complementary ground receiver set used with the Sterling transmitter was the Marconi short-range Mk III tuner, shown in Figure 6.11, which was later followed by a three-valve amplifier, the combination used until the end of the war. The tuner incorporated a Perikon or carborundum crystal and provided a tuning range of 100 to 700 m through two circuits, including one to match the aerial to its tapping on the main tuning inductance [13].

Spark transmitters, however, were limited in their use, for reasons given earlier, and their operational performance deteriorated as the number of ground stations and aircraft fitted with wireless increased. Some improvement was obtained by allocating different sets of operating wavelengths to adjacent squadrons. For example, the three flights of one squadron would use wavelengths of, say, 140, 180 and 220 m, while an

[1] The density of artillery spotting planes in the air above the Western Front towards the end of the war was high. At the time of the Armistice, one plane could be in use for every 400 m of the front [13].

adjacent squadron would use 160, 200 and 240 m. In the longer term, however, the use of CW systems, with their narrower transmission bandwidth, in conjunction with a more selective receiver, offered the only practical solution to the communications problem which was becoming more pressing as the hostilities continued. CW systems, already in use by the army, were initially applied to ground wireless telegraphy communication between brigades, wings and squadrons before being considered for airborne installation. By late 1917 the RFC was able to fit telegraph CW sets into its night bombing aircraft, and was also beginning to experiment with air-to-ground and air-to-air wireless telephony. Transmission from the air was soon considered essential for combat aircraft since it was difficult for a pilot to manipulate a Morse key while flying an aircraft capable of rapid manoeuvres. Telegraphy sets continued to be used in bombers, however, where a separate wireless operator could be carried as part of the crew. The larger T57 and T21 telegraph sets used for this work were manufactured by the Marconi Company and had a range of several hundred kilometres, operating at the then short wavelength of 120 m. They were used primarily for navigation and for the reception of meteorological information and instructions transmitted from the base station.

Aircraft direction-finding was the second major application for airborne wireless equipment. The need for pilots to be able to find their way to their destination and return without disclosing their position was vital in an offensive operation. Experiments started in 1915 with a loop aerial fitted inside the aircraft and capable of being rotated. This enabled aircraft to take a bearing on a ground transmitting station without making a transmission or altering the aircraft's position. However, until the gain of the receiver could be augmented by valve amplifiers a satisfactory service was not achieved. To increase the range a fixed wing loop aerial was developed in place of the small rotating loop. When the aircraft was heading towards the transmitting station, signals of minimum strength were received and the device could be used for homing. To support a direction-finding service a number of high-power wireless telegraphy ground stations were installed towards the end of the war (see also Chapter 8). These were arranged to transmit at pre-arranged times so that bearings could be taken with the direction-finding wireless telegraphy apparatus carried within the aircraft. In effect they provided a 'radio beacon' service with a 5-minute 'fixing' transmission every hour and were used by the new Handley Page long distance bombers which came into service in May 1918.

Towards the end of the war, the variety of wireless telegraphy equipment in used for ground and air communications had been reduced to a small number of basic sets with several variants covering different wavelengths and output power. These are summarised in Table 10.1, which lists the equipment used by the RFC and RAF up to 1918. Spark

Table 10.1   RFC/RAF wireless equipment to 1918

Year	Type	Frequency (kHz)	Power (W)	Purpose*
1916	Rouzet spark TX	650–3000	250	Artillery cooperation
1917	Sterling No. 1 spark TX	1150–3000	30	Short-range artillery cooperation
1917	Sterling Type 52 spark TX	1150–3000	40	Short-range for RNAS
1917	Sterling Type 52a spark TX	1150–3000	40	Long-range artillery cooperation
1917	Tb	500–3000	—	Crystal receiver
1917	Tf	120–2000	—	Valve receiver
1917	T21	120–375	60	Transmitter
1918	Mk III	375–857	—	Receiver crystal + valve
1918	Mk III	428–2500	—	Ground receiver crystal + valve
1918	No. 12 amplifier	—	—	Five-valve intercommunication amplifier
1918	Mk II + Type 7 amplifier	500–750	—	Receiver used in bombers
1918	T57	120–300	75	Airborne and ground valve transmitter
1918	Field portable tonic train	300–150	120	Ground valve transmitter
1918	Th	50–150	—	Seven-valve receiver for D/F

* Airborne use only unless otherwise indicated.

transmitters were still in use, although shortly to be phased out. With a few exceptions these were all based on the successful Sterling No. 1 transmitter, developed in the very early days of the war and still being used in large numbers in 1919. Several versions were produced, some covering longer wavelength ranges, and others augmented in power by the provision of heavy-duty accumulators, or through the use of a.c. generators driven through the aircraft engine. A significant variant was the Type 52a, used for long-range artillery observation; it included an alternator producing 40–50 W of power, which could be coupled to an air-driven propeller mounted on the front of the aircraft. Several of these variants are shown in Figure 10.3.

Wireless telegraphy equipment based on thermionic valves began to make its appearance during 1917 in airborne equipment, initially in amplifiers used to augment the weak output of the crystal detector and Mk III tuner, produced by the Marconi Company. These augmented receivers were produced in several versions (usually covering different wavelength ranges): Types Ta, Tb, Tc and Td. The amplifier designed for this purpose was the Type C, containing three low-frequency valve

Figure 10.3   Four First World War spark transmitters, from left to right: Type 52m, battery and rotary of Type 54b, Rouzet set, Type 54b, Type 52

Source: Public Record Office

amplifying stages. This combination of crystal detector and amplifier was superseded by a multi-valve receiver which used one triode valve as detector followed by two low-frequency amplifying valves, known as the Type Tf (Figure 10.4). This became used extensively for night bombing work in conjunction with the transmitter Type T21. This transmitter (Figure 10.5) used two VT1A valves connected in parallel in an inductively-coupled oscillator circuit providing a nominal power of 60 W [14] and achieved an air-to-ground transmission range of up to 250 km. The T21/Tf installation formed the basis of RAF aircraft communication for several years after the war, together with the Type T57 ground transmitter, the latter operating on long wavelengths of 1000 to 2500 m. Several variants of the Type T57 were developed for service with the RNAS, installed in flying boats and for ground use. For the shorter wavelengths used in ground installations, an ICW set known as the field portable tonic train transmitter was produced by Marconi, and later modified to become the Type T23 transmitter used for aircraft artillery cooperation. This could transmit on sixteen different wavelengths between 350 and 550 m, each of which could be modulated by a choice of ten different tones. This was a valuable feature which enabled the Morse message from a given transmitter to be read through interference when many spotting aircraft were operating over a short length of front. The

Figure 10.4   RAF Type Tf valve receiver, 1917
Source: Public Record Office

Figure 10.5   RAF Type T21 transmitter, 1921
Source: Public Record Office

transmitter had two valves, one as an oscillator with a nominal power of 100 W, and the other acting as a modulator, replacing an earlier version which simply applied a.c. to the transmitting valve so that it produced bursts of oscillation controlled by the rate of the alternator or its interrupter circuit [15, 16]. Towards the end of the war a new receiver, the Type Th, became the standard equipment in the new Handley Page bombers. This was a TRF receiver containing three HF amplifying stages, a valve detector and three LF amplifying stages (the superhet receiver had yet to make its appearance in service equipment), and gave good results once the complex procedure for accurate HF tuning had been mastered [17].

10.2.2 Training telegraphists for air operations

By 1916 the operation of all the airborne equipment was becoming a major responsibility for the pilot or, in larger aircraft, the navigator/ operator. In Britain most of the equipment was Marconi-designed, and to facilitate the supply of suitably trained pilots and operators a training school was opened at the Marconi experimental establishment at Brooklands in Surrey and staffed by the Marconi company. A leading figure at the school was R.D. Bangay, who had been transferred from the American Marconi Company after carrying out a series of air-to-ground tests in 1911 with a prototype spark equipment of his own design. The tests were carried out by a Canadian pilot, J.D.A. McCurdy, flying a Curtiss biplane over Long Island, and the results obtained initiated a period of intensive development of airborne wireless telegraph equipment at Brooklands, alongside its normal training function. Later the training programme at Brooklands was taken over by the military. Qualified Marconi engineers were given commissions, and the school was placed in charge of Major C.E. Price, a Marconi engineer from Chelmsford [2]. The school produced a weekly output of 36 bomber pilots trained to operate the Marconi equipment. At about the same time the RNAS initiated the formation of 'boy mechanics for wireless tuition' in 1915. Entrants at the age of seventeen were trained as aircraft wireless operators, and were promoted as uniformed air mechanics upon successful completion of their probation. The course included instruction in electricity and magnetism and service equipment [9]. This was similar to the Royal Navy training scheme for seagoing telegraphists and wireless mechanics.

By the end of the war the RAF (as the combined RFC and RNAS had by that time become) had six hundred aircraft fitted with various types of wireless equipment and the gun batteries of the army were provided with nearly two thousand ground stations, all capable of communicating with RAF aircraft, and manned by a force of seven thousand officers and wireless operators.

10.2.3 American airborne equipment

The US Navy had demonstrated an interest in wireless equipment carried in their aircraft as early as 1911. Their first application of the equipment was for artillery observation for the US Army, as the British had done, and for the Navy, to increase the 'scouting range of ships'.

A major component used was the Wireless Speciality Apparatus Company's spark transmitter and its associated crystal receiver, the IP76 [18]. The first successful flight using this apparatus was made in a Wright biplane with four flexible copper wires attached under each wing to act as the transmitting aerial. The transmitter included a quenched spark gap, excited by a 500 Hz, 250 W generator, and the receiver, a crystal set, had its weak output augmented by a magnetic amplifier. In the words of S.C. Hooper of the Office of Naval History, Washington, 'The apparatus was slung round the operator's neck with a telegraph key strapped to his right knee and a wire ammeter to his left.' The signals were successfully received at the USS *Stringham* at a distance of over 5 km (and later 27 km) distance. When the United States entered the war in April 1917, orders were placed for 50 experimental sets to be fitted into US Navy aircraft. Several different manufacturers were approached and provided sets. The American Marconi Company contributed its CM295 spark set, the Sperry Gyroscope Company a CS350 vibrator arc set, the De Forest Company an early valve set, the CF118, and a small company, E.J. Simon, a spark transmitter designated by the Navy as the CE615 set. At the time of the experiments only the Simon set was considered effective. This was a 500 W spark set powered by an air-driven propeller generator mounted on the aircraft wing. A trailing aerial was used. The receiver was an early valve set, employing only one valve acting in a regenerative circuit. The entire equipment weighed 45 kg and achieved an operative range of about 270 km. In 1918 spark transmitters were installed in US Navy flying boats and used in overseas locations. They were manufactured by the Radio Telegraph Company in two types: a 200 W rotary spark transmitter, the Type CQ1115, and a 500 W version, the CQ1111, both powered by air-driven generators mounted on the wing. These transmitters were also used by the US Army Signal Corps for ground transmission when a range of over 2000 km was achieved with the 500 W version.

Towards the end of the war the American Marconi Company manufactured large numbers of valve transmitters for aircraft use. These were CW transmitters incorporating two valves, one used as an oscillator and the other as a modulator, so that the transmitter could be used either for ICW Morse transmission or for telephony. Other low-power valve transmitters were made by the General Electric Company and the Western Electric Company for naval aircraft. These were generally installed in flying boats, so size and weight did not pose too great a problem. The Western Electric transmitter used ten 5 W valves, five used in parallel as an oscillator and five as the modulator, and the set used either for ICW Morse or for speech transmission [19]. The US Army Airforce was less concerned with range of transmission, unlike its naval counterparts who needed to maintain contact with aircraft carried on board US ships when they were engaged in offensive missions, and had no requirement for wireless on long-range aircraft in the First World War. Instead, it concentrated experiments on short-range R/T equipment, principally the De Forest CF118 sets and some of the American Marconi and Sperry Gyroscope sets [20]. A major effort to develop R/T equipment for the USAF (and also the newly formed US Tank Corps, which shared the same communication problems), was made by the chief signals officer for the US Army, George Owen Squier. He organised a Radio Division within the Signal Corps, and by 1918 a working airborne system was in production with the first R/T receiver and transmitter (SCR-67 and SCR-68) [21]. While Squier regarded the radiotelephone development as one of the Signal Corps' most creative achievements in the war, later experience with long-range bombing operations, as occurred during the Second World War, demanded W/T contact, and a parallel development of telegraph equipment was later to form an important part of the Signal Corps' activities.

10.2.4 German airborne equipment

German experiments in wireless transmission from the air commenced at about the same time as the American trials. In 1911 Dr Huth of the Telefunken company carried out some balloon transmissions with a spark transmitter and achieved a working range of about 56 km. This success quickly aroused interest in air-to-ground transmission from aircraft, and several engineers from Telefunken and Lorenz worked on equipment installed in Albatross and Wright biplanes. A variety of spark transmitters were employed, and some successful transmissions were made with a French Rouzet set which incorporated a rotary spark gap, then considered the lightest spark set available for aircraft use – although it still weighed 42 kg, a not inconsiderable weight. With this set installed in an Albatross biplane, a communication range of 65 km was obtained using a relatively short aerial. Reception on board the aircraft was made difficult by the noise and vibration of the engine. The vibration disturbed the stability of the crystal detectors, a problem not overcome until carborundum detectors became available in 1915. In an attempt to avoid the noise problem a visual method of reception was tried, consisting of a Saiten galvanometer with the image projected onto a screen, but this was soon abandoned as too difficult to maintain in an operational aircraft.

The military began to make its first active experiments in air-to-ground transmission at Kiel in 1914, when the Torpedo Inspector of Marines, Dr Harrwig, carried out aircraft transmission to ships. Using a D5 biplane he was able to contact a cruiser, the SMS *Magdeburg* at 75 km range using a remarkable *vertical* aerial support mounted on the wings, some 6 m high, which must have limited the aircraft's manoeuvrability quite considerably! Despite these and other experiments, neither the army nor the marines had a wireless system ready for action at the beginning of the war. In February 1915 the first attempts were made to use a biplane for artillery observation on the Western Front by a Lieutenant E. Neumann in conjunction with the Telefunken Company [22]. To spare the artillery observer from having to learn Morse, a Telefunken Morse sending tablet was used, similar to a design described by Samuel Morse in his very early Washington experiments in 1844 (Figure 10.6). This consisted of a series of metal strips, some long and some short, over which a metal pencil contact could be run to generate a Morse character of dots and dashes. A set of such strips, connected to the transmitter, was mounted on the tablet and selected by the observer to transmit his

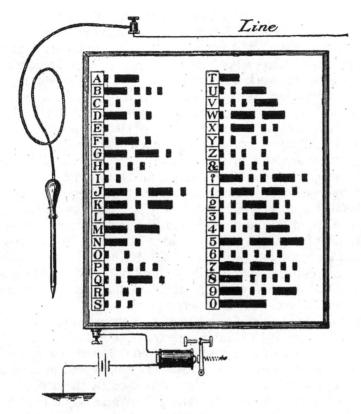

Figure 10.6   A Morse sending tablet
Source: Freebody, 'Telegraphy', 1959

message. The difficulty of ground-to-air reception was solved in 1915 when a carborundum detector was used, followed by an early two-valve amplifier – probably the first use of valve equipment in a military aircraft. This encouraged the military to attempt air-to-air contact, which was successfully achieved between two Albatross biplanes carrying out artillery fire observation flights over the Russian front in August 1915. The equipment used became the basis for a spark set manufactured by Telefunken and Huth as the Series Type AFS 35, a 125 W spark transmitter with a crystal receiver, to be replaced later by an improved Telefunken Type D quenched spark set and the addition of a three-valve amplifier to the receiver, the combination forming the standard equipment for the German Airforce from 1916. As with many airborne transmitters of the period, power for the spark set was obtained from an air-driven propeller generator mounted on the wing.

Equipping the military with this equipment was the responsibility of the energetic Major Bredow, one of the directors of Telefunken. He volunteered as an officer, and later commanded an airbase for the 2nd Army, where he introduced airborne wireless equipment and established a ground receiving station. A new army department for spark telegraphy was formed at Döberitz near Berlin, mainly through Bredow's efforts, and although much work was carried out there, its establishment met with some opposition. Many airmen refused to use the apparatus and one officer was heard to comment, 'One does not wish for the danger of flying without adding to it the possibility of electrocution!' [22]. Despite the opposition, development work proceeded at Döberitz with a two-way wireless system used for artillery observation. Several different types were put into production. For the larger aircraft the Type G station was designed, having 250 W aerial power transmitting over a frequency range of 250–1000 kHz. A smaller Type A was produced by Telefunken as a transceiver carried by the flight leader only, with receivers installed in other aircraft of the flight. Type A had a 70 W transmitting power, and the receiver incorporated a valve amplifier. Both transmitter and receiver were adjusted on the ground for single spot frequency operation, and the send/receive switch connected the aerial to the appropriate unit, and disconnected the Morse key during reception.

By this time, the value of wireless equipment was appreciated by the authorities. In June 1916 Hindenburg and Ludendoff had founded the Fliegerfunkentruppe ('flying wireless company'), with 518 aircraft fitted with wireless equipment, supported by 79 fighter and general aircraft wireless stations and 320 ground stations used by the army. With this large number of stations operating along the Western Front some mutual interference was inevitable, despite the efforts made to maintain a clear separation between the frequencies used by different combat groups. This problem became particularly acute during the French attack at Verdun, when communications within the German forces were seriously hampered. New rules for the use of wireless equipment were imposed, and new types of equipment, operating over a narrower bandwidth, were produced by Telefunken in 1917. These were Type N and Type O transmitters having a radiated power of 70 and 125 W respectively, both using a quenched spark unit sending out an almost CW signal. In the same year, 1917, the manufacturers began to construct airborne sets enclosed in pressed-metal cases containing integral shielding for the different stages, a technique that was to become a hallmark of German wireless construction for aircraft use in all future designs.

In January 1918 Telefunken produced the first design of valve transmitters to be used in aircraft. These were the Type USE 1 series, followed later by a second series, Type ARS 80, both of which used a single-valve, anode-keyed oscillator for the transmitter, operating over the narrow frequency range 500–700 kHz, and a four-valve receiver comprising a grid-leak detector and a three-valve LF amplifier. The ARS 80 was enclosed in a pressed-metal case measuring 330×320×175 mm, and weighed only 10.9 kg (without generator), representing a considerable improvement over the heavy spark transmitters used in the early phase of the war. Both these model series were produced in large numbers in the spring of 1918, and were the last major airborne sets produced by Telefunken for their war effort. In the meantime, Telefunken had been developing a telephony transmitter that could be used by fighter pilots to transmit messages without the aid of the Morse key. This had already been achieved on the ground between Berlin and Nauen as early as 1915, using a valve transmitter, but proved difficult to achieve in the air, and no suitable set was available to the military before the end of the war. Some efforts were also made to devise a 'copy-telegraph' (*Kopiertelegrafen*) for use in aircraft to achieve the same purpose, in the late summer of 1918. The operator in the aircraft was expected to prepare his report in the form of a punched roll of paper (presumably in advance of operations) and feed it into a copy-telegraph sender which automatically transmitted the information to the ground station. This effort was soon abandoned when peace returned, but the experience gained at Telefunken proved applicable to teleprinter development at a much later date.

While the Germans were undoubtedly ahead of the Allies in the design and deployment of wireless equipment at the beginning of the war, they had fallen far behind in the research and application of CW airborne equipment by the end of the conflict. A document found on a German prisoner dated June 1918 makes clear the importance of this lead to their own war effort. This was a letter from corps HQ to army corps staff, located south of the Somme, and refers to the salvage of enemy aircraft:

The enemy has secured a distinct advantage in his successful use in aeroplanes of continuous wave wireless apparatus which possesses great superiority over the spark apparatus in use with us . . . It is of the greatest importance to us to salve further enemy wireless apparatus of this description. In this way millions of money will be saved, as we have not so far been successful in constructing a continuous wave wireless apparatus for aeroplanes, which can work without certain disadvantages [6].

10.3 The inter-war years

The First World War had greatly accelerated the development of aircraft and of airborne wireless equipment. After the war the equipment used in RAF aircraft and ground stations became considerably modified as the experience gained in the war years was assimilated, and industry was able to achieve improvements in performance not possible before 1918. In the next decade spark transmitters were dropped and more effective special-purpose CW transmitters were produced. The performance of aircraft wireless equipment had been improved by the introduction of engine-driven generators, a practice which had been avoided in earlier designs when specially fitted generators were driven by a separate propeller mounted at the front of the aircraft. The drag effect this had on a high-speed aircraft was now considered unacceptable, and transmitters derived their power directly from a generator linked to the engine or from a motor-generator powered by the aircraft's electricity supply.

By 1929, three major transmitter–receiver combinations were in wide use: the T25/R31 set used for W/T and R/T ground and aircraft communication, the T21a/Tf set for W/T communication carried in bombers, and the T19a/R40 set used principally for point-to-point W/T ground communication. The T25 transmitter was a CW valve set. Unlike its wartime equivalent, it avoided the use of a trailing aerial, which was a considerable disadvantage in operational aircraft, and instead worked with fixed aerials. In common with many naval and air transmitters of the day it used two triode valves connected in parallel to achieve sufficient transmitting power. The set had two wavelength ranges, 50–70 m and 80–130 m (6–4.3 MHz and 3.75–2.3 MHz), and was one of the first transmitters to benefit from the new developments in short-wave transmitters for the RAF, discussed in the next section. Its complementary receiver, the R31, was an elaborate 'straight' (TRF) receiver employing seven triode valves and having four tuned RF stages, a grid-leak detector and two audio-frequency stages, and required some skill to tune all four circuits accurately to the operational frequency. The T21a/Tf combination was one of the most effective wireless sets of similar power in existence at that time. The Tf receiver, developed in 1925, with its simple configuration of grid-leak detector and two audio-frequency stages, was easier for the operator to tune for good reception in the air, although some care was needed when making use of its reaction control for effective W/T reception. Good results were, however, generally obtained, and the installation was still in use by the RAF until a few years before the start of the Second World War [10]. For long-distance point-to-point communication at long wavelengths, the T19a/R40 was in use at RAF ground stations. The transmitter consisted of a single-valve oscillator, with a separate valve telephone attachment for using the set for R/T. It was generally employed for W/T, either by direct manual keying or in conjunction with an interrupter for ICW, and in both cases control was by manipulating the bias on the oscillating valve, not only to achieve a safe electrical condition for the operator, but also to obtain a high effective resistance in series with the key at the moment of break to avoid arcing (see Reference 23 for a detailed description of the operation). The T23 and T32 transmitters were designed to meet the needs of aircraft artillery cooperation. They were both valve transmitters producing ICW, which caused far less interference on receivers operating on adjacent wavelengths than did the

Sterling spark transmitter, used previously for this purpose. The T23 was one of the first airborne receivers to be remotely controlled from the dashboard by the pilot, who was able to operate a Morse key from this position [24].

The two areas where aircraft wireless telegraphy had found its greatest use were in air-to-ground communication and in wireless direction-finding, and during the war developments in both these technologies had been rapid. Direction-finding using the Bellini–Tosi aerial for ground stations or a rotating loop fitted to the aircraft had been widely used, but in both cases performance was found to vary with time of day/night, and to be hampered by reflections from the ground or the ionosphere, so that good accuracy was difficult to achieve. In 1919 a major improvement was made for ground stations by introducing the Marconi–Adcock technique of direction-finding [25], and its use was to dominate navigational operations in the RAF for the next half-century.[2]

Successful air-to-ground transmission was difficult to achieve as communication bands became congested – more and more squadron frequencies were needed, and the operational range of aircraft increased. Ground communication between commands was also hampered by a shortage of available transmission frequencies, and an increase in their number was recognised by the air staff as an essential requirement of the expanding service. The existing inter-command transmitters and receivers operated on wavelengths generally between 200 and 264 m, and although narrow-band CW transmission was now in general use, the bandwidths available to the services were limited and congestion was a major problem in the early 1920s. The position was eased considerably in the post-war years as new types of thermionic valve made their appearance, following the introduction of broadcasting in 1923 and the expansion of the radio industry that followed. They made possible the extension of reliable wireless communication to much shorter wavelengths, a factor which was to have a most significant effect on the operation of the air service in the next two decades.

10.3.1 *Use of shorter wavelengths*

By 1923 experimenters had found that the shorter wavelengths gave improved communications, used less power and provided more frequencies

---

[2] In experiments carried out in 1917 with loop aerials on the ground for reception, it was found that serious errors were caused by reception of the horizontally polarized component of the wave radiated from the aircraft aerial (at that time invariably a trailing aerial). Frank Adcock, then working with Marconi Ltd, proposed that the horizontal element in the receiving aerial should be dispensed with and the D/F aerial constructed of two pairs of vertical elements, aligned with the four points of the compass, and connected to the receiver input in the same way as the four inputs of the Bellini–Tosi system. This technique proved successful in reducing errors, and was applicable over a wide range of frequencies.

within the limited bands designated for military use.[3] This advance originated from work carried out by an enthusiastic band of radio amateurs
in various parts of the world. One such amateur was Flight Lieutenant
Durrant, then stationed in Iraq in the early 1920s, who used his own
design of wireless equipment to maintain regular contact with various
other amateurs, including a colleague in the RAF, Flight Lieutenant
Rodney, who had constructed his short-wave apparatus at Cranwell,
the premier RAF training station in England. The two officers soon
discovered that their home-made equipment could carry wireless telegraphy traffic under conditions where the existing official long-wave
RAF wireless telegraphy organisation failed. By 1926 they were in daily
communication with RAF stations in Palestine, Iraq, India, Baghdad,
Ismailia, Malta, Cairo and Amman, and had occasional contact with
Australia. Their short-wave sets enabled semi-official communications to
be maintained regularly between RAF headquarters at Cairo and units
in Aden, Khartoum and Bir Salem. The situation was appreciated by the
Air Ministry, which noted in an internal report that, 'As regards economy, a record kept during the last three months of messages sent between
various points, shows that these would have cost about £300 per month if
they had been sent by cable' [26]. It is interesting to note, from the same
source, that the Admiralty had also been assisted by the use of amateur
equipment in the same way in their communications with central China
during their 1926 operations in Hankow. Shortly afterwards this new
system of short-wave communication was placed on a service footing.
The several amateur RAF short-wave stations were made official and
established in Palestine, Malta, Egypt and India, in addition to Durrant's
initial experimental station in Iraq and other stations in Britain, using
the short-wave band between 35 and 37 m. A final accolade came in June
1926 when direct short-wave telegraphic communication was instituted
on a daily basis between RAF Cranwell and the Royal Australian Air
Force (RAAF) in Melbourne.

   The experimental short-wave equipment initially installed in all these
stations was of local construction and privately owned by the service
radio amateurs, except for the transmitting valves and power generators
which were contributed by the RAF authorities, so there was considerable variation in design. A service design to replace these transmission
sets was put into production in 1926, following the work of Durrant,
who had been posted back to the United Kingdom as an instructor at the

---

[3] The acute shortage of available bandwidth in the MF spectrum for the military resulted from the
competing requirements of public broadcasting having developed at a time before the use of wireless
transmission by military and civil aircraft became general. By the mid-1920s the popularity of
broadcasting had created for it an entrenched position in the frequency spectrum from which it
could not be removed. Various international conferences on the allocation of frequencies were set up
in the 1930s, and the right of exclusiveness for aircraft frequency bands was agreed, but these bands
were still not enough for the growing air requirements.

Technical and Wireless School at RAF Kidbrooke, where he supervised the construction of 12 transmitters, 12 receivers and 12 wavemeters. For a short time this became the standard Type A short-wave equipment used by the RAF. In 1931 this system was replaced by an improved version manufactured at the Royal Aircraft Establishment, Farnborough, (RAE); the transmitter became known as the Type T27, which covered the 16–60 m (5–18 MHz) short-wave band, and to facilitate service operations the master oscillator was arranged to provide six crystal-controlled spot frequencies. Initial tests between RAF Malta and the Air Ministry in London, established the techniques of frequency selection for different times of day, later to be used for overseas working throughout the command organisation, using a wavelength of 23 m in daylight hours and 42 m at night.

10.3.2 *The 1929 development programme*

In 1927 the RAF decided that all its aircraft should carry two transmitters, one for W/T and a second for R/T, and three receivers, one of which was to be used for direction-finding. The operational and maintenance problems imposed by this policy made the design of a suitable general-purpose transceiver capable of fulfilling all three needs the most urgent equipment problem at that time. A review by the Air Staff in 1928 found that 'the installation of all the individual sets required at the time was so complicated and cumbersome as to interfere with other duties of the aircraft crew' [27]. Unfortunately the number and variety of operational requirements meant that no single piece of equipment could be developed to meet all three needs. The RAF's 1929 wireless development programme recommended a number of separate developments which would go some way towards solving the problem:

1   An aircraft transmitter combining W/T and R/T using a fixed aerial (the trailing aerial, often as long as 70 m, was to be retained for emergency use only) for bombers, with a range of 480 km for W/T and 8–24 km for R/T and having up to 50 alternative frequency channels.
2   An aircraft receiver with a fixed aerial to correspond to the above transmitter and combined with it if possible.
3   An aircraft transmitter for army cooperation having combined CW, ICW and R/T using a fixed aerial, to cover the 100 120 m band (2.5–3.0 MHz) with a W/T range of 80 km to and from a mobile ground station and 40 km for R/T.
4   An aircraft receiver with a fixed aerial to correspond to the above transmitter and combined with it if possible.
5   An aircraft transceiver for single-seater fighters using a fixed aerial and a choice of 30 channels of R/T with a range of 56 km to the

ground station and 8 km air-to-air, operating over a waveband of 50–70 m (4.3–6 MHz).

Other recommendations were made for naval and ground station equipment, the latter including the replacement of Bellini–Tosi direction-finding equipment with the more accurate Marconi–Adcock system, a process that commenced with the fitting of Adcock equipment at Croydon, Lympne, Pulham and Bristol in 1937 [29, 30].

The 1929 programme was to lead to a new series of transmitters and receivers, several of which were for short-wave transmission and hence inherently supporting a large number of transmission channels within the allotted military wavebands. This was in fact the beginning of the process of changing the general transmission policy in the RAF, complete by the late 1930s, from predominantly MF and HF to VHF operation, most specifically for airborne equipment. The new aircraft systems consisted of a general-purpose HF/MF set, the Type T1083/R1082, an older set, the Type T21c/Tf, used for bombers, and an R/T transceiver, the Type TR9, working on VHF, for fighters. For ground point-to-point communication the T77/R1084 combination was the principle set used for command and Air Ministry communication for both W/T and R/T transmission. The Type T77 transmitter, introduced in 1936, consisted of four triode valves used as master oscillator, power amplifier, modulator and sub-modulator for speech. When used for W/T the power amplifier was keyed, leaving the master oscillator in continuous oscillation to maintain good frequency stability. A similar but higher-power set, the T1087/R1084 combination, was used for general station and aircraft communication work, together with the T1083/R1082 used as a ground set [31]. Some earlier systems were still in use, including the R68 receiver, especially adapted for direction-finding, and the Type T27, an MF transmitter used in mobile equipment.

10.4 War in the air, 1939–45

The significant improvement in the performance of military aircraft since 1919, and their potential for dominating any future conflict, made their control and communication with ground stations and one another a fundamental need rather than simply a desirable attribute. Fortunately this aspect of aircraft development in the late 1930s was well recognised by the Air Staff, and as a result of earlier reviews of communication requirements – particularly as regard to short-wave transmission and channel requirements, and the decision to carry wireless equipment in all aircraft of the RAF despite the additional weight and power requirements – placed Britain in a fairly good operational position at the commencement of hostilities in 1939. The problem was one of supply of equipment in the quantities required, and it was to be several years into the war before this was fully met.

Both Britain and the United States became adept at mobilising scientific talent rapidly in order to devise the new techniques needed, and to improve existing models, whereas German industrial policy was to establish a rapid standardisation of equipment for all branches of the armed services to obtain maximum production before hostilities commenced – a policy based on the expectation that the war would be a short one. Although this gave Germany sound, well-engineered units for wireless communication equipment, the Allied policy of continuously improving equipment to meet the changing pattern of military activity eventually achieved a superiority of performance, and this was a vital factor in the Allies' eventual success.

10.4.1 British airborne equipment

In Britain three sets of airborne equipment had evolved out of the 1929 programme: a general-purpose T1083/R1082 HF-MF transceiver; the older T21c/Rf set still in use for bombers, army cooperation and coastal defence; and the TR9 VHF R/T transceiver set for fighters. The TR9 equipment became vital in early engagements of the war, particularly the Battle of Britain, not only because of its good communications features, but also because the shift to higher frequencies enabled it to operate in an almost 'interference free' environment where a wide choice of narrow-band transmission frequencies became available [32]. It was later developed into a highly stable system offering over a hundred communication channels in the frequency band 100–124 MHz that was allocated to fighter command.

The requirement for the production of point-to-point wireless equipment in the United Kingdom was initially less urgent, since by 1936 such communication had been relegated to a standby role, with much of the W/T and teleprinter communications carried out over landlines. Landline communication was well advanced, with all bomber stations connected by at least two landlines, and this requirement was quickly extended to fighter command stations. To some extent this was also the case for overseas communications via the submarine cable network, with W/T communication acting as a standby alternative [10]. This latter came into its own during the later stages of the war when a large number of additional telegraph and teleprinter channels were required, both at the home base and for overseas locations. The technique developed by the Post Office in cooperation with the American Telephone & Telegraph Company was to operate a number of Morse or teleprinter channels in each of their wireless communication links which embodied several single sideband transmission channels, used initially for multi-channel speech transmission. Eventually the RAF used a six-channel speech frequency system, each channel incorporating a number of teleprinter circuits, on a number of their overseas wireless links in place of high-speed Morse transmissions, with considerable effect.

Arrangements were made in the late 1930s for quantity production of the essential aircraft equipment, noted above, in time for the expected commencement of hostilities in 1939. The general-purpose sets developed during the inter-war years, were in turn to be modified or replaced in the following few years, but the principle of establishing a small number of sets to cover a wide range of operational requirements was fairly consistently followed with advantage during the war period. The T1083/R1082 installation was the principal airborne equipment available to the RAF in 1939. The transmitter consisted of a master oscillator valve and a power amplifier valve, the frequency bands being covered in four ranges by plug-in coils: 136–500 kHz, 3–6 MHz, 6–10 MHz and 10–15 MHz. A switching unit provided the selection of CW, ICW or R/T transmission, with appropriate connections made in conjunction with the receiver, R1082, so that 'listening through' was possible [33]. The R1082 was a TRF receiver containing five valves: a screened-grid valve RF amplifier, a grid leak detector and three resistance–capacitance amplifying stages. A diode shunted across the aerial input circuit protected the receiver against excessive pick-up from the transmitter. Alternative aerial and anode inductances for the RF stage could be plugged in to cover a range of frequencies between 111 kHz and 15 MHz, and a reaction coil controlled the sensitivity via a potentiometer mounted on the front panel. This panel also carried separate tuning capacitors for these aerial and anode coils. The several panel controls, all affecting frequency setting, made the receiver difficult to tune accurately. This difficulty, together with the need to change coils in both the transmitter and the receiver for most changes in operational frequency, made the T1083/R1082 combination not entirely satisfactory for the operational squadrons.

A replacement was under discussion as early as 1935/36, but was nowhere in sight when hostilities commenced in 1939. Eventually, in the early 1940s, a general mobile and aircraft bomber transceiver was evolved by the Marconi Company: the T1154/R1155, which was manufactured and used in large quantities up to the end of the war.[4] The transmitter, Type T1154, (Figure 10.7a), known to all RAF wireless operators as the 'fruit machine' from its preponderance of gaily coloured control knobs, enabled 24 pre-selected frequencies to be set quickly in the transmitter over a frequency range between 200 kHz and 10 MHz, with the transmission capable of switching between CW, MCW and R/T. The transmitter contained a triode master oscillator driving two pentode valves connected in parallel as a power amplifier. A second triode valve

[4] Over 80 000 T1154/R1155 sets were manufactured by Marconi during the war years.

(a)

(b)

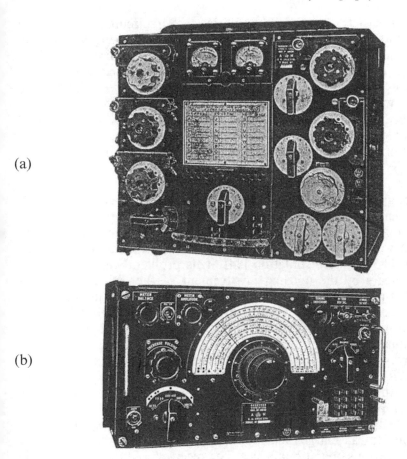

Figure 10.7   RAF Type T1154/R1155 HF-MF transmitter and receiver:
(a) T1154, (b) R1155
Source: RAF Signals Manual AP1186, 1944

provided a audio frequency tone source at 1.2 kHz, when the transmitter
was switched to MCW, or used as a modulating stage for R/T. The power
supply for the transmitter was derived from two rotary transformers
providing HT and LT indirectly from the aircraft a.c. supply system. The
LT generator was used in common with the receiver, supplying both its
HT and LT requirements. This was a welcome departure from the previous practice of requiring LT accumulators for the wireless equipment,
carried within the aircraft, which had required considerable
maintenance.

The Type R1155 receiver (Figure 10.7b) was the most advanced
airborne British military set yet seen in service. It embodied a sensitive

ten-valve superhet receiver with additional RF circuits for use in a direction-finding mode, using a loop aerial. Two other aerials were carried in the aircraft: a short fixed aerial, and a long trailing wire which could be let out when the aircraft was in the air. Five frequency ranges, continuously tunable over the MF and HF bands between 75 kHz and 18.5 MHz, were available. The high sensitivity of the receiver gave it a performance which made it also a valuable adjunct to mobile units, and it was much used in vehicles and small boats. From 1940 the T1154/R1155 combination was being fitted to five types of aircraft, the Wellington, Blenheim, Hampden, Whitley and Hudson, and it was also being used extensively as a ground mobile installation. The rate of fitting was never enough in the first few years of the war and caused not a little correspondence between Bomber Command, the Air Ministry and the Minister for Aircraft Production, when the lack of the new Marconi equipment was blamed for losses arising from wireless failure using the existing T1083/R1082 installation [34]. This problem was eased in the early war years by the unusual involvement of the civilian producers directly at operational RAF stations, where the W/T equipment was being installed in Bomber Command aircraft by Marconi staff, a task which also involved fitting an additional generator to supply the extra power required by the new installation [10].

The corresponding ground equipment, designed somewhat earlier, was the T1087/R1084 installation, operating on the higher frequencies now being used throughout the service, replacing the earlier T77/R64 equipment developed in 1934. The R1084 was one of the first superhet receivers used by the RAF and was available by 1937 for ground communications. It was a large rack-mounted receiver conceived as a flexible unit to suit various operational roles. The receiver contained 13 valves in a complex circuit which provided the facility to switch from superhet to TRF operation when required. The range of frequencies covered extended from 120 kHz to 20 MHz and was achieved by inserting plug-in coils. These were contained in an accompanying set of storage cases – 14 cases containing the 84 coils required! The detailed handbook noted that 'since the receiver is capable of a very high degree of selectivity some experience is necessary before the full advantages are realised', and included a set of instructions for configuring all the control knobs, trimmers and coil changes available (including a choice of two intermediate frequencies of 40 and 180 kHz) [33]. Although complex and requiring some skill to set up, it was used for much of the war, and as late as 1943 it was considered that 'although this receiver is obsolete it is the only one we have which is suitable for field use and we must continue to use it until something better is available' [35]. Later ground station communication receivers were less complex, and many made use of the R1080, a six-valve superhet of commercial manufacture, or one of several United States receivers received under the wartime Lend–Lease arrangements.

Point-to-point transmission using W/T between commands and headquarters, when landlines were not available, was carefully arranged to avoid aircraft frequency bands used strictly for air-to-ground communication. In addition, the siting of aerial equipment for ground point-to-point installations was now made well away from the airfield to avoid interference with local airborne operations. A major piece of equipment used at this time was a Marconi medium-range, short-wave transmitter, the SWB8, the first of a line of service equipment for fixed and mobile point-to-point communication, often installed in a vehicle for mobile use by Army as well as RAF units. It first saw service in France in 1939 when a fleet of three converted Green Line buses containing two SW8B transmitters, accompanied by several heavy lorries and a 60 kW power trailer, formed a mobile station to provide a W/T service between the Air Ministry and Paris. This service had much in common with the Golden Arrow service used for army mobile communications in th First World War (see Chapter 8), and was in fact dubbed the 'Silver Arrow' by its users. The reliable high-speed Morse telegraph system installed enabled a rate of 150 wpm to be maintained over the 24-hour period preceding the fall of France in 1940 when no other form of official communication existed [36].

From the 1940s, spot frequency control by the use of plug-in crystals was beginning to enhance the performance of the new equipment. Developments in VHF equipment, derived from the short-wave experiments of the 1930s, had advanced to the stage where several W/T and R/T transceivers for airborne use were now in use. One of these was the Type TR1133 transceiver, which could be switched to any of four spot frequencies selected by remote control from the pilot's dashboard. Its counterpart on the ground was the T1131/R1132, which had a similar frequency range. Other transceivers were the TR1143, used for aerial reconnaissance, and the TR1196, a replacement for the TR9 fighter communication set, and which was used only for R/T. Although by the mid-1940s most aircraft transmitters were fully crystal controlled and capable of at least four changes of frequency, easily effected in operational flight, this was not sufficient for the expanding needs of the RAF in 1944, by which time there were many squadrons in operation, all needing individual sets of communication channels (see the footnote on p. 354 and Reference 10). This gave rise to a new problem, the need for crystal holding economy (a problem shared with army communications and discussed earlier in Chapter 8). In one solution a reference crystal technique was used which relied on a series of harmonics generated by a crystal considered as a reference standard, thus giving a very large number of available crystal-controlled frequencies [37].

10.4.2 American airborne equipment

In 1944 the available W/T equipment of the RAF was augmented by United States sets, supplied under Lend–Lease. This also included certain British designs which had been manufactured in quantity in the United States. Much of the ground HF and MF equipment used by the Allies during the war for communication with aircraft was of US manufacture. Prominent among them were the RCA transmitter RT4332, known in the RAF as the Type T1179, and the SCR274-N, developed early in 1941 for the US Navy and Airforce. The latter was used both as a ground command set and in modified form for aircraft, and was manufactured in large numbers for the Pacific theatre of war. The RAF used a different set for this purpose, also manufactured in the United States, the SCR 522 transceiver, which operated on a higher frequency band.

RCA communication receivers used in large numbers by the RAF were the AR77 and AR88. The AR77, renamed Type R1188 in the RAF, was used as a replacement for the Type R1084 receiver, which was in demand as a mobile receiver mounted in vehicles. The AR77, a modern communications receiver developed for commercial and radio amateur use, included features such as automatic volume control, noise limiting and a best frequency oscillator for CW reception. It was configured as a ten-valve superhet receiver having exceptionally good selectivity for W/T through the use of a crystal filter circuit [38]. For airborne use the US Air Corps adopted the SCR522 transreceiver, much used by the British RAF. This covered the VHF frequency band 65–156 MHz and was a crystal-controlled set having a choice of four spot frequencies within this band. A long-range airborne transmitter also used in quantity during the early part of the war was the SCR 187, transmitting W/T or R/T on the high-frequency band from 1–12 MHz over a range of up to 3200 km. For airborne reception the US Air Corps installed the RCA BC-224-A, an eight-valve superhet receiver, again including a crystal bandpass filter to enhance selectivity for W/T reception. A Collins Type 18Q HF transceiver was adopted by the Corps for mobile use. Some units met a special need, such as the RCA triple diversity HF receiver, used on important long-distance high-speed automatic W/T circuits. This category also included a special-purpose transmitter, the Type T1333 which began to be carried in all bomber and coastal command aircraft in the later years of the war. This was a 'dinghy' emergency transmitter, radiating on the distress frequency of 500 kHz. It sent out a continuous SOS signal and could be keyed to transmit messages, although a second operator would then be needed since the power for the transmitter was produced by the hand generator, visible in Figure 10.8. The two-valve transmitter could deliver 5 W of power to an aerial raised by a hydrogen-filled balloon, to give a range of about 160 km [33]. These Morse transmissions were generally detected by naval stations situated around Britain's coasts and by

Figure 10.8   Lifeboat Type T1333 transmitter
Source: RAF Signals Manual AP1186, 1944

the Post Office direction-finding stations, the information was passed to naval headquarters, where the position of the dinghy was determined and rescue arranged. Some idea of the scale of these air–sea rescue operations may be seen from the statistics recorded by the Post Office between 1st January and 30th August 1943, probably the most active period in the air war, when no fewer than 1011 airmen were rescued from the sea [39].

10.4.3 German airborne equipment

Although the Versailles Peace Treaty precluded the German authorities from recreating the Luftwaffe after the First World War, a certain amount of planning continued after 1918 within the German Government in anticipation of an eventual military force, and this included the development of airborne and ground wireless equipment. Germany had been permitted under the treaty to retain a Defence Ministry, which provided the opportunity to retain a nucleus of Luftwaffe personnel to establish a General Staff for this purpose. Further, the Paris Air Agreement of 1926 allowed Germany complete freedom in the sphere of civil aviation [40]. The Deutsches Lufthansa (DLH) equipment already in use for civil aircraft was the starting point for military development. Equipment such as the Telefunken Spez 205F and 257, both containing 70 W transmitters and operating on a suitable frequency range of 230–1000 kHz, would serve as a basic design for a variety of military aircraft wireless equipment. The planning was initially disguised in terms of that required for an aviation information agency or flying school (*Vekehrsfliegerschule*). The flying schools were popular and encouraged by the Defence Ministry through the extensive use of gliders to train pilots and in the membership of Germany's main air society, the Deutsche Luftsportverband, which had more than 50,000 members by the end of the 1920s. The military nature of this planning became publicly apparent in 1932 when the firm Lorenz AG began the development of three wireless systems for military aircraft [22]:

1   Model A, a 100 W medium-frequency transceiver for close reconnaissance aircraft using a fixed aerial and a frequency range of 600–1667 kHz.
2   Model B, a high-frequency receiver for fighter aircraft operating in the frequency range 2500–3750 kHz.
3   Model C, a 100 W low-frequency transceiver for bomber and larger aircraft using a trailing aerial and a frequency range of 310–600 kHz.

Of these, Model A was constructed for testing as the Type S1 transmitter and Type E1 receiver, both of which could be modified to produce Model C as Types S2 and E2. A combined HF/LF transceiver was also constructed by Telefunken for the new military force as the Spez 636F, operating at the lower aerial power of 40–70 W (in effect combining Models B and C). A year later, in 1933, Major Wolfgang Martini, who had operational experience in the First World War, was appointed as Chief of the Luftnachrichten Verbindungswesen, an organisation charged with the task of recreating the military wireless plan for the Luftwaffe which was gradually taking shape.[5] He was later to become General der Luftnachrichtentruppe, in charge of all Luftwaffe signals and radar throughout the 1939–45 war.

An important change in nomenclature took place for Luftwaffe wireless equipment in 1933. Model A became FuG 1 (Funk-Gerät No. 1) and an alphanumeric series, prefixed by FuG, was established for all the military aircraft wireless equipment subsequently produced. Table 10.2 lists the more important of the German aircraft wireless installations used from 1933 to 1945. The FuG 1 was fitted into the first Heinkel combat aircraft, the He46 and He126, and the Model C, which became FuG H, was installed in the He45. Model B appeared as the FuG 4, fitted to

---

[5] Some of the instructions included in their equipment handbooks were remarkably warlike. Accompanying the detail for taking direction-finding readings using a Morse transmission was the warning, *Nicht länger als unbedingt nötig auf Taste drücken. Feind hört mit und peilt!* ('Don't press the key too long, or the enemy will take a bearing on your transmission!').

Table 10.2 Luftwaffe airborne equipment, 1933–45

Year	Type	Frequency (MHz)	Power (W)	Purpose*
1933	FuG 1	0.6–1.667	100	Close reconnaissance
1933	FuG 3	0.3–0.6		
		3.0–6.0	40–70	Bomber transmitter
1934	FuG 6	2.5–3.75	20	Fighter transceiver
1934	FuG 16	38.5–42.5	35	Bomber transceiver (used with EZ6 receiver)
1934	FuG 17	42.1–47.9	35	Army cooperation transceiver (used with EZ6 receiver)
1936	EZ6	0.3–0.6		
		3.0–6.0	—	Superhet receiver
1939	FuG 7	2.5–3.75	20	Dive-bomber transceiver
1939	FuG 10	0.3–0.6		
		3.0–6.0	40–70	Fighter transmitter
1942	E52	1.5–25.0	—	Ground standard receiver
1942	FuG 10K3	6.0–18.0	70	Command liaison
1942	NS4	53.5–61.0	1–2	Emergency dinghy transmitter

* Airborne use unless otherwise indicated.

fighters as a receiver only, initially a four-valve TRF receiver. Other models followed from Telefunken and Lorenz, including for the first time a telephony set, the FuG 6, operating in the frequency band 2.5–3.75 MHz. This used a fixed aerial and was supplied with power from a rotary transformer connected to the aircraft supply battery. Lorenz redesigned its civilian Lufthansa 20 W long-wave set for the Luftwaffe (still a clandestine organisation in 1934) under the guise of equipment for a training aircraft. This became the FuG XX, a 25 W set which, together with similar models from Telefunken, became the forerunner of a series of transceiver sets with which the Luftwaffe was equipped in 1939, namely the FuG 10 to FuG 18.

The FuG 10 and its derivatives were used mainly for W/T work. The transmitter was a three-valve set in which the first valve operated in a highly stable Colpitts oscillator circuit, and the other two valves were connected in parallel as a power amplifier. A schematic diagram of the transmitter is shown in Figure 10.9, which illustrates the method of 'listening through' by the operator of the signal being transmitted. The complex switching control shown would also be connected to a second transmitter and receiver, if these were carried, to provide alternative transmission and reception channels. Keying was made operational at a suitable setting of the switching unit by removing an inhibiting bias control at the power amplifying valve. An installation in a bomber would have at least two FuG 10 sets, one for MF and the other for HF working,

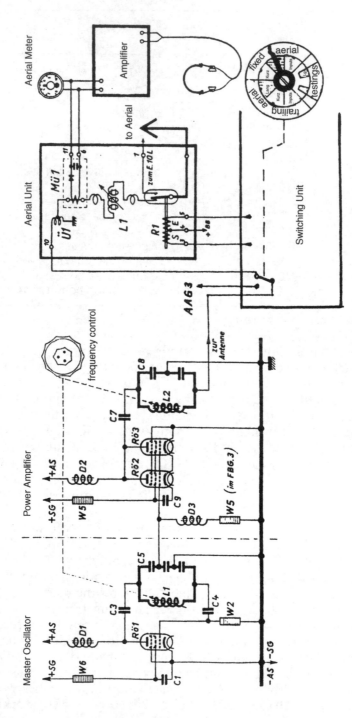

Figure 10.9   Schematic diagram of the German FuG 10 transmitter
Source: Trenkle, 'Bordfunkgeräte vom Funkensender zum Bordradar', 1986

with a possible third for direction-finding. This is seen in the assembly of units from a military aircraft shown in Figure 10.10. Both fixed and trailing aerials would be available, with the former switched between sending and receiving equipment. The switching unit, connected to all these units and situated close to the pilot/operator, would enable CW, ICW or R/T to be chosen at a selected operational frequency. The power output for the FuG 10 transmitter depended on the aerial and frequency in use. With a fixed aerial this varied between 15 and 50 W, depending on frequency, reaching its full design power of 70 W only when the trailing aerial was used [41]. Two frequency bands were provided covering the MF and HF bands of 300–600 kHz and 3–6 MHz, although separate transmitter and receiver units were generally installed when coverage of both bands was needed. The receiver design, Type EZ6, an eight-valve superhet, was incorporated in the FuG 10KL from the earliest days of the war, unlike the corresponding British aircraft receiver at the time, the R1082, a TRF receiver, replaced only in 1942 by the R1155 superhet receiver. A simpler design was however incorporated in the Stuka dive-bombers, which were the major offensive aircraft in the early days of the war. This was the FuG 7 transceiver, used as an R/T set providing one

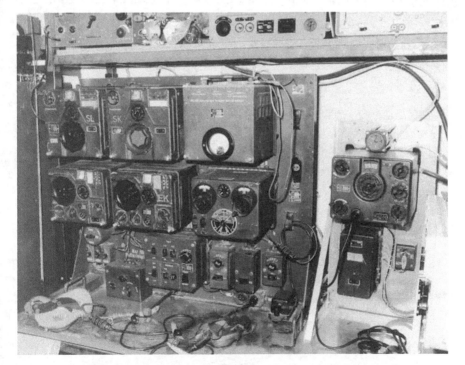

Figure 10.10   Assembly of units forming a complete FuG 10 aircraft installation
Source: A.O. Bauer, Foundation Centre for German Communications and Related Technology, Diemen, The Netherlands

spot frequency in the band 2.5–3.75 MHz, but it did not compare in performance or design to the British TR9 set described earlier. In general the German designs, and particularly in their receivers, were all of single-valve type. For the receiver this was the RV12P2000, a small specially designed HF pentode which served as diode, triode or pentode in both RF and LF circuits. It had a moulded base with side pin connections and a top cap grid connection. In use it was inserted inverted into its holder and then screwed into position (Figure 10.11), an arrangement both mechanically sound and ideal for maintenance and replacement purposes.

All the Luftwaffe units were constructed in a pressed aluminium frame, similar to the design of the German Army equipment already discussed in Chapter 8, a technique which had been used since the 1930s for civil aviation work. While crystal control was not applied until the closing years of the war, the rigid mechanical construction and attention to suitable component types (including the use of negative-coefficient capacitors to correct for temperature change), together with anti-backlash drives, led to good frequency stability so that 'click-stop' frequency setting mechanisms could be used [42]. As noted with the corresponding German Army units, the rigid mechanical structure made

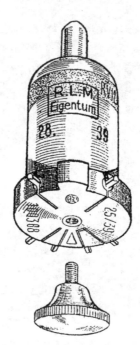

Figure 10.11   Interchangeable HF Pentode, type RV12P2000, used in Luftwaffe airborne receivers

design changes to meet different operational requirements difficult to apply, and for this reason the flexibility and performance of the aircraft wireless units fell behind the Allied equipment towards the end of the conflict. A feature of German wireless equipment produced for all their services was the complex mechanical design, inherent in many of the Luftwaffe designs discussed above. In a remarkable development in 1942 Telefunken and Hesche produced a design of the FuG 10K3 set for a fighter aircraft in which the length of the fixed transmission aerial was mechanically changed by motor control to suit the transmission frequency selected, the adjustment automatically carried out until the aerial current reached its maximum. A sophisticated aircraft set, the FuG 15, began to appear in 1944 which included a number of design features not previously seen in operational sets. This was a VHF system operating in the frequency range 37.7–47.7 MHz, having crystal control for a selected spot frequency, and with the facility to switch the transmitter modulation between AM or FM. The associated receiver was a double superhet having some of its RF stages and its first local oscillator shared with the transmitter through the send/receive switch [43]. In general, however, the Luftwaffe did not rely on crystal control, and when it did it used frequencies in the lower end of the VHF band, up to 47.7 MHz, and only towards the end of the war did it venture into designs in the UHF band, with frequencies of 100 MHz or more, and then not in large quantities.

In some of the larger aircraft a separate R/T set, the FuG 16, was used for short-range air-to-ground telephony communication. This was designed by Lorenz and covered the VHF frequency range 38.5–42.5 MHz. The transmitter and receiver dials were calibrated in frequency and carried click-stops to four alternative frequencies. A similar transceiver unit, the FuG 17, was used by army cooperation aircraft and could be operated as a W/T set using CW or ICW at the slightly higher frequency range 42.1–47.9 MHz. At a later stage in the war the Luftwaffe started to carry a portable dinghy emergency transmitter in its bomber aircraft. This was a simple MO–PA device powered by an (unsmoothed) vibrator power supply, fed by eleven 2 V accumulators, and radiated 1 or 2 W of modulated power into a short vertical aerial on a frequency between 53.5 and 61.0 MHz. Unlike its British/American counterpart it did not include a hand-driven generator, and the battery supply in good condition would last for 2–3 h of intermittent transmission.

10.5 RAF wireless training

The scale of training required for wireless personnel in the Second World War was immense. Time was short, and the 'need to know' about new communication techniques and equipment was far greater for more people than in the previous major conflict. In the First World War much training was carried out by civilian wireless schools, particularly by the Marconi Company, and this practice was continued for supplying the extra wireless operators required by the merchant navy in the inter-war years as a result of the changes in marine regulations. The rapid expansion of the RAF in terms of aircraft, airfields and bases at home and overseas during the first few years of the Second World War required urgent action to train the various types of wireless personnel required. The service needed ground wireless operators, slip-readers, wireless operator/air gunners, wireless mechanics and wireless operator mechanics, cipher operators, teleprinter operators and signal officers, not to mention other competing posts in the field of radio communication and radar, also required on the same time-scale. The wireless equipment fitted in ground bases and aircraft changed dramatically as the war progressed, which modified the instruction of operation and maintenance technicians, so that a need for supplementary training for those already working added to the burden of training requirements. In the following the author is indebted to the Air Historical Branch of the MOD for details of the course content and scale of the training carried out for the RAF in wireless work during the Second World War [44].

At the commencement of the war in September 1939 there were only two signals schools available for the RAF, at Cranwell and Yatesbury, responsible for all areas of signals training. A rapid expansion of training locations, some of them new RAF establishments, plus Post Office schools and civilian training centres, met the shortfall until September 1940, by which time four main RAF wireless training centres were established: the No. 1 Electrical & Wireless School, Cranwell; the No. 2 Electrical & Wireless School, Yatesbury; and the No. 3 Electrical & Wireless School, Compton Bassett; with a combined initial training given at the No. 10(S) Recruit Centre at Blackpool, later to become No. 13 Radio School. This last was important in providing an initial training capacity for ten thousand airmen and airwomen, with intakes of up to a thousand a week. Some elementary information in wireless and electrical theory was included, but the main role for Blackpool was in general recruit training and Morse instruction. This occupied some twenty weeks, of which twelve weeks were spent in learning to send and receive code in stages up to a speed of 20 wpm and in RAF W/T procedural instruction. Tests took place at the upstairs rooms of Burton, the tailors, with little opportunity for a retake in the event of failure. Those successfully completing the course and securing the coveted wireless operator brevet were sent to one of the three main wireless schools to continue their training in their chosen specialisation. Ground wireless operators were transferred to No. 3 Electrical & Wireless School at Compton Bassett, although this was also used for part of the training for wireless operator/air gunners (WOAGs). At Compton Bassett the twelve-week course included further instruction in RAF W/T procedure, and detailed lectures and practical work on some of the ground point-to-point and aircraft wireless equipment, described earlier in this chapter. Some wireless operators went on to complete a fairly lengthy course of twenty-six weeks as a 'slip reader'. The development of high-speed W/T in the services had evolved around the concept of automatic reception of Morse code at rates beyond the capability of aural reception. Instead the message was recorded visually as inked marks on a long paper tape or 'slip', and read by the operator in this different form. The course would also include some instruction in dealing with enciphering messages for transmission and preparing them for fast automatic transmission. Other operators, in particular those from the WAAF, were sent on a ground direction-finding course at No. 1 Electrical & Wireless School at Cranwell before joining operational units. There the techniques and procedures required for operating Marconi–Adcock direction-finding equipment were taught. In order to reduce the pressure on the Signals Schools (the Electrical & Wireless Schools were renamed as Signals Schools in 1942), some of the WAAF wireless operators received their initial operating and Morse training at Post Office centres, which supplemented the RAF schools during 1941–42. Those trained ground wireless operators destined to become aircrew as WOAG, or later as W/Op(Air), required a further period of training at a gunnery school and a certain amount of air experience in operating wireless equipment in flight. In early 1942 the minimum time for a WOAG to pass through the various stages of training was eighty-one weeks, and in some cases two years before joining an operational squadron. This was caused to some extent by a bottleneck in the air gunnery schools, a problem which remained for much of the war. At a later stage of the war all wireless operator training for WOAG was given at No. 2 Signal School at Yatesbury, with the No. 3 Signal School at Compton Bassett confined to ground operator training for the RAF and WAAF.

The maintenance of ground and aircraft wireless equipment (excluding radar) was the province of wireless operator/mechanics (WOM) or later wireless mechanics (W/Mech), for which Morse experience was not required, and their training was initially carried out at No. 1 Signal School at Cranwell. The pre-war source of recruits for WOM was the aircraft apprentices who were trained both in operating and in maintenance on a three-year course. This was reduced to two years at the outbreak of war. A nine-month conversion course from W/Op to WOM was also in place. This too was reduced, to thirteen weeks, with entries at the rate of 125 per week. A trade of Wireless Mechanic was introduced at Cranwell in 1940 which catered for semi-skilled personnel from civilian life, where the incumbents did not need a knowledge of Morse code (e.g. Post Office employees or radio servicemen). For this trade a certain amount of *ab initio* training was commenced at eight technical colleges under the auspices of the Board of Education, with later training at Cranwell reduced accordingly. Eventually twenty-seven technical training colleges were involved in W/Mech instruction, with a potential training population of eight thousand airmen and airwomen.

The W/Mech and WOM courses were designed to give a very thorough basic instruction in the theory and practice of wireless reception and transmission in order that later special courses in service equipment, operational at the time, could be followed. These latter were often given at command level, and consisted of training in new forms of equipment as they appeared in operational aircraft. This included maintenance and location of faults, a major part of the wireless mechanics' duties, where the basic instruction in theory proved invaluable. At the commencement of the war many aircraft still needed to carry accumulators, for the T1083/R1082 and other early sets, and consequently the maintenance and charging of accumulator banks became an important task. Replacement of low-voltage supplies in aircraft with rotary converters demanded a certain amount of electrical maintenance instruction. These tasks were in addition to the main requirement to understand and maintain the operation of ground receivers and transmitters. This became more complex as receiver development in the RAF introduced the use of the superhet, and the transmitters employed crystal control and a remote switching capability between W/T, MCW and R/T. The need to continually update course material and establish short retraining courses was a responsibility carried out extremely effectively during the war years, and played a significant part in maintaining effective wireless communications.

References

1   RALEIGH, W.: 'The war in the air', vol. 1 (Oxford University Press, Oxford, 1922)
2   BAKER, W.J.: 'A history of the Marconi Company' (Methuen, London, 1970)
3   MEISSNER, A.: 'Die deutsche Elektroindustrie in den Kriegsjahren' (Telefunken, Berlin, 1919)
4   'Aircraft radio', RAF Signal History vol. 3 (1945), p. 6, Doc. AIR10/5557, Public Record Office, Kew, Surrey
5   VINCENT-SMITH, T.: 'Wireless in the RFC during the War', *The Electrician*, 1919, **83**, pp. 445–7
6   'History of wireless telegraphy in the RNAS, RFC and RAF in the First World War' (Air Ministry, 1920), Doc AIR1/2217, Public Record Office, Kew, Surrey
7   'Transmitter Rouzet wireless telegraphy' (Air Ministry, 1916), Doc. AIR10/216, Public Record Office, Kew, Surrey
8   ORME, R, Major: 'Radiotelegraphy and aviation', *in* 'Yearbook of wireless telegraphy and telephony' (Wireless Press, London, 1920), pp. 988–94
9   'RAF signals history – signal branch policy and organisation', vols 1 and 2 (1958) Doc. AIR10/5554, Public Record Office, Kew, Surrey

10 HONAN, J.J.: 'Aircraft wireless sets – the Sterling transmitter', *Wireless World*, May 1919, **7**, pp. 101–3; August 1919, **7**, 278–81
11 VYVYAN, R.N.: 'Wireless over thirty years' (Routledge, London, 1933), p. 137
12 HONAN, J.J.: 'Aircraft wireless sets – crystal receiver', *Wireless World*, January 1920, **75**, pp. 605–8
13 'Handbook for transmitters Types T57 and T21' (Air Ministry, 1921), Doc. AIR10/854, Public Record Office, Kew, Surrey
14 'Portable transmitter Tonic Train 120W' (1919), Doc. AIR10/140, Public Record Office, Kew, Surrey
15 'Handbook for transmitter Type T23' (1921), Doc. AIR10/969, Public Record Office, Kew, Surrey
16 'Particulars of wireless apparatus in use in the RAF, December 1918' (1918), Doc. AIR10/149, Public Record Office, Kew, Surrey
17 HOYT TAYLOR, H.: 'Radio reminiscences – a half century', Report, Office of Naval Research, Washington, DC, 1920, pp. 110–11
18 JOHNSON, T.: 'Naval aircraft radio', *Proceedings of the Institute of Radio Engineers*, 1920, **8**, (2), pp. 6–7
19 CLARK, G.H.: 'Radio in war and peace' (MIT Press, Cambridge, Mass., 1914)
20 CLARK, P.W.: 'Early impact of communication in military doctrine', *Proceedings of the Institute of Electrical and Electronics Engineers*, 1976, **64**, (9), pp. 1407–13
21 TRENKLE, F.: 'Bordfunkgeräte vom Funkensender zum Bordradar' (Bernard u. Graefe Verlag, Koblenz, 1986)
22 'Transmitter ground Type T19' (Air Ministry, 1921), Doc. AIR10/35, Public Record Office, Kew, Surrey
23 'Transmitter Type T23', Air Publication 952 (Air Ministry, 1923), Public Record Office, Kew, Surrey
24 KEEN, R. 'Wireless direction finding' (Iliffe & Sons, London, 1938), chap. 7
25 'History of private short-wave W/T experiments in the RAF' (Air Ministry, 1926), Doc. AIR5/455, Public Record Office, Kew, Surrey
26 'RAF signal history', vol. 3, 'Aircraft radio' (1958), p. 589, Doc. AIR10/555 (see also AM File S23185), Public Record Office, Kew, Surrey
27 'Wireless technical development programme' (Air Ministry, 1930), Doc AIR 2/1426, Public Record Office, Kew, Surrey
28 'Wireless development programme, 1936–37' (Air Ministry, 1936), Doc AIR2/2803, Public Record Office, Kew, Surrey
29 'Wireless development programme for 1928–1929' (Air Ministry, 1929), Doc AIR2/1244, Public Record Office, Kew, Surrey
30 'W/T point-to-point communication' (Air Ministry, 1941), Doc AIR2/5058, Public Record Office, Kew, Surrey
31 'VHF in the RAF' (Air Ministry, 1945), Doc AIR2/2946, Public Record Office, Kew, Surrey
32 'RAF signals manual' (Air Ministry, 1943), chap. 5, Doc AIR10/2296, Public Record Office, Kew, Surrey
33 Letters between the Minister and Bomber Command, *in* 'Marconi Wireless' (1941), Doc. AIR19/241, Public Record Office, Kew, Surrey

34 'Requirements for main items of ground wireless equipment for year 1944' (1943), Doc. AIR2/5506, Public Record Office, Kew, Surrey
35 'Wireless telegraphy communication' (Air Ministry, 1939), Doc AIR2/3136, Public Record Office, Kew, Surrey
36 HELLER, D.M., and STENNING, L.C.: 'Reference crystal-controlled VHF equipment', *Journal of the Institution of Electrical Engineers*, 1947, **94**, (3A)
37 'The United States RCA Type AR77 receiver' (Air Ministry, 1940), Doc. AIR10/3125, Public Record Office, Kew, Surrey
38 'Post Office report on wireless telegraphy 1939–1945' (1945), POST 56/142, British Telecommunications Archives
39 'Rise and fall of the German Air Force', Air Ministry Pamphlet 248 (1945), Doc. AIR41/10, Public Record Office, Kew, Surrey
40 Document issued 'for the troops' by R.L.M., General and Chief of the Luftnachrichtenschule (Luftwaffe Information Office), Saale, 1939
41 EDWARDS, C.P.: 'Enemy airborne radio equipment', *Journal of the Institution of Electrical Engineers*, 1944, **91**, (3A), pp. 44–66
42 'FuG 15 transmitter/receiver' (1945), Doc. AIR 14/3618, Public Record Office, Kew, Surrey
43 'RAF training 1939–44', Air Ministry Document 60<#>844–1, January 1945, pp. 96–116

Chapter 11

Epilogue

On 31st December 1997 a group of telegraphists gathered at Land's End Radio Station (2359Z) at midnight to witness the end of distress and commercial telegraph operations in the Morse code on the frequency of 500 kHz for all the United Kingdom coast stations, a service which had been in operation since the days of spark telegraphy communication. Four CQ (calling all stations) shut-down messages were sent, and hand-sent Morse interchanges took place with operators over the whole of Europe. The French Coast Guard had officially terminated their 500 kHz telegraph service a year earlier with a poignant final Morse message, 'CQ, CQ, CQ, this is our last cry before eternal silence'. Elsewhere silence had already been imposed by the US Coast Guard, which ceased Morse operations in March 1995, following the US Navy and Marines which ceased Morse transmission a few months earlier, while in Australian waters the use of Morse lingered on until 1st February 1999, the date set by the International Maritime Organisation for the final and worldwide demise of the Morse code for distress transmissions.

11.1 The demise of Morse

The surprising thing about the use of Morse for all forms of wireless contact, including distress calls, was its longevity. From 1837 to 1910 it reigned supreme, and only began to be supplanted by other and more efficient codes and techniques when the volume of traffic began to over-take the ability of operators and their equipment to convert information into a coded form at the required speed. The search for faster methods of code transmission dominated the development of established telegraph systems in Britain from about the time that the various telegraph companies were amalgamated to form a national service in 1870, culminating in a major and coordinated review of the possibilities for change through a High Speed Telegraph Committee, established by the Postmaster General in 1913.

11.2 High-speed telegraphy

Before this date, however, some progress had already been made in improving the efficiency of the telegraph line by duplexing and quadrilexing to increase the number of channels carried on a single pair of wires, and also in the automatic transmission of pre-recorded Morse signals at a speed faster than could be achieved by manual operation. The Wheatstone system had been developed to a high degree of efficiency to work at a speed of 300 wpm duplex and 400 wpm or more for simplex operation [1]. Early experiments were being carried out with keyboard perforators in which the punches derived their power from a pneumatic source or from electromagnets. Preparing paper tape 'off line' proved a successful technique for transmitting code at high speed, but reception was still a difficult operation, requiring skilled operators to transcribe the Morse 'slip' from the output of the receiving inker device. An improvement in overall efficiency was achieved in 1908 when the Morse slip was read by an operator who could simultaneously produce a gummed paper slip by means of a specially adapted typewriter for directly affixing the typed message to the telegraph form for onward transmission [2]. This system was carried further in the Post Office lines from London to Edinburgh, when a Creed instrument was employed for reperforating the tape automatically upon reception for retransmission to another telegraph station at a speed of 150 wpm; or by feeding the tape, again automatically, into a printer to produce a gummed paper tape output printed in Roman characters for affixing to the telegraph form for delivery to the addressee. Although rapid, these schemes were cumbersome and required an army of sending and receiving telegraphists to operate the complex equipment.

On Continental Europe new methods were being applied, and the Baudot system was being installed by the French authorities to operate the London–Paris triple duplex circuits at a cumulative rate of 180 wpm for the six 30 wpm channels in service. It was the consistently good performance of this circuit that led the Postmaster General in 1913 to institute a far-reaching inquiry into high-speed telegraph operation. This was not confined to Morse code systems, and by including a comparison of Morse and non-Morse systems in terms of efficiency and cost, the way forward for the British telegraph service was made clear. The decisions reached by this committee were to have a beneficial impact on the standardisation of telegraph systems throughout Europe in the next two decades.

The committee formed by the PMG consisted of Sir John Cavey MIEE, John Lee and A.B. Walkley from the Telegraph Branch, W.M.

Mordey, A.M. Oglivie, W. Slingo and G.E. Wood, under the chairmanship of Captain Norton (later to become Lord Rathereedan). The committee held eight meetings in 1913 and 1914 and published its findings fully in 1916, having released its interim conclusions two years earlier [3]. The types of equipment under consideration were automatic printing systems, typically the Wheatstone–Creed systems, and the new multiplex devices which were beginning to be put into use by several European postal authorities and were already well established by Western Union in the United States (although American systems were not specifically examined by the committee). A list of the systems which were considered is given in Table 11.1, with comments on each of them made by Post Office engineers at the time [4]. Several working conclusions emerged from this study:

1   the superiority of multiplex systems over automatic,
2   the improvement obtained by the use of a type keyboard,
3   the value of Baudot over Morse codes,
4   printers dealing with a complete page of information were preferable to printed slips, even if the latter were gummed to facilitate onward delivery.

In addition, the committee stated specifically that 'The Wheatstone system, even with the addition of the Creed accessories, is not considered by the committee as suitable for commercial telegraph work', and commented on 'the correction of errors [with the Wheatstone system] becomes a matter of considerable difficulty'. These conclusions were far-reaching and spelt the end of the dominance of the Wheatstone–Creed system, which had been in use by the Post Office for several decades. In 1914 multiplex techniques were in their infancy, and the techniques of frequency-division multiplexing (FDM) had yet to be satisfactorily employed to the extent to which they were later used in telegraph and telephone circuits. Automatic systems had reached their limit in terms of complexity and were proving too costly to install and operate. The value of the Baudot five-unit code had been shown to be superior to Morse code (excluding the operation of submarine cables), and this was later to lead to the use of binary code supplanting Baudot for long-distance working. The change to page printing for received information, now taken for granted, was only beginning to happen.

11.3 Baudot and the new codes

The Baudot system was the first of the non-Morse codes to be used successfully in a commercial telegraph service. It reached its zenith in multiplex systems employing at least four (and often six or eight) telegraph transmissions along a single line. Jean Maurice Émile Baudot was

Table 11.1 Systems considered by the High-speed Telegraph Enquiry of 1913

System	Type*	Speed (wpm)	Channels	Comment
American telegraph	Keyboard	40	Single	An American invention by A. D. Cardwell. Too costly for short lines. Page printing
Baudot multiplex Printing telegraph	Five-key Perforator	30/channel	2–12	The basic invention for many multiplex machines
Creed & Bille Receiver/perforator	Perforator	150 (receive only)	Single	Used to translate the Wheatstone Morse slip into printed characters and provide perforated slip for onward transmission
Delaney multiplex Telegraph	Printer from received slip Morse keys	20/channel	4	Bad working and technical difficulties
Gell telegraph Perforator	Keyboard Perforator	70–80	Single	Best keyboard perforator but costly Used as a Wheatstone perforator
Hughes printing Telegraph	Piano keyboard	30	Single	Too complex and costly. Slow
Kleinschmidt Telegraph perforator	Keyboard Perforator	70–80	Single	A good machine in use by the P.O. Used as a Wheatstone perforator
Kotyra Telegraph perforator	Three-key Perforator	70	Single	Best 3-key perforator
Morkrum printing Telegraph	Keyboard	60	Duplex	Too slow for main lines. Costly Page printing
Murray multiplex Printing telegraph	Keyboard	40/channel	8	Based on the Baudot system Page printing

Potts printing Telegraph	Keyboard	—	Single	Perforations in page form for retransmission below printed message
Siemens teleprinter	Keyboard	20	Single	Line printing telegraph Small, compact
Siemens automatic Printing telegraph	Keyboard	160	Duplex	In use by P.O. between London and Liverpool producing a printed slip
Telewriter	Facsimile	18	Single	Combined sender and receiver recording messages in handwriting Complex and unreliable
Western Electric Multiplex telegraph	Keyboard	40/channel	4	In user by P.O. between London and Manchester producing a printed slip
Wheatstone automatic	Three-key	400	Single	Labour of transcription from Morse slip. Difficult to correct errors

* Keyboard = Typewiter keyboard, three-key = keyboard with only 3 keys, five-key = keyboard with only 5 keys, piano = keyboard arranged as in a piano.

an employee of the French telegraph administration in the 1870s. He realised that with printing telegraphs, such as the Hughes or House printers of the day, the line is idle most of the time, apart from the brief intervals when a current would initiate the transmission of a single letter. Baudot planned to replace this inefficient system with one of the first applications of time-division multiplexing, (TDM) used for telegraphy. In his system he connected the line to a rotating contact, which he named a distributor, which made brief contact with a series of sectors (from four to eight). Five such rotating contacts were used to send a group of positive and negative signals, making up a single letter for transmission (Figure 11.1). Each operator was allocated a single sector and, at an appropriate time, pressed the five keys of the keyboard in the correct order to transmit a single letter of the message, before the distributor moved on from that operator's sector to give the next operator an opportunity to transmit a letter. This nascent TDM system and its accurate operation depended on the distributor at the transmitting end keeping exactly in step with the distributor at the receiving end, and the operator sending the letters of the message only at times when the contacts pass over their allocated sector. This could be achieved at a uniform speed of 30 wpm by strictly observing the 'cadence' or rhythm of the system when the distributor gave the operator the use of the line. The method required the operator to memorise a five-unit code in order to manipulate five piano-type keys in their correct order, a specialised task for which Morse training would be of no value. The five-unit code is

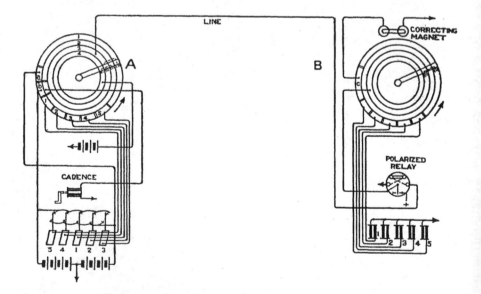

Figure 11.1   The Baudot transmission system
Source: Fleming, 'Principles of Electric Wave Telegraphy', 1909

shown in Figure 11.2, each letter being formed by some permutation of positive or negative currents sent to the line in quick succession. To advise the operator when to commence the keying sequence, a device called a 'cadence tapper' gave a slight sound just in advance of the time when the distributor reached the operator's allocated sector.

Although this appears a complex and difficult system to put into operation, not least from the skill it demanded of the operator, it quickly outreached automatic Morse code systems, which proved ill-adapted to multiplex operation. However, the Baudot system carried with it a major defect in the operation of the five-position keyboard, demanding a skill that could not be employed on any other machine and was difficult to learn. The solution to this problem was found in The United States, where the typewriter was being pioneered as a business tool. Alternative coded transmitting equipment was sought, based on the typewriter keyboard and a page-printing mechanism. Various attempts were made to design such a system, the most successful by British inventor, Donald Murray, who had devised a TDM system for the British Post Office in 1903, following the ideas suggested by Baudot. His system, the Murray multiplexer and printing telegraph, transmitted four messages simultaneously along one wire, the operator using a modified typewriter keyboard which made a group of holes for each letter in a moving paper tape, passing this automatically through a transmitter which in turn read

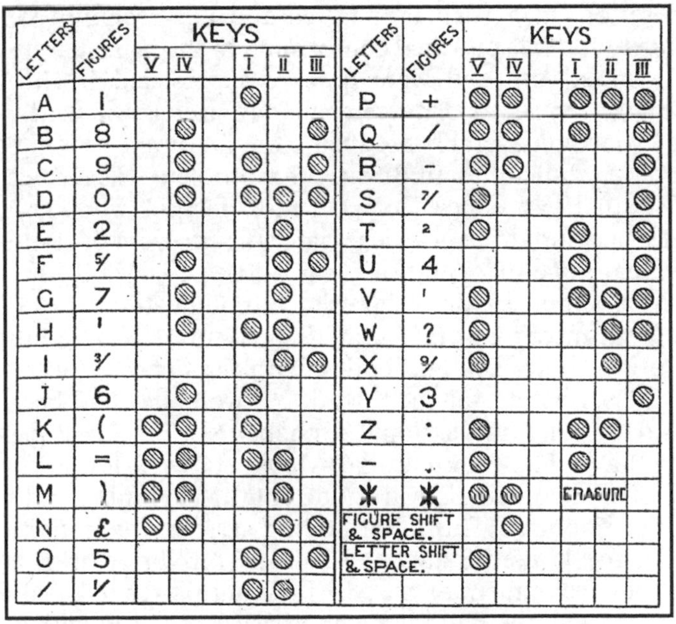

LETTERS	FIGURES	V	IV	I	II	III	LETTERS	FIGURES	V	IV	I	II	III
A	1			◎			P	+	◎	◎	◎	◎	◎
B	8		◎			◎	Q	/	◎	◎	◎		◎
C	9		◎	◎		◎	R	–	◎	◎			◎
D	0		◎	◎	◎	◎	S	7⁄	◎				◎
E	2				◎		T	2	◎		◎		◎
F	5⁄		◎		◎	◎	U	4			◎		◎
G	7		◎		◎		V	'	◎		◎	◎	◎
H	'		◎	◎	◎		W	?	◎			◎	◎
I	3⁄				◎	◎	X	9⁄	◎		◎		
J	6		◎	◎			Y	3					◎
K	(	◎	◎	◎			Z	:	◎		◎	◎	
L	=	◎	◎	◎	◎		—	.	◎		◎		
M	)	◎	◎			◎	✳	✱	◎	◎	ERASURE		
N	£	◎	◎		◎	◎	FIGURE SHIFT & SPACE.			◎			
O	5			◎	◎	◎	LETTER SHIFT & SPACE.		◎				
/	⁄			◎	◎								

Figure 11.2   The five-unit Baudot code
Source: Fleming, 'Principles of Electric Wave Telegraphy', 1909

the tape and sent the recorded signal as a series of positive and negative current impulses along the line, as did the Baudot machine. Printing at the receiving end could be directly onto a telegraph form for immediate transmission to the addressee. There were two remarkable features of Murray's system:

1   The operator could use a fairly conventional keyboard, for which experience may well have been gained from using an ordinary business typewriter.
2   The transmitters were entirely automatic, and were started and stopped in response to the paper tape, which included control characters at the beginning and end of each five-unit code letter.

The system initiated the use of a very compact and successful five-level code, based on the Baudot code, which went on to become a vital feature of later coding techniques for electronic data systems, such as the teletypewriter and the early digital computers.

11.3.1 Keyboard machines

The Murray system was the first of a number of multiplexers, all of which used a typewriter-like keyboard, operated at speeds of 30–60 wpm, and were capable of sending at least four messages simultaneously along one telegraph wire. A few keyboard instruments were constructed for Morse operation, in particular the Gell and Kleinschmidt perforators, where a paper tape was punched conforming to the Wheatstone code, but these had limited use and were quickly superseded by Murray-type machines. The keyboard arrangements for alphabetic and numerical operation varied considerably between manufacturers, causing considerable operational difficulties. In Britain, this problem was resolved in 1921 by A.C. Booth, Post Office Engineer-in-Chief [5], who designed and implemented a standard. A few years later the problem of standards over a much wider range of operations was addressed by the International Consultative Committee for Telegraph Communication (CCITT), which at its inaugural meeting in Berlin in 1926 commissioned a study of the standardisation of the five-unit codes, with the expectation of reaching an agreed international standard. By 1929 two codes had been adopted for universal use: The International Alphabet No. 1 for multiplex systems (Figure 11.3) and The International Alphabet No. 2 for start–stop systems in which control impulses are added at the beginning and end of the five character impulses. This latter coding system was initially put into operation in the United States and Europe for the Morkrum typebar instrument, a development of the Murray multiplexer referred to earlier. Much later, in 1966, the Baudot code itself was replaced with a seven-bit digital code, the American Standard Code for Information Exchange (ASCII), which allowed 128 different coded letters or symbols to be used,

NUMBER OF SIGNAL	LETTER CASE	FIGURE CASE	No. OF IMPULSES					REMARKS
			1ST.	2ND.	3RD.	4TH.	5TH.	
1	A	I						
2	B	8						■ INDICATES POSITIVE CURRENT
3	C	9						
4	D	O						
5	E	2						
6	F	SEE NOTE 1.						
7	G	7						□ INDICATES NEGATIVE CURRENT
8	H	+						
9	I	SEE NOTE 1.						
10	J	6						
11	K	(						
12	L	=						
13	M	)						NOTE 1. AT THE DISPOSAL OF EACH ADMINISTRATION FOR ITS INTERNAL SERVICE.
14	N	SEE NOTE 1.						
15	O	5						
16	P	%						
17	Q	/						
18	R	–						
19	S	.						
20	T	SEE NOTE 1.						NOTE 2. FOR PAGE PRINTING INSTRUMENTS
21	U	4						
22	V	' (APOSTROPHE)						
23	W	?						
24	X	, (COMMA)						
25	Y	3						
26	Z	: (COLON)						
27	CARRIAGE RETURN	CARRIAGE RETURN						
28	FRESH LINE SEE NOTE 2	FRESH LINE (SEE NOTE 2)						
29	LETTER BLANK (SPACE)	LETTER BLANK (SPACE)						
30	FIGURE BLANK (SPACE)	FIGURE BLANK (SPACE)						
31	⋇ (ERROR)	⋇ (ERROR)						
32	INSTRUMENT AT REST	INSTRUMENT AT REST						

Figure 11.3   The International Alphabet No. 1
Source: Fleming, 'Principles of Electric Wave Telegraphy', 1909

as compared with 32 for the Baudot code. This became a standard data interchange for use in digital computers.

11.3.2 The teleprinter

In 1924 the American Telephone & Telegraph Company (AT&T) introduced a printing system using a typewriter keyboard, which they named the teletype, and which became widely used for business communication. Each keystroke of this new device produced a series of five-unit Baudot coded electrical impulses which could be sent down the line and accepted by a second teletype machine to be converted back into print. Later, when computers came into use, this was supplanted by the ASCII code for teletype control, with similar operational results.

With the introduction of the teletype (known as the teleprinter in Europe), working on-line at a reasonable speed and allied with an FDM system, already in use with the Post Office (see Section 3.5), telegraphy in Britain took on a new form. Hand-keying in Morse code virtually disappeared for the internal terrestrial network, and Wheatstone and other automatic systems were phased out. Submarine cable operation and long-distance lines still needed the conventional telegraph key operators, and the military services made extensive use of Morse for security purposes. The teleprinter provided a direct complement to the office typewriter, and with the introduction of a version that reproduced messages in a page format, rather than on a ribbon of paper, the instrument took its place alongside the telephone – and was later able to use the same transmission equipment. In Britain this took the form of the Post Office Model No. 7 teletypewriter, which frequently made use of 'phantom' circuits enabling a teleprinter circuit to be carried on existing telephone cables. In 1932 this became a common adjunct to system working in which, under the name of the Van Rysselberghe system, had been used on Continental Europe for a number of years (Section 3.5). Business users made much of the system, which was later developed to become the telex service, described below.

11.3.3 Military use of the teleprinter

The military was not slow to take advantage of the standardisation of teleprinter systems. Almost a thousand private teleprinter circuits were employed by Bomber Command alone during the 1939–1945 war, and large numbers were also used by other commands and by USAF bases in Britain [6]. They were generally carried within eight-channel groups in voice frequency systems, a method developed by the Post Office for the public service. The terrestrial teleprinter connections were supplemented in 1943 with a network of VHF transceiver installations, created through cooperation between the RAF and the US Army during the war. This consisted of a series of installations each carrying one R/T and one teleprinter circuit, to connect with the US Tactical Air Command and several RAF Group centres in Continental Europe and in southern Britain.

In the British Army, Morse training was reduced as non-Morse methods of communication became established, being progressively replaced by teleprinter instruction, although the Royal Armoured Corps and Royal Artillery still used Morse operation for some purposes. The transmitters in the Army Chain of communicating systems, discussed earlier, were upgraded in the post-war years to include carrier frequency-shift equipment, and the existing stations converted to single sideband (SSB) operation with a view to maintaining a network service using speech and teleprinter channels only [7].

Although the Royal Navy had first considered the problem of maintaining teleprinters at sea somewhat doubtful, tests were carried on HMS *Royal Oak* in 1937 using Creed teleprinters on the bridge and various locations on board ship. These tests were successful, and led to the widespread use of teleprinters on larger ships and aircraft carriers, where unambiguous messages could be conveyed in the noisy conditions of active service when telephone conversation could be difficult [8]. The conversion of mobile wireless communication in the Royal Navy to non-Morse coding was initiated in the late 1950s, and transmission of paged teleprinter information became the norm until the adaptation of binary (digital) transmission in the 1980s and 1990s. This now included facsimile transmission, which employs a highly redundant code making error detection and correction a much more reliable operation than in the five-unit Baudot code used initially for teleprinter data transmission.

11.4 Telegrams, telex and the telephone

Changes were also taking place in non-Morse communication in the public usage of the telegram service, with its transmission increasingly effected more by Baudot code than by automatic Morse methods. At one time the prime method of rapid and efficient message transmission for business and the public, the telegram was on the decline. In its peak year (1919) some 82 million messages were transmitted over the inland telegraph network. By the end of 1969 this was down to 8 million, despite being assisted since early in the century by the transmission of telegram messages through the use of the telephone to the nearest Post Office – a method termed 'phonogram working',[1] which reduced the number of retransmissions and saved delay, particularly in the conveyance of the printed telegram by messenger directly to the recipient. The phonogram service grew rapidly in the United Kingdom from just before the onset of the First World War in 1913, when an average day's traffic was about three thousand messages, to five times this number in 1928, when the work was transferred from the telegraph operator to the telephone service [9]. Business users, however, were already making much use of the telephone connection for passing messages using a teletypewriter system and, in 1932, a new fast message system, the telex service was inaugurated, which largely supplanted the telegram for business subscribers.

11.4.1 The telex service

The telex (teleprinter exchange) service was based on the use of existing telephone exchanges and the telephone network, operating as a private teleprinter network for the benefit of industry. It enabled teleprinters situated at either end of a line to exchange messages through the typewriter keyboard. Both speech and teleprinter signals could be carried out on the same connection, though not simultaneously. This service was introduced first in large towns, such as London, Birmingham, Liverpool and Manchester, between individual business enterprises, and was later extended to cover the entire country. The method became popular for

---

[1] 'Mailgram' in the United States.

business in 1936 when a limited service was opened with Continental Europe. In the following year the number of subscribers reached 250, most of whom used the service to transmit short messages, 'printer-grams', as a cheaper alternative to the telegram service (a role today taken over by e-mail). One final change in the service was inaugurated in 1937, when the standard carrier frequency for the service over international telephone lines was raised from 300 to 1500 Hz by international agreement, resulting in a marked improvement in reliability for the service.

For inland working, however, it was the subscriber telephone that dealt automatic Morse transmission the *coup de grâce*. As early as 1876, when the early experiments of Alexander Graham Bell in the United States were proving a commercial success, the telephone was seen as a likely successor to the telegraph. The utility of the telephone, and the advantages it possessed over the ABC and Morse sounder methods of the day, soon became recognised by the Post Office in Britain, which began to supply telephone instruments on terms similar to those adopted for ABC telegraphs. At first the service provided lacked public confidence, and telephone instruments were fitted alongside the telegraph sets, switches being provided at the subscriber's premises, and also at the Post Office, whereby either instrument might be connected to a single circuit line, the idea being that, should the telephone fail, the telegraph could be resorted to. The first telephone exchange was established in 1879, and several companies competed to provide a telephone service for different areas of the country. With the consolidation of the various telephone companies by the British Government in 1912 to form part of the British Post Office, the network developed rapidly so that by 1931 over 11 million km of local lines had been established and over 1.5 million km of 'trunk' lines connected large exchanges, with a similar explosive growth for telephone lines in the United States [10]. Carrier systems supporting multiple speech transmissions came into operation in the 1930s, a development which also aided the widespread use of teleprinters, which were able to use the same transmission and multiplex system as the telephone network. The reliability and economy of the carrier frequency telephone service led in 1933 to the conversion of practically the whole of the inter-urban telegraph networks to carrier operation, freeing for the telephone network the telegraph cables thus released [11].

The telegraph service, now mainly served by teletypewriters and assisted by phonogram working, began to be run at a loss after the Second World War. The fall in the use of telegrams was arrested for a short time in the 1950s when an anomaly in the charging system, not previously noted in operation, was recognised by the public at large. Since 1893 a little known 'telephone letter service' had been available whereby a message could be sent via the telephone operator at the ordinary rate charged for a three-minute call, irrespective of the actual time needed for the sending operator to read the message and for the telephone operator at the receiving end to transcribe it, the message then being delivered locally as a letter. The total charge for, say, a 30-word message sent in this way, could be as little as one-ninth of the normal telegram charge. This fact was made known to the general public through the press in 1954, at a time when the telegram charges were being increased by the Postmaster General. The *Daily Mail* commented, 'Phone letters – If they put another shilling on the telegram get your own back by using the phone express-letter service instead . . . it's cheap and efficient'. This loophole was closed by the withdrawal of the service in 1954, and the fall in telegram usage continued [12].

The service continued to operate at a loss for a further number of years but was retained mainly on account of its value in an emergency, a need which, 'concerns no one very often, but many people very occasionally, and which, when it does concern them, matters very much' [13]. As the telephone coverage reached a very wide section of the public, the familiar yellow telegram envelopes containing the good news and the bad were discontinued in 1982, together with the acceptance of international telegrams, being replaced directly by a telephone message to subscribing premises.

11.4.2 Telephony by submarine cable and satellite

There were still two areas where for a time, no satisfactory alternative to telegraphy existed. These were in long-distance military applications, where a high level of security was needed, and the use of submarine cables, which could not convey telephone signals except on the shortest routes.

The military had a need for rapid and secure communication which would reduce or eliminate manual intervention. A method used by the United States forces during the Second World War relied on Creed equipment and the Baudot code, a system which had also been adopted by the RAF. This was the fully automatic 'tape relay system'. Punched paper tapes carrying the message were rapidly reproduced at the relay station, and automatically transmitted at high speed to the next station in the chain – a development which proved very suitable for teleprinter operation [14]. In Britain a similar project was initially applied across the Army Chain transmitter network, where the links between the stations were formed through wireless transmission. For tactical operations, complex multiplex systems were later developed using microwave transmission and very high transmission bandwidths, as discussed previously in Chapter 8. A large number of telephone and teleprinter channels could now be conveyed in complete military security, making use of digital techniques for control and correction of transmission errors.

For civilian purposes several cross-Channel submarine cables had been able to convey telephone conversations since the beginning of the century and, on several of these, phantom circuits were incorporated to work teleprinter lines [10]. The transmission of cable-carried telephone conversations across the Atlantic was a commercially viable proposition, but this had to await the laying of the first transatlantic telephone cable, TAT1, in 1956. This made use of coaxial cable to carry a broader bandwidth carrier signal than had been possible with earlier solid core systems, and incorporated submerged electronic amplifiers to counter attenuation with distance. These amplifiers also contained circuits for equalising the attenuation and phase of the signals to reduce characteristic distortion to a minimum, enabling effective use to be made of the increased bandwidth so obtained [15]. The new submarine cable could carry 36 simultaneous telephone channels and its success led to demand for yet more telephone channels to be carried by cable under the oceans. Increasing the bandwidth (and hence the number of speech channels that could be carried) could be achieved by increasing the diameter of the coaxial cable, but a limit was quickly reached. At the same time as the transatlantic telephone cables were being installed, satellite transmission was being put into operation and beginning to gain ground as a powerful competitor to cable systems, despite the inevitable quarter-second delay which accompanied speech transmission via satellite introduced by the much longer transmission path. In 1962 the United States launched its first communications satellite, Telstar, which permitted VHF bandwidths to be used, enough to support 480 simultaneous speech channels. The later generation of Intelsat satellites in the late 1980s carried several thousand two-way speech conversations and a number of television channels. In the meantime a new technique for transatlantic submarine cable transmission was put into effect. This employed optical fibres instead of copper conductors, allowing a greatly increased bandwidth for multiple telephone conversations across the Atlantic. The new fibre optical system, TAT8, provided a total of 7680 channels, a figure which in the 1990s was exceeded by a factor of 14 by making use of direct optical amplification instead of conversion into and out of an electronic form [16]. Transmission coding in all these newer submarine or satellite communication systems had by now become a purely digital operation, supplanting all other encoding techniques. The adoption of digital encoding for all forms of transmitted information in long-distance communication completed a process initiated by Morse over 150 years earlier.

11.5 The digital revolution

After the end of the Second World War much new technology became available, leading to the creation of the 'information' industry, now understood as the communication of data, audio and speech signals as well as visual signals (television and facsimile) by digital means. Frequencies in terrestrial carrier-phased systems, principally the telephone network, had been increased by replacing copper wires with coaxial cable and, more recently, by microwave links and fibre optic cable, the latter incorporating optical amplifiers. And in satellite systems the wide bandwidth available meant that transmission speeds were limited only by the capabilities of the terminal equipment used. Information transmission began to share the same type of digital encoding for many different purposes, a process which led to a number of significant advantages, not least that of being accepted by computers, used to control information transmission from the 1960s. Methods of accepting telephone conversations, carried on the FDM telephone network, for transmission over digital data lines were developed by the use of modulator/demodulator (modem) equipment of ever increasing speed and reliability, and the transmission of large quantities of data (files) over the network, revolutionised the acquisition and retrieval of written information.

The alliance between computers, the telephone network and the use of modems, initiated in the 1970s, produced a unique form of personal and business communication which we now know as the Internet. Like the telegraph network, the Internet allows people to communicate across great distances by the establishment of suitable connections between numerous separate networks. We have seen how, towards the end of the nineteenth century, businessmen, administrators and members of the public became entranced by their newly found ability to communicate swiftly with one another through the medium of telegraph networks, connected for overseas transmission to an extensive submarine cable network which crossed all the world's oceans. Now, as we enter the twenty-first century, a similar enthusiasm may be found within the same sections of the community for the ease with which communication may be established by e-mail and the World Wide Web through the Internet. Other comparisons may be made: the control of errors by repeating messages for the second time over the network – a practice common with the early mechanical telegraphs, now referred to as ARQ (automatic repeat request), a digital technique in which the retransmission of a block of digital characters is automatically carried out when required; message enciphering practised with commercial and military Morse code transmissions; and priority tagging, necessary when military telegraph lines are shared with civilian organisations. All these features are found in current digital transmission systems. The remarkable developments in 'communication at a distance', commencing with the invention of the telegraph in the eighteenth century, can clearly be recognised as leading directly to the present equally remarkable growth in digital global communications – the wheel of technology has indeed turned full circle!

References

1 'Telegraphy', *Post Office Electrical Engineers' Journal*, 1956, **49**, pp. 166–72
2 MARTIN, J.: 'Systematic Wheatstone working', *Post Office Electrical Engineers' Journal*, 1911, **3**, p. 322
3 WOOD, G.E.: 'Telegraphs: high speed committee report 1913–1914' (1916), POST 30/3589B, British Telecommunications Archives
4 'Report of Committee appointed to consider the question of high-speed telegraphy', *Post Office Electrical Engineers' Journal*, 1914, **10**, pp. 1–25
5 BOOTH, A.C.: 'Telegraph keyboard perforators', *Post Office Electrical Engineers' Journal*, 1921, **14**, pp. 72–9
6 ROBBINS, A.G.: 'The bomber lines', *Post Office Electrical Engineers' Journal*, 1945, **38**, pp. 105–8
7 NALDER, R.F.H.: 'History of the Royal Corps of Signals' (Royal Signals Institution, London, 1958)
8 'Fitting of ships with teletypewriter equipment' (1937), Doc. ADM 220/1908, Public Record Office, Kew, Surrey
9 JONES, J.S.: 'Changes at the Central Telegraph Office during the last fifteen years', *The Telegraph and Telephone Journal*, 1928, **15**, pp. 52–4
10 CRUICKSHANK, W.: 'Telegraphy and telephony, 1929–1931', *Journal of the Institution of Electrical Engineers*, 1931, **70**, pp. 153–6
11 SMITH, R.P.: 'The inland telegraph service: the introduction of modern machinery and methods', *Journal of the Institution of Electrical Engineers*, 1933, **72**, p. 189
12 'Proposals for abolition of the telephone letter service' (1954), POST 33/6061, British Telecommunications Archives
13 'Report of the Advisory Committee on the Inland Telegraph Service 1958' (Post Office/HMSO, London, 1958)
14 'Shore-based communication in the U.S. Navy: the tape relay station' (1954), Doc. AS330/43, pp. 45–50, British Library
15 MORRELL, P.O.: 'A recent development in telegraph repeaters for submarine cables', *Post Office Electrical Engineers' Journal*, 1945, **38**, pp. 34–8
16 BARNES, S.R, DEVOS, J., GABLE, P.M., and LE MOUEL, B.: '150 years of submarine cable systems', *Alcatel Telecommunications Review*, 1997, 1st Quarter, pp. 55–68

Index

ABC Telegraph 40, 42, 43, 61, 79
Abyssinian War 117
Admiralty 4–6, 8, 103, 309, 322
aerials 186, 236–8
African campaigns 288–9
air-to-ground communication 367
air-sea rescue 376–7
airborne wireless
    1914–18 350–65
    1939–45 370–83
    before 1914 348–50
    inter-war years 365–70
airborne wireless equipment
    British 353–9, 366–75
    German 362–5, 377–83
aircraft direction-finding 355
airship 348–9
Alexander, William 32–3
Alexander's electric telegraph 32–3
Alexanderson, Ernst Fredrik Werner
    202, 218
Alexanderson alternator 202, 218
Allied army wireless sets 296–301
    type 297
Allied navy wireless equipment 331–6
    type 332–3
alternator 201–2, 217–18
amateur radio movement 260–2, 345
American airborne wireless equipment
    360–2, 376–7
American Civil War 66–7, 110–15
American Marconi Company 216,
    242–3, 361
American Morse Code 55–6
American Railway Association 48
American Standard Code for
    Information Exchange 396
American Telegraph Company 64, 67,
    110, 115
Ampère 25

amplifier 227–8
Apparatus Carrier Telephone 287
Arco, Georg von 213
Army School of Signals 305
Army signalling 128–31
    time coding 129
Army telegraph training 126
Army wireless
    Allied equipment 296–301
    before 1914 267–8
    First World War 276–84
    inter-war years 284
    Second World War 285–303
Army wireless sets 284–5, 290–1,
    296–301
    German 301–3
    No. 10 290–1
arrival curve 148–9
artillery spotting aircraft 351–4, 362
astatic galvanometer 25, 27–8
astatic operation 35
Atlantic Telegraph Company 147, 155
audion 227–8
Australia 171–4
Austro-French conflict 115
auto-alarm devices 252–3
automatic recorders 87
automatic repeat request 403
aviation, civil 257–260

Bacon, Francis 28
Bain, Alexander 62
Bain printing telegraph 62, 87
Bangay, R. D. 360
Barretter 217
Battle of Jutland 316–7
Baudot, Jean Maurice Emile 391, 394
Baudot code 391, 394–5
Baudot printer 90
Baudot system 390–1, 394–5

beam aerials 236–8
Beardslee, G. W. 113
Bellini-Tosi direction finder 243, 269–70
Beresford, Major 90
Bewley, Henry 137
BF (British Field) set 279
binary code 28, 47–8, 52
Boer War 122–6, 170, 267, 314
    armoured train 125
    railway blockhouse 124
bombers 355, 359, 370
Braaten code 261
Braun, Professor Ferdinand 188, 190–1
Braun-Siemens system 210–11
Brazil-Paraguay war 116
Brett, J. W. 147
bridge duplex method 83
Bright, Charles Tilston 63, 157, 164
Bright's bells 63
British & Irish Magnetic Telegraph
        Company 74, 77
British airborne wireless equipment
        353–9, 366–75
    short wavelength 367–9
    type 356
British Australian Telegraph Company
        172
Brown's relay 317, 319
'bug' key 68
Bull, William 38
buzzer 274

'C' Telegraph Troop 117, 119
Cable & Wireless Ltd 239, 340–1
cable capacitance 148, 159
cable carts 120, 124
cable code 150, 239
    double current 239
cable-laying ships 154–6, 175, 292, 340–2
cable plough 106
cable resistance 148–9, 159
cablegram 48
cadence tapper 395
Calahan, Edward 86
Campbell, G. A. 152
canal towpaths 79
Canrobert, General 108
Cardew, Major Phillip 120, 126
cavalry 268–9
central battery working 84
Central Telegraph Station, London
        69–72, 76
Centralized Wireless Scheme 330
Chappé, Abraham 8

Chappé, Claude 6–7, 20
Chappé's telegraph system 6–9, 14, 17
Chinese Telegraph Dictionary 170
cipher books 48
ciphers 6, 48, 52, 111
Clark, Josiah Latimer 72, 106
'C.M.' 20–1
coast stations 240–2
code tables 28, 30, 36–7
codes 47–8, 247–8, 352–3
    binary 28, 47–8, 52
    digital 402
    Morse 48
    Schilling 48
    Steinheil 48
coherer 184–6
Collins, Perry 142
commercial semaphore systems 13–14
commercial telegraph companies 57–81
    Britain 69, 72–81
    developers 58
    United States 57–9, 62, 64–5
computers 403
Confederate Telegraph Company 111
Continental Morse code 244
convoy protection 330–1
Cooke, William Fothergill 29–32, 36–7,
        44
Cooke and Wheatstone telegraph system
        32, 34–5, 42, 44
Cornell, Ezra 66, 135
Coxe, J. R. 23–4
Creed printer 84, 90, 390
Crimean conflict 14, 103–8
cross-Channel wireless transmissions 188
crystal detector 191
crystal rectifier 225
curbing the signal 150–1
CW transmission 208–9, 212
    United States 216–20
Cymophone portable telegraph
        apparatus 241

Dalhousie, Lord 95
Davis, Jefferson 111
Davy, Edward 32
De Forest, Lee 226–8
Defence Teleprinter Network 286
Denmark 14
Depillon semaphore system 8
desert conditions 119–20
detection devices 225–7
dial telephones 40, 42–7
differential relay 82–3

digital communications 403–4
diode detector 227
Direct United States Cable Company 156
direction-finding 243, 269–70, 272, 325,
    355, 367, 370
    training 385
dirigible 348–50
distress signals 248–53, 256, 376, 389
    wartime 327
Dolland, John 3
double-current working 88–9
double frequency effect 192, 194–5
Duddell, William 200, 232
duplex telegraphy 82–3
    wireless 210
Durrant, Flight Lieutenant 368–9
Dyer, Harrison Gray 22, 24

earth return 22
East Africa 283
Eastern Telegraph Company 147,
    169–71, 176
Edison, Thomas 86, 225
Edison effect 225
Egyptian campaigns 119–22
Electric & International Telegraph
    Company 74–5
electric arc 200–1
electric synchronous telegraph 21
Electric Telegraph Company 32, 36–7,
    72–3, 105–6
electrical renewer 32
electrochemical devices 22–4, 26
electrochemical telegraph 22–3
electromagnet 51–2
electromagnetic telegraph 25–6
electromagnetic wave theory 183
electrostatic devices 20–2, 26
Elwell, Cyril F. 219
exchange switchboards 81
Exchange Telegraph Company 80–1

Faraday, Michael 135
Fardeln, William 44
Fardeln's dial telegraph 44–5
Fessenden, Professor Reginald A. 193,
    217–18
fibre optical systems 402
Field, Cyrus 147
field lines 105, 110, 112
field telegraph equipment 112–13
fire beacons 3
First Ashanti War 117–18
First World War

at sea 314–24
    cable operations 323–4
    in the air 350–65
    trench warfare 272–81
    wireless direction-finding 269–72
five-needle telegraph 34–5
flag signalling 18
Fleming, John Ambrose 225–6
Fleming, Sir Sandford 175–6
Fleming diode 226–7
four-channel cable 288
Franco-Prussian War 116
Franklin, Charles Samuel 234–5
French valve 229
frequency division multiplex system 85,
    287
frequency multiplication 202–3
frequency-multiplying transformer 212
'fruit machine' 372–3
Fuller, Algernon Clement 274
Fullerphone 274–5, 287

Gale, Professor Leonard 52, 55
Galletti, R. 197–8, 200
Galletti's multiple spark gap transmitter
    198–200
Galton, Captain Douglas 160
Gamble, Reverend John 4, 12
Gamble's mobile semaphore system
    12–13, 103
Gauss, Carl 25
GBR 231, 234, 240
Gell perforator 396
General Electric 218, 243
German airborne wireless equipment
    362–5, 377–83
    type 379
German army wireless sets 301–3
    type 303
German navy wireless equipment 337–40
    type 338
Germano-Austrian Telegraph Union 33
Gintle, J. G. 82
Giovanni 22
Gisborne, Francis N. 142, 172
Globotype telegraph 88
'Golden Arrow' 289–90, 298
Goldshmidt generator 203
Grant, General Ulysses S. 111
Great Northern Telegraph Company
    147, 170
Greenwich Mean Time 75
Grout, Jonathan 16
Grove, William Robert 57

Grove battery 57
gutta-percha 77, 88, 91, 93, 135–8
Gutta-Percha Company 137, 139, 167

Hagenuk 337
hard valve 228
hatchment telegraph 34–5, 37
Head, Sir Francis Bond 69
headphones 191
Heaviside, Oliver 151
heliograph 108, 125
heliostat 108, 125
Hell-Schreiber system 240
Helsby Wireless Telegraph Company 241
Henley, William 164, 166
Henry, Joseph 31, 52
Hertz, Heinrich 183–4
heterodyne method 229, 329
Heurtley, E. S. 152
Heurtley magnifier 152–3
HF alternator 201–2, 217–18
HF pentode 382
high-power transmitters 231–2
High Speed Telegraph Committee 1913
    390–3
high-speed telegraphy 390–1
Highton, Edward 97
Holyhead-Liverpool telegraph 13
Hooke, Robert 4
House, Royal Earl 61
House printing telegraph 61–2
Housekeeper seal 230
Hughes, David 64
Hughes automatic printing telegraph 79,
    89

Imperial and International
    Communications Ltd 239
Imperial Wireless Scheme 232–7
India 94–9, 110, 162–8
    cable routes, Britain-India 162–8
Indian Mutiny 96, 108–10
inductive communication 60
inductive field transmission 184
Institution of Electrical Engineers 127–8
insulation 134–8
insulation resistance 159–60
Intelsat 402
inter-war years 284, 327–30, 365–70
international agreements 254–6
International Alphabet No. 1 396–7
International Alphabet No. 2 396
International Convention on Telegraphy
    254–5

International Flag Code 18
International Morse Code 55–6
international Q code 247–8
International Telecommunications
    Union 256
International Telegraph Union 84, 93
Internet 403
ionosphere 235, 237, 260–1
Italian conflicts 115

Jackson, Captain Henry Bradwardine
    260, 308–10
jigger 190

Kendall, Amos 58–9
keyboard machines 391–3, 396
keying at high power 215
Kleinschmidt perforator 396

latex 135–7
leakage 152
Leclanché cell 84
Lepel, Baron Egbert von 197
Lepel spark gap 193, 197
Li Hung-chang 147
Lieben-Reisz relay 228
life-saving at sea 248–54
lifeboats 253–4
lightning protector 96–7
lightning strainer 97
Lincoln, Abraham 111–12
line tapping 113, 124
Link radios 300
loading 151–2
Lodge, Sir Oliver Joseph 184–6, 189
Lodge-Muirhead Company 206
Lodge's coherer 184–5
London & District Telegraph Company
    74, 77–8
London-Portsmouth line 8, 10, 12, 18
loop set 279
Lorenz 337–9, 378

Macdonald, Lieutenant-Colonel 5
'Maggie' 191–2, 207
magnetic blowout 192
Magnetic Telegraph Company 57–8
mailgram 67, 399
man-packs 300–1
Marconi, Guglielmo 186–90, 204, 308–9
Marconi-Adcock technique of direction-
    finding 367, 370
Marconi Company 204–10, 226, 233–6,
    241

Marconi high-power stations 208–10
Marconi International Marine Company 204, 240–1, 252
Marconi magnetic detector 191–2, 207
Marconi short-range multiple tuner 207–8
Marconi spark transmitters 206–7
  quenched 207
Marconi system 206–8
Marconi timing disc 194
Marconi uniform 244, 246
marine galvanometer 160
maritime wireless communication 240–3
  life-saving 248–54
  newspapers 247–9
  telegrams 246–7
  training 244–6
maritime wireless telegraphy 205–7
Martin, Horace G. 68
Maxwell, James Clerk 183
McCallum, David 88
McClellan, General George B. 112–14
Meissner, Alexander 228
merchant navy 244–6
Mesopotamia 281, 283
minesweeping 341–2
mirror galvanometer recorder 153–4, 161
modem 403
Montgomerie, Dr William 135
Morkrum typebar instrument 396
Morse, Professor Samuel F. B. 17, 51–2, 59
Morse code 48–9, 52, 54–6, 69, 150, 304–5
  American 55–6
  demise 389
  International 55–6
Morse equipment 47, 52–3, 55, 57
Morse inker 87
Morse key 52, 67–8, 97
Morse recorder 72, 87
'Morse Relay' patent 57
Morse sending tablet 362–3
Morse sounder 63, 69, 96–7
Muirhead, Alexander 82
multiplex systems 391–2
Murray, Donald 395
Murray, Lord George 4
Murray multiplexer and printing telegraph 395–6
Murray system 4–6
Myer, Major A. J. 113

Napoleonic Wars 5–6

National Electric Signalling Company (NESCO) 217, 219
nationalisation 65, 73–4, 81
Nauen high-frequency alternator station 213–14, 315–16
Naval wireless telegraph network 322–3
navigation 243, 258, 269
Navy wireless
  1914–18 314–24
  1939–45 329–40
  Allied navy equipment 331–6
  before 1914 308–14
  inter-war years 327–30
  shore stations 324–7
  tunes 310–11
Navy wireless equipment 317–22, 327–9
  German 337–40
  type 321, 328
needle telegraph 24–38
  five-needle 34–5
  single-needle 69
  six-needle 28–30
  two-needle 35–6, 38
netting 295, 299
New York and Mississippi Valley Printing Telegraph Company 65–6
Newall and Company 107, 140
newspaper companies 59–60, 75, 80
newspapers at sea 247–9
Nile expedition 119–21
nocturnal telegraph 12
Norderney cable 80
Norman Report 234

O'Etzel, Major Franz August 14, 91
O'Rielly, Henry 58–9
O'Rielly Contract 59
O'shaughnessy, William Brooke 94–7
Oersted 24–5
oscillator 228–9
overhead wires 90, 93

paper tape 87, 390
  gummed 390
Parker, John R. 16–17
Pasley, Lieutenant-Colonel 12
Pederson Tikker 201, 227
Pender, John 161, 167, 169–71
Peninsular War 103
pentode 382
permalloy 152
Phelp's printer 89
phonogram 399
pneumatic tube network 72–3

polyurethane 137
Popham, Sir Home Riggs 8
Popham's semaphore line 8, 10–12
Popov, Aleksandr Stepanovich 185–6,
   312
portable keyboard Morse instrument 113
Porthcurno telegraph school 168
Post Office 73–4, 81, 89, 126–7, 286,
   325–6
Potsdam-Koblenz semaphore line 14–16
Poulsen arc 200–1
Preece, Sir William 184, 187
printergram 400
printing telegraph 86
Prussia-Bohemia war 116

quadruplex working 82

R. S. Newall and Company 107, 140
radiation field transmission 184
Radio Corporation of America 220, 242
Radio Society of Great Britain 260, 262
   National Field Days 262
radio telephony 184
radiogram 246
RAF Civilian Wireless Reserve 262
RAF wireless equipment 356–9,
   366–75
   short wavelength 367–9
   type 372–5
RAF wireless training 383–6
   training centres 384
railroads 59–60, 65–6
railways 31–41, 44, 47
   India 97–9
   safety 35–9
recording telegraph 86–7
rectifying detector 217
relay 55, 57, 82–3, 166, 228
repeater 152–3
repetitive strain injury 68
Reuter, Baron Julius 80
Reuter's Telegram Company 74, 80
Reuters 240
Robinson, William 38
Robinson's 'Wireless Electric Signals'
   38–9
Roebuck, George 4
Ronalds, Sir Francis 21, 24
roof-to-roof wiring 77–9, 81
Round, Henry Joseph 269–70
Rouzet transmitter 351–2, 357, 362
Royal Air Force 350
   see also RAF

Royal Aircraft Establishment 369
Royal Engineers 105–6, 117, 119–20,
   126–7, 130
   Postal Corps 130
   telegraph training 126
Royal Engineers Telegraph Battalion: *see*
   Telegraph Battalion
Royal Flying Corps 350–1
Royal Naval Air Service 350–1
Royal Naval Wireless Auxiliary reserve
   261–2, 342
Royal Society of Arts 11, 135
rubber 135
Russian Navy 312–13
Russian telegraph network 103–4
Russo-Japanese war 268, 312–13

Salvá, Don 21–2
satellites 402
Schilling, Baron Pawel Lwowitsch 25,
   138
Schilling's telegraph 25, 27–30
   six-needle 28–30
School of Army Signalling, Bangalore
   110
Schweigger, J. S. C. 25
Schweigger galvanometer 25
Scudamore, Frank Ives 73, 128
Second World War 285–305, 329–40,
   370–83
   African campaigns 288–9
   Allied army wireless equipment
      296–301
   Allied navy wireless equipment
      331–6
   cable ships 340–2
   cables 292–3
   German army wireless equipment
      301–3
   German navy wireless equipment
      337–40
   line working 286–8
   RAF wireless equipment 371–5
   RAF wireless training 383–6
semaphore systems 6–18
   Denmark 14
   England 8, 10–14, 18
   France 6–8
   Prussia 14–16
   Russia 14, 18
   Sweden 14
   United States 16–17
separator 126
shore stations 315–16, 324–7

short wavelength systems 235–8
shutter systems 4–6
  England 4
  Sweden 5
Siberian Telegraph 142–3, 147
Sibley, Hiram 66
Siemens, Charles William 128
Siemens, Werner von 44, 104, 137
Siemens & Halske 91, 93, 96, 103, 138,
  165, 210
Siemens' automatic writing system
  104–5
Siemens' dial telegraph 44–7
Siemens' Morse equipment 47
signal regeneration 176–7
signal shaping 151
Signals Schools 385
Silver Arrow 375
Simpson, General 108
single-needle telegraph 69
siphon recorder 154
six-needle telegraph 28–30
Slaby, Professor Rudolf 210
Slaby-Arco system 210
slip readers 385
Smith, R. 24
Smith, Willoughby 160–1
Society of Telegraph Engineers 127–8
soft valve 228
Sömmerring, Samuel Thomas 22
Sömmerring's electrochemical telegraph
  22–3
South America 156–8, 170
South Foreland Lighthouse 187–8
Southwestern Telegraph Company 110
Spanish civil war 1838–40 115
spark gap 189–90, 192–200
  quenching 192–3, 195, 197–8
  short 194–7
  synchronous 193
spark resonator 184
spark transmitter 206–7, 256
spiral four 288
Stager, Anson 112
Standard 92 Code 67
Standard Cipher Code 48
Stearns, J. B. 82
Steinheil, Professor Karl August 32,
  36
Steinheil's recording telegraph 33,
  86–7
Sterling spark transmitter 351,
  353–4, 356
stock ticker 86

submarine cable 107–8, 134–78
  Africa 170
  armoured 140, 164
  Atlantic 147–8, 154–6
  Australia 171–4
  Caribbean 157–8
  Committee of Inquiry 160–1
  English Channel 138–40
  Far East 168–70
  India 162–8
  Irish Sea 140–1
  Mediterranean 141
  North America 142
  Pacific 174–8
  South America 170
  telephony 402
submarine cable-laying technology
  158–61
  cable recovery 158–9
  cable storage 161
submarines 310–11, 315, 317–8, 334,
  340
  aerials 340
superheterodyne receiver 259, 329
Sweden 5, 14
synchronous rotary disc discharger
  194
synchronous spark gap transmitter
  193
syntonic wireless telegraphy 189
syntony 189

tank wireless set 302
tape relay system 401
Telcon (Telegraph Construction &
  Maintenance Company) 167,
  169–70, 175
Telefunken 202–3, 211–13, 216, 258–9,
  277, 302, 339, 364–5
  growth 211
  wireless sets 258–9
Telefunken quenched spark ship
  installation 212
telegrams 48, 81, 238, 246–7, 255,
  399–401
  shilling 81
Telegraph Act of India 95
Telegraph Acts of 1868–69 73–4, 204
Telegraph Battalion 97, 119–20, 122,
  126–7
telegraph relay 32
telegraph stamps 97
telegraph train 113
telegraphiste 77–8

telephone 40, 42–7, 84, 125, 398, 400–2
    satellite 402
    submarine cable 402
telephone letter service 400
teleprinter 397–8, 400
    military use 398–9
telescope 3
teletype 397
teletypewriter 299, 396, 398
telex 399–400
Telstar 402
Tesla, Nikola 193
thermionic valve 224–31
    amplification 227–8
    detection 226
    oscillation 227–9
    transmission 229
Thomson, Elihu 192
Thomson, G. P. 151
Thomson, Professor William 148, 150,
    153–4, 156
time ball 75
time-division multiplexing 394
Todd, Charles 171–2, 174
training
    air operations 359–60
    British Army 126, 304–5
    British Navy 309, 342–5
    direction-finding 385
    RAF 383–6
    wireless operators 244–6, 269, 342–5,
        359–60, 383–6
transatlantic wireless transmission
    190–1
transmission rate 83
Transvaal War 119
Treaty of Six Nations 64
trench set 277, 279–80
trench warfare 272–81
Treuenfeld, Herr von 116
Trinity House 187
triode oscillator 229
triode valve 226–8
tuning 189–90
two-needle telegraph 35–6, 38

U-boats 319, 321, 339–40
ultra-audion 229
underground telegraph lines 90–4
    Europe 92
    Germany 91–2
underwater cables 95
    see also submarine cables
undulator 154

United Kingdom Electric Telegraph
    Company 74, 78–9
United States Military Telegraph
    Department 111
United States Signal Corps 110–11
Universal Private Telegraph Company
    74, 79–80

Vail, Alfred 53, 55
valve transmitter 229–31
    Army 284–5
    external anode 230–1
Van Rysselberghe telephone system 84,
    125, 398
Varley, C. F. 88
vibrating sounder 120
Vibroflex key 68
Voice Frequency Telegraph System 294
voltaic pile 22

Walker, Charles 138
walkie-talkie 301
war at sea
    1914–18 314–24
    1939–45 329–40
war in the air
    1914–18 350–65
    1939–45 370–83
war on the ground
    1914–18 268–84
    1939–45 285–303
warships 310, 330
Washburn, Senator C. C. 65
Waterlow, Sydney 77
Watson, Lieutenant Barnard Lindsay
    13
Watson's General Telegraph Association
    13
wavemeter 335
Webber, Captain Charles 128
Weber, Wilhelm 25
Wellington 103
Western Union Company 110, 115
Western Union Telegraph Company
    65–7
Wheatstone, Charles 31–2, 37, 40, 42,
    44
Wheatstone automatic system 72–3,
    83–4, 87–9
Wien, Professor Max 195–6
Wien's short spark gap 195–6
Wilson set 279
wireless mechanic 385–6
wireless operator/air gunner 384–5

wireless operator/mechanic 385–6
wireless ship 294
Wireless Telegraph & Signal Company
    187
wireless telegraphy 184, 189
Wireless Telegraphy Act 254
wireless training 244–6, 269, 342–5,
    359–60, 383–6

X code 247

Yakobi, B. S. 86

Z code 247
Zeppelin 270–1, 349–50
Zimmerman telegram 324
Zulu War 117, 119

Printed in the USA
CPSIA information can be obtained
at www.ICGtesting.com
JSHW011508221024
72173JS00005B/1242